Fundamentals of Arithmetic

McGraw-Hill Basic Skills System

Tools for Learning Success

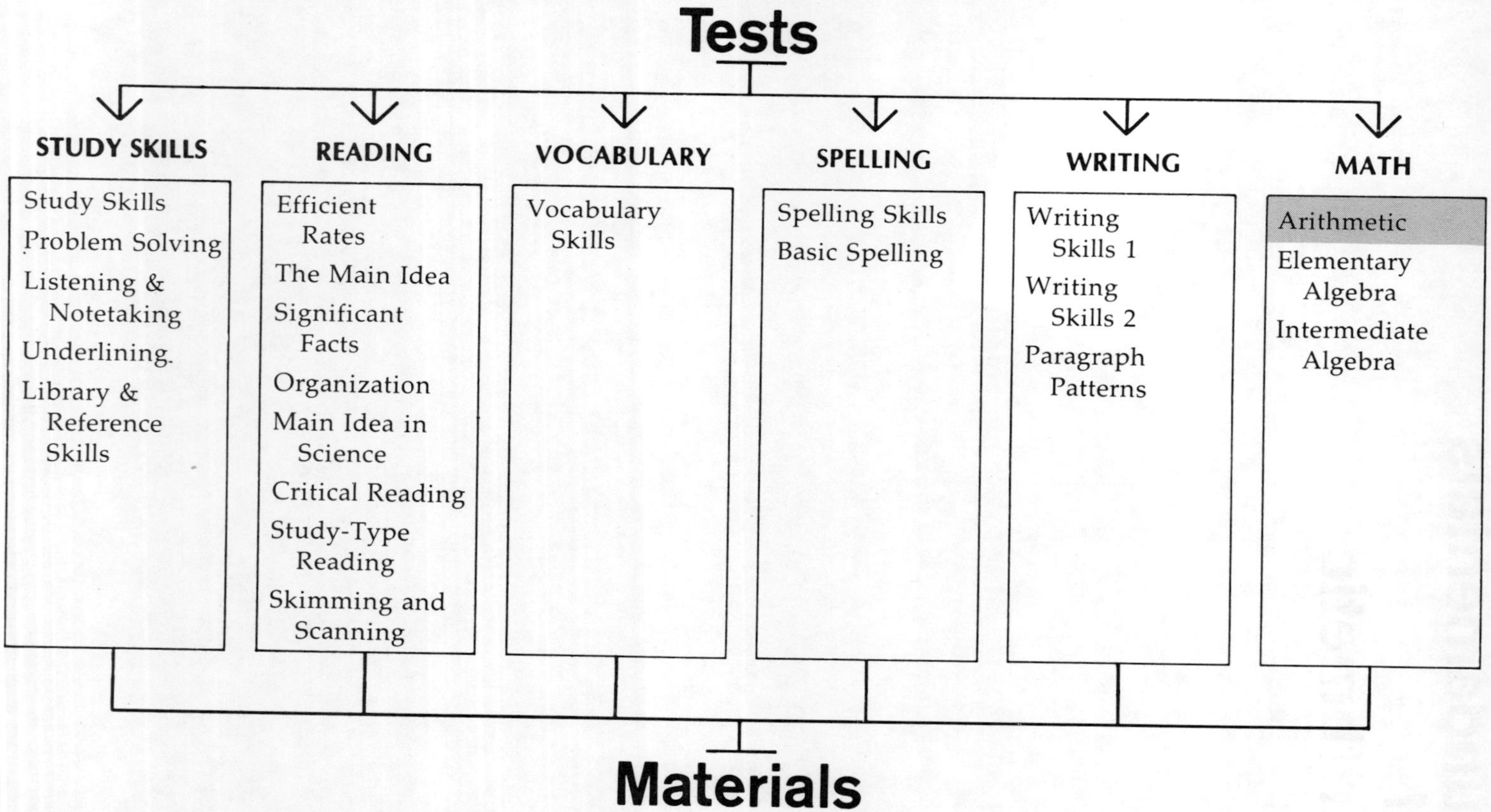

Editor's Introduction

This book is one of three parts in a series devoted to instruction in mathematics skills. This mathematics series, in turn, is part of a larger system of instructional materials—the McGraw-Hill Basic Skills System: Tools for Learning Success. Designed at the University of Minnesota Reading and Study Skills Center, the Basic Skills System is aimed at college-bound high school students, and junior college and college students who need to improve those skills necessary for academic success. The system consists of *tests* to determine instructional needs and *materials* designed to meet those needs, plus instructor's manuals to explain the tests and materials and the relationship between them. The purpose of the *tests* is to find out what instruction a student needs in basic skills, and the purpose of the *materials* is to give him that instruction. Each student gets what he needs without wasting time on unnecessary tasks.

Six basic skill topics—study skills, reading, vocabulary, spelling, writing, and mathematics—are covered, and two tests (A and B forms) are provided for each topic. Subscales on the tests are matched to accompanying instructional materials: thus a student with a low score on one or more subscales gets instruction in the corresponding skill. The second form of the test may be used to evaluate progress after instruction.

The materials in the Basic Skills System have been field-tested and revised to provide the best possible results. While most of the materials are self-instructional programs, the tests, of course, are designed for supervised administration. These testing instruments have been carefully developed and standardized by California Test Bureau. The latest research techniques and procedures have been utilized to ensure the highest possible validity and reliability.

The instructional materials are designed to be used separately, if desired, and can be purchased as single units. Most of the materials are suitable for adoption as textbooks in such basic skill courses as Freshman English, Communications, How to Study, Vocabulary Development, and Remedial or Developmental Mathematics. Individualized diagnosis and instruction are optional in such settings.

Alton L. Raygor
Consulting Editor
University of Minnesota

Fundamentals of Arithmetic

A Program for Self-instruction

MICHAEL ERAUT
Centre for Educational Technology
University of Sussex, England

McGRAW-HILL BOOK COMPANY
New York St. Louis San Francisco Düsseldorf London
Mexico Panama Sydney Toronto

Fundamentals of Arithmetic
A Program for Self-instruction

07-051401-1

7890VBVB79876543

McGraw-Hill Basic Skills System:
Design Director: Barbara Bert
Production Supervisor: Adam Jacobs
Editing Supervisor: Antonia Stires
Editor: James Wallace

The author developed this book in connection with an urban education project involving approximately 200 marginally prepared and specially enrolled freshman students at the University of Illinois at Chicago Circle, while he was on the staff of the Office of Instructional Resources.

Preface

I began this book because I believed that the problem of freshman mathematics could be considerably simplified by specifically catering for variations in student achievement rather than trying to compromise with them. It is virtually impossible to adapt traditional courses to such a wide range of student needs; so I set out to design a self-instructional course which students could start at any point. The next problem was how to help the students decide which parts of the course they needed, and this led me to the rather complex system of diagnostic tests described on page ix.

The first version of each chapter was tried out with a very small group of students and carefully discussed with them so that I could use their experience and advice to produce an improved version. Sometimes I had to do this two or three times before I was satisfied. Then the testing with larger numbers of students began, and this led to two further revisions. I also had four mathematics professors independently examine the material in detail, and I have incorporated most of their suggestions. Nearly all the testing was carried out at the University of Illinois at Chicago with students scoring in the lowest 20% on the Mathematics Placement Test. Students placed higher would still derive considerable benefit, but the aim of the testing was to make sure that the book was good enough to teach even the weakest students. The results obtained with the last version prior to publication show good evidence of its effectiveness with these students; and it has been improved since then. These results were obtained with nearly 80 freshmen at the University of Illinois:

Average score on diagnostic tests at beginnings of chapters	28%
Average score on final tests at ends of chapters	89%

The time taken varies greatly from one student to another. In general the shorter sections take 10 to 40 minutes and the longer sections 30 to 90 minutes to complete; but some students take most of the sections and some take very few.

Michael Eraut

To the Student

THE PURPOSE OF THIS BOOK

This book was written because not all students are the same. You differ in the way you learn, in your knowledge of mathematics, and in your ability to apply your knowledge to problems in science, social science, and technology. These differences make it particularly difficult to cater for your needs when you first arrive in college. There may be three or four mathematics courses you can take, but a course represents a lot of work. You won't want to take Math 100 when you only need the last 2 weeks of it; yet the last 2 weeks of Math 100 may be essential background for Math 101. If Math is one of your weaker subjects, you may even find that Math 100 starts at a level too high for you and assumes that you know things which you can only hazily remember from high school and never properly understood. If your first college course is too easy, you get bored; if it is too difficult, you get frustrated. The purpose of this book is to solve that problem.

The book has two main components, diagnostic tests and the self-instructional program. The purpose of the tests is to make sure that (1) you understand the fundamentals and (2) you skip all the sections of the book you don't need. Then you won't waste any time. The program itself will provide you with learning sequences in the form of numbered questions and answers called "frames." These have been thoroughly tested and revised so that you can learn to a very high degree of proficiency all the material in the book that you don't already know.

Since you will probably be meeting much of the material for the second time, this book attempts to approach most of it from a fresh viewpoint and to make it meaningful to students who have taken either a modern or a traditional course in high school. At the same time it tells you exactly where you are going and why. It leads you up to rules through sequences of examples, often allowing you to deduce them for yourself first, and it makes sure that you are convinced of their validity. But it does not attempt many proofs. The overall purpose is to find out what you need, to give it to you as quickly as possible, and to encourage an understanding of the fundamental principles in addition to developing a proficiency in applying them.

HOW TO USE THE TESTS

Each chapter in this book has a diagnostic test at the beginning and a final test at the end. The chapters are divided into sections, each with a pretest at the beginning and a posttest at the end. You mark all the tests yourself, using the answers provided. This may seem to you like a lot of tests even though some of them are very short, but the point of the tests is to help you find your best way through the book. It is impossible to adapt to your individual needs without them. You will understand this better when you are halfway through the book and have a better grasp of how the system works. Meanwhile let us follow Edward, a typical student, in his progress through a chapter. Edward has some previous knowledge but is not fully proficient. The diagram below shows his path through the chapter and is followed by an explanation of how the testing system allocated it.

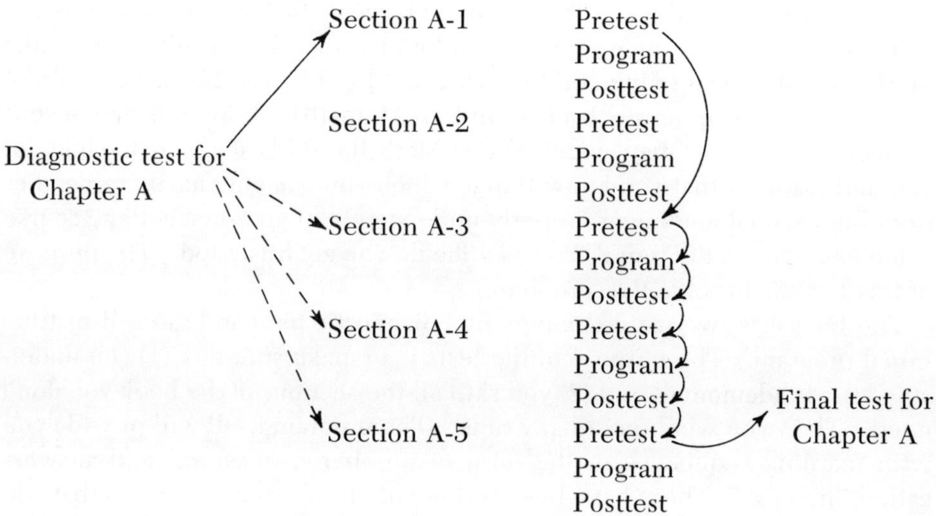

Edward began by taking the diagnostic test and marked it himself as follows (✓ indicates an error)

1. ✓	4. ____	6. ✓	9. ✓
2. ____	5. ____	7. ____	10. ✓
3. ____		8. ✓	

He then followed the instructions below and circled Secs. A-1, A-3, A-4, and A-5:

	section	page
If you made any mistakes in Questions 1 to 3, circle Sec. A-1.	(A-1)	000
If you made any mistakes in Question 4 or 5, circle Sec. A-2.	A-2	000
If you made any mistakes in Questions 6 to 8, circle Sec. A-3 *and* Sec. A-4.	(A-3) (A-4)	000 000
If you made any mistakes in Question 9 or 10, circle Sec. A-5.	(A-5)	000

This means that the diagnostic test has assigned Edward to Secs. A-1, A-3, A-4, and A-5 (but not A-2).

Edward now takes the pretest for Sec. A-1 and gets it all correct (his wrong answer to Question 1 of the diagnostic test was a careless mistake which he didn't repeat). This enables him to skip Sec. A-1.

Since the next section assigned is A-3, Edward takes Pretest A-3. He gets two out of six, and so he stays with A-3, works through the program, and takes the posttest. He gets six out of six on the posttest (the typical student *does* learn nearly all the material!).

Then Edward takes the pretest for Sec. A-4 and gets none of them right. He works through the program for A-4 and gets five out of six on the posttest.

When he takes the pretest for Sec. A-5, he answers all the questions correctly, and so he skips Sec. A-5. He couldn't have passed this test at the beginning, but now that he has worked through two sections he has remembered some of his high school work (and probably understands it better).

Edward is now ready to take the final test at the end of the chapter. He gets nine out of ten because he made a careless mistake.

The tests also serve two other purposes. They help you to see in advance just what you are going to learn, and they help you to check up afterward on your success. A diagnostic test tells you what a chapter contains, and a pretest tells you what a section contains. Then the posttest and the final test help you to check that you have learned everything. You should make very few mistakes on these as the book has been well tested and thoroughly revised.

Do not worry if you do not remember everything in this introductory section. You will receive detailed instructions as you work through the book. The object of this section is to give an overview. But make sure that you follow the instructions exactly. If they say that you need to go through a program, they are right. This has been carefully tested because I did not want to give students any unnecessary work. But during the testing of this book, those students who failed to follow the instructions nearly always did worse than the others.

HOW TO USE THE PROGRAMS

Each section of this book is a self-instructional program, which has been carefully designed to lead you through the material efficiently and effectively. The content is broken down into a series of steps, or frames, and each frame contains one or more questions and the appropriate answers. The material is arranged like Frames 1 to 4 on the next page.

1 Which section pretest was skipped by Edward in the diagram on page x?

Sec. A-2

2 Why was he allowed to skip this pretest?

Because he got both Questions 4 and 5 of the diagnostic test correct.

3 Edward was also able to skip the frames for Secs. ______ and ______ because he passed the section pretests.

A-1 A-5

4 He worked through the frames for Secs. ______ and ______.

A-3 A-4

Nearly all students find it useful to cover the answers to each frame while they are working on the questions. An easy way to do this is to use the mask card provided with this book and to position it with its top against the line in the middle of the frame. After you have studied the frame, write out your answer in full on a separate piece of paper. Then move the mask down the page to the next line, exposing the answer, and compare your own answer with the one given in the book.

You are unlikely to be wrong more than about 10% of the time, but when you are, make sure that you can understand how the answer in the book was derived. If you don't do this, you will find that later material begins to get very difficult because you are not sure of the fundamentals. In any case your performance on the posttest at the end of the section will tell you if you have any serious misunderstandings and need to review.

Michael Eraut

Contents

Editor's Introduction *iii*
Preface *vii*
To the Student *ix*

CHAPTER 1 OPERATIONS WITH SMALL NATURAL NUMBERS **1**

1-1 Addition of Numbers under 100 3
1-2 Subtraction of Numbers under 100 14
1-3 Multiplication of Numbers up to 10 24
1-4 Simple Division 32
1-5 Some Operations with Three-digit Numbers 37

CHAPTER 2 MULTIPLE OPERATIONS WITH NATURAL NUMBERS **48**

2-1 The Commutative Principle 50
2-2 The Associative Principle and the Use of Parentheses 55
2-3 The Distributive Principle Applied to Multiplication 62
2-4 The Distributive Principle Applied to Division 70

CHAPTER 3 OPERATIONS WITH LARGE NATURAL NUMBERS **80**

3-1 Addition of Large Numbers 82
3-2 Subtraction of Large Numbers 94
3-3 Multiplication of Large Numbers by One-digit Numbers 106
3-4 Long Multiplication 117
3-5 Division of Two-digit Numbers by One-digit Numbers 124
3-6 Division of Large Numbers by One-digit Numbers 131
3-7 Long Division 145

CHAPTER 4 NEGATIVE NUMBERS **162**

4-1 Definition and Addition of Negative Numbers 164
4-2 Multiplication and Division with Negative Numbers 175
4-3 Subtraction of Negative Numbers 183
4-4 Multiple Operations with Negative Numbers 189

CHAPTER 5 INTRODUCTION TO RATIONAL NUMBERS 201

5-1 Definition and Graphing of Rational Numbers 203
5-2 Equivalent Rational Numbers 212
5-3 Simplification of Rational Numbers 223
5-4 Mixed-numeral Forms 232

CHAPTER 6 OPERATIONS WITH RATIONAL NUMBERS 243

6-1 Multiplication of Rational Numbers 245
6-2 Division of Rational Numbers 255
6-3 Addition and Subtraction of Fractions 263
6-4 Addition and Subtraction of Mixed Numerals 273
6-5 The Lowest Common Denominator and Multiple Additions 279
6-6 Multiple Operations 289

CHAPTER 7 INTRODUCTION TO DECIMALS 299

7-1 Definition of Decimals and Conversion to Fractions or Mixed-numeral Forms 301
7-2 Multiplication or Division of Decimals by Multiples of 10 308
7-3 Metric Units 316
7-4 Conversion of Fractions to Decimal Form and Introduction to Infinite Decimals 321

CHAPTER 8 OPERATIONS WITH DECIMALS 329

8-1 Addition and Subtraction of Decimals 330
8-2 Multiplication by Integers and Division by Decimals 335
8-3 Multiplication by Decimals 341

CHAPTER 9 APPROXIMATIONS 349

9-1 Approximation by Rounding Off 350
9-2 Significant Figures 356
9-3 The Use of One-figure Approximations 362

9-4 Error and Proportional Error in Using Approximations 369
9-5 The Use of Two-figure Approximations to Evaluate Large Expressions 376

CHAPTER 10 POWERS OF 10 383

10-1 Multiplication and Division of Positive Integer Powers of 10 385
10-2 Multiplication and Division of All Integral Powers of 10 391
10-3 Conversion of Decimals into Standard Exponent Form 396
10-4 The Use of Powers of 10 in the Multiplication and Division of Decimals 404
10-5 Addition and Subtraction of Numbers in Standard Exponent Form 413

CHAPTER 11 SIMPLE EQUATIONS 421

11-1 The Addition Transformation with Numerals 423
11-2 The Multiplication Transformation with Numerals 428
11-3 The Addition Transformation with Variables 435
11-4 The Solution of Equations with Integral Coefficients 443

CHAPTER 12 RATIO AND PROPORTION 452

12-1 The Use of Proportion in Interpreting Tabulated Data 455
12-2 The Use of Ratios in Interpreting Tabulated Data 462
12-3 Multiple Ratios 469
12-4 Direct Proportionality 475
12-5 Inverse Proportionality 482

CHAPTER 13 PERCENTAGES 493

13-1 Definition and Conversion to Decimals or Fractions 497
13-2 The Use of Information in Percentage Form 504
13-3 Conversion into Percentage Form and Problems with Percentages 509
13-4 Percent Change 517

Chapter 1 Operations with Small Natural Numbers

The natural numbers consist of 0 and the counting numbers 1, 2, 3, 4, ... Since operations with small natural numbers form the basis of all arithmetic, they are extremely important. In order to do arithmetic problems with any speed you must memorize a number of facts about small whole numbers; but it is important for you to have a chance to discover these facts for yourself before memorizing them. Otherwise you might think they were bogus.

The chapter is divided into five sections. Section 1-1 deals with the addition of numbers between 0 and 99. Section 1-2 deals with the subtraction of numbers between 0 and 99. Section 1-3 deals with the multiplication of numbers between 1 and 10, and Sec. 1-4 with division facts that arise from these multiplications. Section 1-4 also covers multiplications and divisions involving 0. Then Sec. 1-5 extends the work of the previous sections to cover three-digit numbers.

In order to find out which sections, if any, you need to work through, take the diagnostic test for this chapter. If you make no mistakes on this test, you can skip the rest of the chapter and go straight on to Chap. 2. If you do make mistakes, the instructions tell you which sections you need to do. You will get a second chance to skip each section when you take the section pretest.

DIAGNOSTIC TEST 1

1. $24 + 59 =$ ______
2. $35 + 45 =$ ______
3. $0 + 18 =$ ______
4. $68 + 26 =$ ______
5. $75 - 39 =$ ______
6. $60 - 21 =$ ______
7. $93 - 67 =$ ______
8. $55 - 55 =$ ______
9. $6 \times 7 =$ ______
10. $9 \times 9 =$ ______
11. $9 \times 5 =$ ______
12. $56 \div 8 =$ ______
13. $4 \div 4 =$ ______
14. $3 \times 0 =$ ______
15. $0 \div 6 =$ ______
16. $439 + 293 =$ ______
17. $518 - 198 =$ ______
18. $873 - 569 =$ ______
19. $80 \times 5 =$ ______
20. $630 \div 7 =$ ______

When you have finished the diagnostic test, correct it yourself using the answers on page 45. Then check all the questions you got wrong on the list below:

1. ______	5. ______	9. ______	16. ______
2. ______	6. ______	10. ______	17. ______
3. ______	7. ______	11. ______	18. ______
4. ______	8. ______	12. ______	19. ______
		13. ______	20. ______
		14. ______	
		15. ______	

If you made no mistakes at all, go straight on to Chap. 2 on page 48. But if you made any mistakes (even if it was only *one* mistake), follow these instructions:

	section	page
If you made any mistakes in Questions 1 to 4, circle Sec. 1-1.	1-1	3
If you made any mistakes in Questions 5 to 8, circle Sec. 1-2.	1-2	14
If you made any mistakes in Questions 9 to 15, circle Secs. 1-3 and 1-4.	1-3	24
	1-4	32
If you made any mistakes in Questions 16 to 20, circle Sec. 1-5.	1-5	37

You should now work through the sections whose numbers you have circled, taking them in order. The sections begin on the pages indicated.

Section 1-1
Addition of Numbers under 100

Do this section if you made a mistake in Questions 1 to 4 of Diagnostic Test 1. Begin by taking the pretest below, which will show you how much of Sec. 1-1 you already know and what you will be able to do by the time you have finished it.

PRETEST 1-1

1. $7 + 8 =$ _____
2. $0 + 4 =$ _____
3. $13 + 35 =$ _____
4. $45 + 39 =$ _____
5. $30 + 62 =$ _____
6. $23 + 0 =$ _____
7. $39 + 48 =$ _____
8. $15 + 65 =$ _____

When you have finished the pretest, correct it yourself using the answers on page 45. Then check all the questions you got wrong on the list below:

1. _____	4. _____	7. _____
2. _____	5. _____	8. _____
3. _____	6. _____	

If you made any mistakes (even if it was only *one* mistake), read the introduction and go through the program.

If you made no mistakes at all, skip Sec. 1-1, because you already know its material, and go on to the next section whose number you circled on page 2. If no further numbers were circled, take Final Test 1 (page 44).

Introduction

In this section you will learn to add numbers between 0 and 99 (it is assumed that you can count). The fact that numbers are used for measuring as well as for counting will help you discover the basic facts of addition.

Equipment

You will need a 12-inch ruler and a set of cardboard strips in this section. Instructions for making the set of cardboard strips follow.

Instructions for Making Cardboard Strips

You will need 10 strips each of the following lengths: 1, 2, 3, 4, 5, 6, 7, 8, 9, and 10 inches. Each strip should be 1 inch wide. Altogether you will need five pieces of cardboard 10 inches long and 10 inches wide and one piece of cardboard 10 inches long and 5 inches wide. You can then cut out your strips by using patterns like these below:

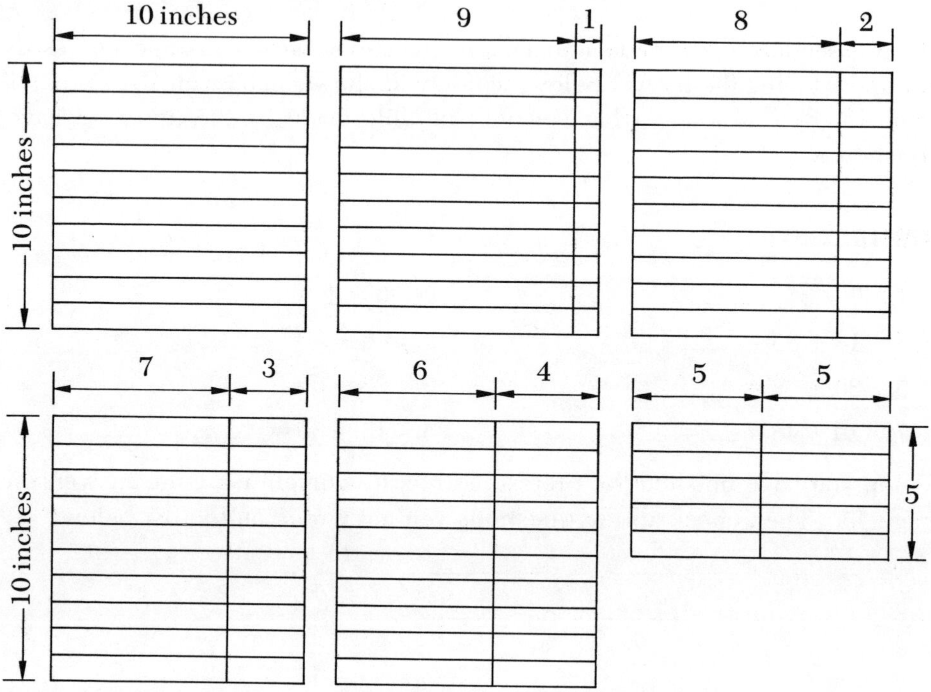

When you have cut out the strips, label each one according to its length, writing 10 on each of the 10-inch strips, 9 on each of the 9-inch strips, etc.

FRAMES FOR SECTION 1-1

1 Use counting to answer the following questions:

a. How many circles are there in square *A*?

b. How many circles are there in square *B*?

c. How many circles are there altogether?

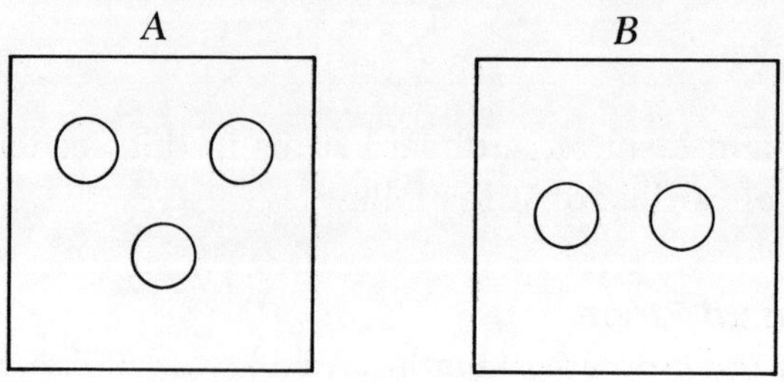

---

a. 3 *b.* 2 *c.* 5

---

2 In Frame 1 we used counting to answer three questions and we found that when 2 circles are *added* to 3 circles, we obtain ______ circles. We could also use counting to find out how many oranges we get when we add 2 oranges to 3 oranges, but we know very well that the answer will be ______ oranges.

---

5 5

---

3 We have done this kind of counting so often in our lives that we know that we always get the same result. Whenever we add 2 objects to 3 objects, we get ______ objects. We call this a *fact of addition* and write it in shortened form as $2 + 3 =$ ______.

---

5 5

---

4 Knowing *addition facts* can save us a lot of counting time. See how many of these addition facts you can complete without doing any counting. Then check your answers:

a. $5 + 7 =$ ______ *b.* $4 + 6 =$ ______

c. $9 + 8 =$ ______ *d.* $8 + 3 =$ ______

e. $7 + 9 =$ ______ *f.* $8 + 4 =$ ______

g. $7 + 6 =$ ______ *h.* $9 + 9 =$ ______

i. $7 + 8 =$ ______ *j.* $5 + 9 =$ ______

---

a. 12 *b.* 10 *c.* 17 *d.* 11 *e.* 16 *f.* 12 *g.* 13 *h.* 18 *i.* 15 *j.* 14 If you answered all these questions correctly without counting, skip to Frame 16. If you had any difficulty or made any mistakes or had to use counting because you couldn't remember an answer, continue with Frame 5.

---

5 There is an even easier way than counting for finding out addition facts. It depends on the way we use numbers for measurement. On a ruler, for example, we find a series of marks, and neighboring marks are always the same distance apart.

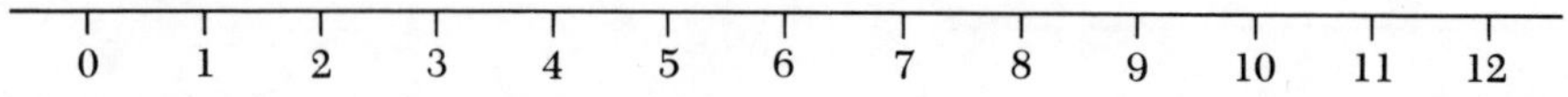

Now suppose we cut a strip of cardboard 2 inches long and place it next to the ruler in position *A*.

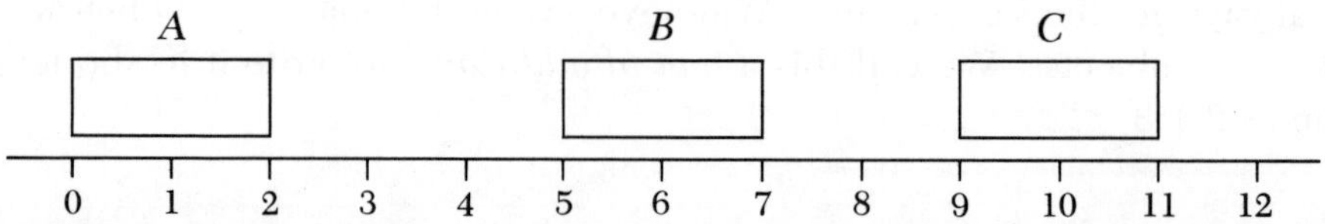

Its left end is opposite the mark 0, and since the strip is 2 inches long, its right end will be opposite the mark ______. If we move the strip to position *B*, its left end will be opposite the mark ______ and its right end opposite the mark ______. And in position *C*, its left end will be opposite the mark ______ and its right end opposite the mark ______.

2 5 7 9 11

6 By using a strip of cardboard and a ruler we can find out some useful addition facts. In Frame 5, observing the strip in position *A* told us that $0 + 2 = 2$. Observing the strip in position *B* told us that $5 + 2 =$ ______, and observing the strip in position *C* told us that ______ $+ 2 =$ ______.

7 9 11

7 Now take one of your strips of cardboard labeled 3 and check its length with a ruler. Put it in positions *E*, *F*, and *G*.

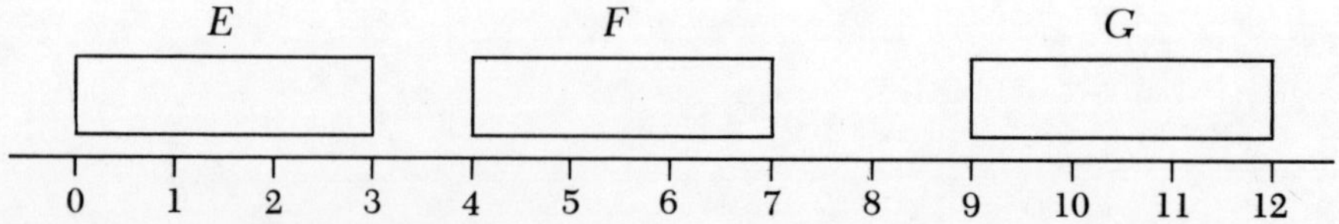

This gives you these addition facts:

a. ____ $+ 3 =$ ______

b. ____ $+ 3 =$ ______

c. ____ $+ 3 =$ ______

a. $0 + 3 = 3$ *b.* $4 + 3 = 7$ *c.* $9 + 3 = 12$

8 Now take one of your 3-inch strips of cardboard (labeled 3) and one of your 5-inch strips (labeled 5). Place them on a ruler as shown:

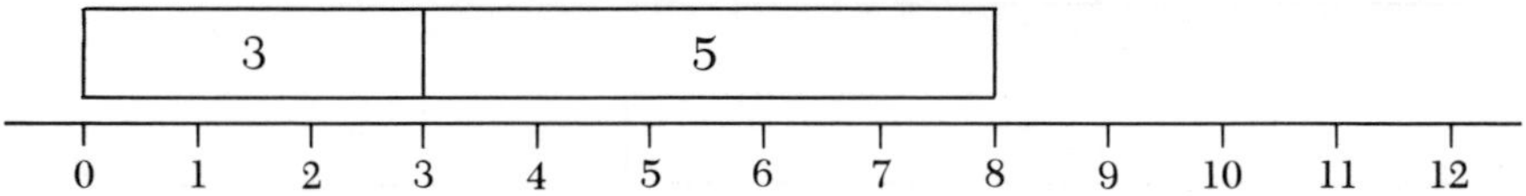

This gives us the addition fact 3 + 5 = ______. Another way to find this fact doesn't even use a ruler. Find a third strip to match the two you already have, as shown below:

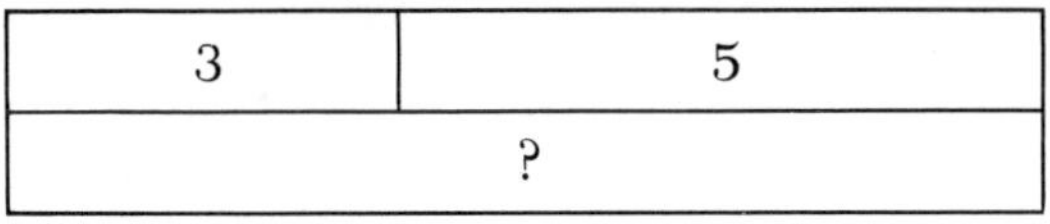

What is the label of this third strip?

---

8 8

---

9 Take one of your 2-inch strips of cardboard (labeled 2) and one of your 4-inch strips (labeled 4). Place them end to end, as shown below, and find a third strip to match them.

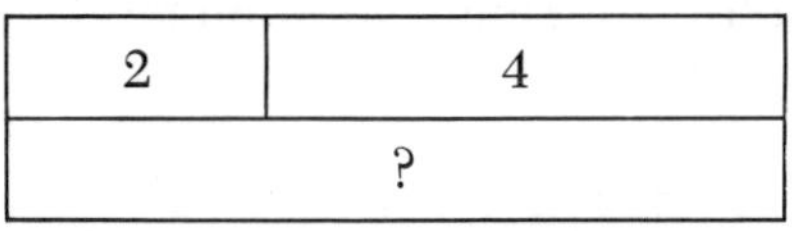

a. What is the label of this third strip?

b. The addition fact you have discovered is ______ + ______ = ______.

---

a. 6 *b.* 2 + 4 = 6

---

10 In the same way, select three strips of cardboard to demonstrate the addition fact 4 + 5 = ______. The strips you used were ______, ______, and ______.

---

9 4 5 9 Your strips should have looked like this:

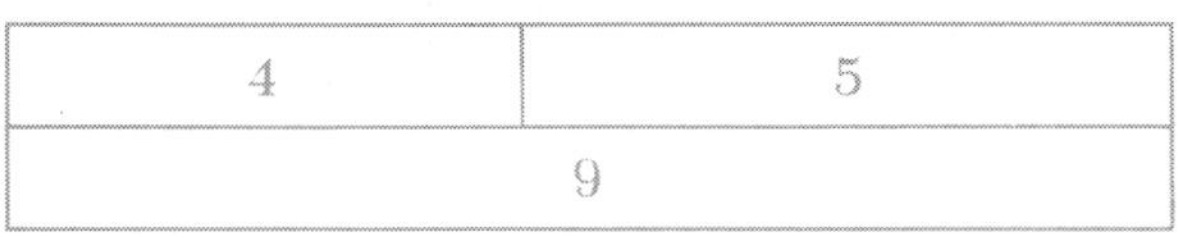

---

11 Since our number system is based on 10, we can use a 10 strip and a ______ strip instead of a 13 strip to demonstrate that 10 + ______ = 13.

---

3 3

---

12 To find the answer to $7+9$, we can use a 10 strip and one other strip as shown below.

7	9
10	?

a. The fourth strip is labeled ______.

b. The addition fact we have discovered is $7+9=10+$ ______ $=$ ______.

a. 6 *b.* 6; 16

13 Use four strips of cardboard, including a 10, to complete the addition fact $5+8=$ ______. You used strips ______, ______, ______, and ______.

13 5 8 10 3 Your strips should have looked like this:

5	8
10	3

14 Use your cardboard strips to complete the following addition facts:

a. $7+5=$ ______ *b.* $6+4=$ ______

c. $8+9=$ ______ *d.* $3+8=$ ______

e. $9+7=$ ______ *f.* $4+8=$ ______

g. $6+7=$ ______ *h.* $9+9=$ ______

i. $8+7=$ ______ *j.* $9+5=$ ______

a. 12 *b.* 10 *c.* 17 *d.* 11 *e.* 16 *f.* 12 *g.* 13
h. 18 *i.* 15 *j.* 14

15 In order to do more complicated additions, you need to memorize all the addition facts for numbers up to 10. The easiest way to do this is the following:

1. Pick out any two of your cardboard strips.
2. See if you know what strip is equal in length to the two you picked.
3. If you can't remember the answer, find it by using some of your other strips.
4. Memorize the answer if you didn't already know it.
5. Repeat steps 1 to 4, using various pairs of strips, until you have memorized all the answers and no longer need the help of the strips.
6. When you have finished all the above steps, answer the questions below:

a. $7 + 4 =$ ______ *b.* $9 + 3 =$ ______

c. $6 + 5 =$ ______ *d.* $8 + 8 =$ ______

e. $5 + 9 =$ ______ *f.* $3 + 7 =$ ______

g. $8 + 5 =$ ______ *h.* $7 + 8 =$ ______

---

a. 11 *b.* 12 *c.* 11 *d.* 16 *e.* 14 *f.* 10 *g.* 13 *h.* 15 If you made any mistakes, return to the beginning of Frame 15 and repeat the procedure.

---

16 *a.* If I have a pile of 5 oranges and I add 7 more, how many do I then have?

$5 + 7 =$ ______

b. If I have a pile of 7 oranges and I add 5 more, how many do I then have?

$7 + 5 =$ ______

c. Does the order in which you add 7 and 5 affect the result?

---

a. 12; 12 *b.* 12; 12 *c.* no

---

17 *a.* If I have a pile of 5 oranges and I add 0 oranges, how many do I then have?

$5 + 0 =$ ______

b. If I have a pile of 0 oranges and I add 5 oranges, how many do I then have?

$0 + 5 =$ ______

c. Does the order in which you add 0 and 5 affect the result?

---

a. 5; 5 *b.* 5; 5 *c.* no

---

18 *a.* $4 + 9 =$ ______ and $9 + 4 =$ ______

b. $8 + 0 =$ ______ and $0 + 8 =$ ______

c. Does the order in which you add any two numbers affect the result?

---

a. 13; 13 *b.* 8; 8 *c.* no

---

19 *a.* Two-digit numbers whose first digit is 1 can be expressed as the sum of 10 and a one-digit number. Thus $12 = 10 +$ ______, and $15 =$ ______ $+ 5$.

b. Similarly two-digit numbers whose first digit is 2 can be expressed as the sum of 20 and a one-digit number. Thus $22 = 20 +$ ______, and $28 =$ ______ $+ 8$.

c. And two-digit numbers whose first digit is 7 can be expressed as the sum of 70 and a one-digit number. Thus $73 = 70 +$ ______, and $76 =$ ______ $+ 6$.

---

a. 2; 10 *b.* 2; 20 *c.* 3; 70

---

20 *a.* $24 =$ ______ $+ 4$ *b.* $35 = 30 +$ ______ *c.* $10 + 7 =$ ______

$44 =$ ______ $+ 4$ $66 = 60 +$ ______ $50 + 1 =$ ______

$74 =$ ______ $+ 4$ $98 = 90 +$ ______ $80 + 3 =$ ______

---

a. 20; 40; 70 *b.* 5; 6; 8 *c.* 17; 51; 83

---

21 Let us see what happens when you count numbers 10 at a time instead of 1 at a time:

a. *Counting 1 at a time*

$0 + 1 = 1$
$1 + 1 = 2$
$2 + 1 = 3$
$3 + 1 = 4$
$4 + 1 =$ ____

b. *Counting 10 at a time*

$0 + 10 = 10$
$10 + 10 = 20$
$20 + 10 =$ ____
__ $+ 10 =$ ____
__ $+ 10 =$ ____

---

a. 5 *b.* 30; 30; 40; 40; 50

---

22 Just as adding 3 can be represented as adding 1 three times, adding 30 can be represented as adding 10 three times.

$0 + 1 = 1$	$0 + 10 = 10$
$1 + 1 = 2$	$10 + 10 = 20$
$2 + 1 = 3$	$20 + 10 = 30$
$3 + 1 = 4$	$30 + 10 = 40$
$4 + 1 = 5$	$40 + 10 = 50$

Thus $2 + 3 =$ ______, and $20 + 30 =$ ______.

---

5 50

---

23 In general, two-digit numbers ending in 0 can be added like one-digit numbers. So $60 + 30 =$ ______ because $6 + 3 =$ ______. Similarly $40 + 10 =$ ______, and $20 + 20 =$ ______.

---

90 9 50 40

---

24 Two-digit numbers that do not end in 0 can be added by separating the digits, adding the first and second digits separately, and recombining:

$24 = 20 + 4$
$31 =$ __ $+ 1$
$24 + 31 =$ __ $+ 5 =$ ______

---

30 50 55

---

25 *a.*

$$\begin{array}{r} 43 = \_\_\_\_ + 3 \\ 16 = \_\_\_\_ + 6 \\ \hline 43 + 16 = \_\_\_\_ + 9 = \_\_\_\_ \end{array}$$

b.

$$\begin{array}{r} 32 = \_\_\_\_ + 2 \\ 37 = \_\_\_\_ + 7 \\ \hline 32 + 37 = \_\_\_\_ + \_ = \_\_\_\_ \end{array}$$

---

a. 40; 10; 50; 59 *b.* 30; 30; 60 + 9 = 69

---

26 Use the same method to complete the following additions:

a. $66 + 21 = \_\_\_\_$ *b.* $23 + 75 = \_\_\_\_$

c. $40 + 39 = \_\_\_\_$ *d.* $41 + 57 = \_\_\_\_$

e. $60 + 14 = \_\_\_\_$

---

a.

$$\begin{array}{r} 66 = 60 + 6 \\ 21 = 20 + 1 \\ \hline 66 + 21 = 80 + 7 = 87 \end{array}$$

b.

$$\begin{array}{r} 23 = 20 + 3 \\ 75 = 70 + 5 \\ \hline 23 + 75 = 90 + 8 = 98 \end{array}$$

c.

$$\begin{array}{r} 40 = 40 + 0 \\ 39 = 30 + 9 \\ \hline 40 + 39 = 70 + 9 = 79 \end{array}$$

d.

$$\begin{array}{r} 41 = 40 + 1 \\ 57 = 50 + 7 \\ \hline 41 + 57 = 90 + 8 = 98 \end{array}$$

e.

$$\begin{array}{r} 60 = 60 + 0 \\ 14 = 10 + 4 \\ \hline 60 + 14 = 70 + 4 = 74 \end{array}$$

---

27 *a.*

$$\begin{array}{r} 55 = \_\_\_\_ + 5 \\ 28 = \_\_\_\_ + 8 \\ \hline 55 + 28 = \_\_\_\_ + 13 = \_\_\_\_ \end{array}$$

b.

$$\begin{array}{r} 66 = \_\_\_\_ + 6 \\ 29 = \_\_\_\_ + 9 \\ \hline 66 + 29 = \_\_\_\_ + \_ = \_\_\_\_ \end{array}$$

---

a. 50; 20; 70; 83 *b.* 60; 20; 80 + 15 = 95

---

28 Use the same method to complete the following additions:

a. $35 + 47 = \_\_\_\_$ *b.* $26 + 38 = \_\_\_\_$

c. $49 + 17 = \_\_\_\_$ *d.* $52 + 29 = \_\_\_\_$

---

a.

$$\begin{array}{r} 35 = 30 + 5 \\ 47 = 40 + 7 \\ \hline 35 + 47 = 70 + 12 = 82 \end{array}$$

b.

$$\begin{array}{r} 26 = 20 + 6 \\ 38 = 30 + 8 \\ \hline 26 + 38 = 50 + 14 = 64 \end{array}$$

c.

$$\begin{array}{r} 49 = 40 + 9 \\ 17 = 10 + 7 \\ \hline 49 + 17 = 50 + 16 = 66 \end{array}$$

d.

$$\begin{array}{r} 52 = 50 + 2 \\ 29 = 20 + 9 \\ \hline 52 + 29 = 70 + 11 = 81 \end{array}$$

---

29 Complete the following additions:

a. 23 + 58 = ______ *b.* 42 + 50 = ______

c. 67 + 19 = ______ *d.* 33 + 26 = ______

a. 81 *b.* 92 *c.* 86 *d.* 59

You have finished the program for Sec. 1-1. Now take the posttest and see how much you have learned.

POSTTEST 1-1

1. 9 + 6 = ______ 2. 0 + 5 = ______

3. 14 + 52 = ______ 4. 25 + 58 = ______

5. 20 + 73 = ______ 6. 58 + 0 = ______

7. 47 + 29 = ______ 8. 75 + 15 = ______

When you have finished the posttest, correct it yourself using the answers on page 45. Then check all the questions you got wrong on the list below:

1. ______ 4. ______ 7. ______

2. ______ 5. ______ 8. ______

3. ______ 6. ______

If you compare your performance with that on Pretest 1-1, you will see how much you have improved. If you made more than one mistake on this posttest, review the material and take the pretest again. Don't move on to the next section until you can get all the questions right.

Now proceed to the next section whose number you circled on page 2. If no other sections were circled, take Final Test 1 (page 44).

Section 1-2
Subtraction of Numbers under 100

Do this section if you made a mistake in Questions 5 to 8 of Diagnostic Test 1. Begin by taking the pretest below, which will show you how much of Sec. 1-2 you already know and what you will be able to do by the time you have finished it.

PRETEST 1-2

1. $17 - 9 =$ _____
2. $8 - 0 =$ _____
3. $25 - 14 =$ _____
4. $7 - 7 =$ _____
5. $53 - 29 =$ _____
6. $70 - 22 =$ _____
7. $95 - 35 =$ _____
8. $56 - 37 =$ _____

When you have finished the pretest, correct it yourself using the answers on page 46. Then check all the questions you got wrong on the list below:

1. _____ 4. _____ 7. _____

2. _____ 5. _____ 8. _____

3. _____ 6. _____

If you made any mistakes (even if it was only *one* mistake), read the introduction and go through the program.

If you made no mistakes at all, skip Sec. 1-2, because you already know its material, and go on to the next section whose number you circled on page 2. If no further numbers were circled, take Final Test 1 (page 44).

Introduction

In this section you use cardboard strips to discover some basic facts of subtraction. You then apply these facts to the subtraction of numbers between 0 and 99. You also find that for natural numbers, that is, 0, 1, 2, 3, 4, . . . , subtraction of a larger number from a smaller number is not defined.

Equipment

You will need a 12-inch ruler and a set of cardboard strips. Instructions for making the strips are on pages 3 and 4.

FRAMES FOR SECTION 1-2

1 Use counting to answer these questions:

a. How many circles are there in the whole box?

b. How many circles are there in the left half of the box?

c. How many circles would remain if we emptied the left half of the box?

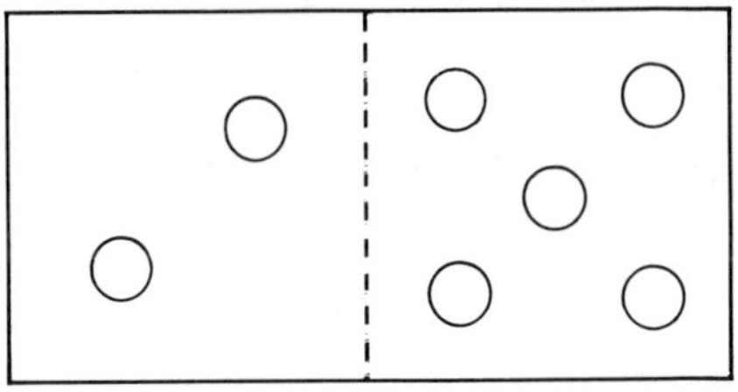

---

a. 7 *b.* 2 *c.* 5

---

2 In Frame 1 we used counting to answer three questions, and we found that when 2 circles are taken from 7 circles, ______ circles remain. We could also use counting to find out how many apples remain when we start with 7 apples and eat 2 of them; but we know very well that the answer will be ______ apples.

---

5 5

---

3 We have done this kind of counting so often that we know that we always get the same result. Whenever we take 2 objects from 7 objects, we are left with 5 objects. We call this a *fact of subtraction* and write it in numerical form as $7 - 2 =$ ______.

5

---

4 Knowing *subtraction facts* can save us a lot of counting time. See how many of these subtraction facts you can complete without doing any counting. Then check your answers.

a. $13 - 9 =$ _____ *b.* $9 - 1 =$ _____

c. $11 - 5 =$ _____ *d.* $17 - 8 =$ _____

e. $12 - 7 =$ _____ *f.* $15 - 8 =$ _____

g. $14 - 6 =$ _____ *h.* $18 - 9 =$ _____

i. $11 - 7 =$ _____ *j.* $16 - 9 =$ _____

---

a. 4 *b.* 8 *c.* 6 *d.* 9 *e.* 5 *f.* 7 *g.* 8 *h.* 9
i. 4 *j.* 7

If you answered all these questions correctly without counting, skip to Frame 18. If you had any difficulty, or made any mistakes, or had to use counting because you couldn't remember an answer, continue with Frame 5.

---

5 There is an easier way than counting for finding out subtraction facts. It depends on the way we use numbers for measurement. On a ruler, for example, we find a series of marks, and neighboring marks are always the same distance apart.

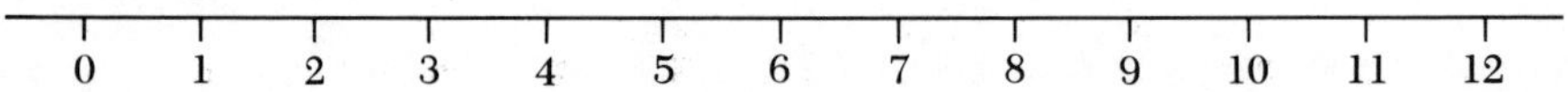

Now suppose we cut a strip of cardboard 3 inches long and place it next to the ruler in position *A*. Its right end is opposite the mark 3, and since the strip is 3 inches long, its left end will be opposite the mark _____.

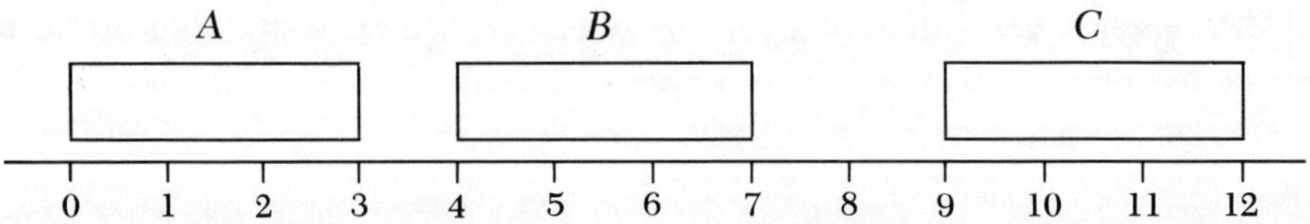

If we move the strip to position *B*, its right end will be opposite the mark _____ and its left end opposite the mark _____. And in position *C*, its right end will be opposite the mark _____ and its left end opposite the mark _____.

---

0 7 4 12 9

---

6 By using a strip of cardboard and a ruler, we can find out some useful subtraction facts. In Frame 5, observing the strip in position A told us that $3-3=0$. Observing the strip in position B told us that $7-3=$ ______, and observing the strip in position C told us that ______ $-3=$ ______.

---

4 12 9

---

7 Now take two of your strips of cardboard labeled 4 and check their length with a ruler. Put them in positions E and F.

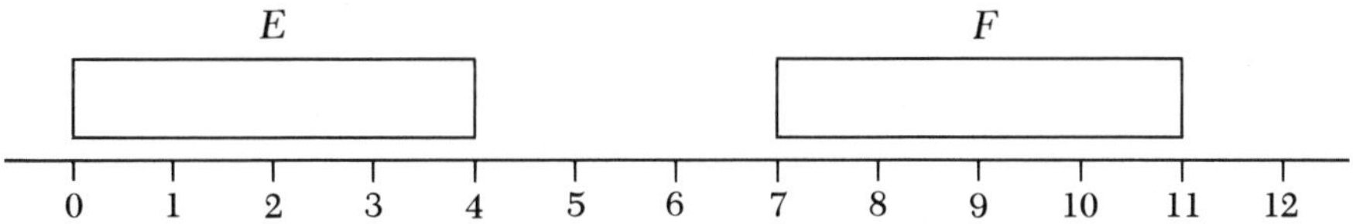

This gives you two subtraction facts:

______ $-4=$ ______ and ______ $-4=$ ______

---

4 0 11 7

---

8 Now take one of your 9-inch strips of cardboard (labeled 9) and one of your 6-inch strips (labeled 6). Place them on a ruler as shown:

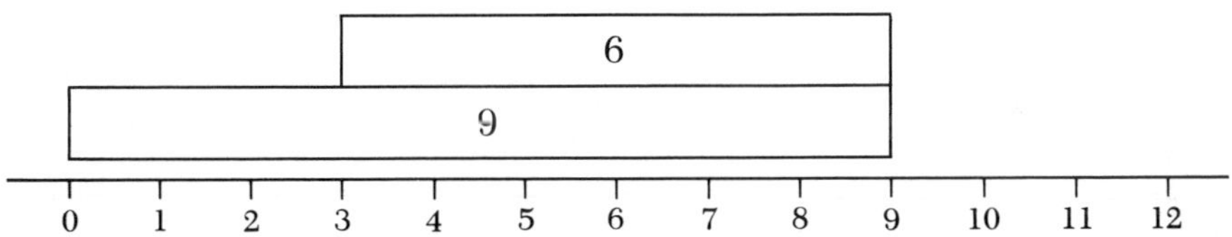

This gives us the subtraction fact $9-6-$ ______. Another way to find this fact doesn't even use a ruler. Find a third strip to add to the 6-inch strip in order to match the 9-inch strip, as shown below:

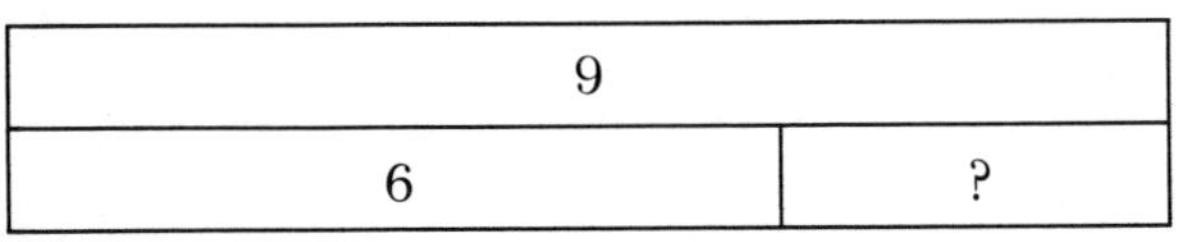

What is the label of this third strip?

---

3 3

---

9 Take one of your 8-inch strips of cardboard (labeled 8) and one of your 5-inch strips (labeled 5). Place them side by side as shown below and find a third strip to complete the match.

8	
5	?

a. What is the label of this third strip?

b. What subtraction fact have you discovered?

_____ − _____ = _____

---

a. 3 *b.* $8 - 5 = 3$

---

10 Use three strips of cardboard to complete the subtraction fact $6 - 2 =$ _____. You used strips _____, _____, and _____.

---

4 6 2 4

---

11 Your strips in Frame 10 should have looked like this:

6	
2	4

This combination of strips gives us two addition facts and two subtraction facts:

$4 + 2 =$ _____ $2 + 4 =$ _____

$6 - 4 =$ _____ $6 - 2 =$ _____

---

6 6 2 4

---

12 Notice how closely connected addition and subtraction are. Find the length of the missing strip below

7	
2	?

and give the four addition and subtraction facts which we can deduce.

a. _____ + _____ = _____ *b.* _____ + _____ = _____

c. _____ − _____ = _____ *d.* _____ − _____ = _____

---

a. $2 + 5 = 7$ *b.* $5 + 2 = 7$ *c.* $7 - 5 = 2$ *d.* $7 - 2 = 5$

---

13 Since our number system is based on 10, we can use a 10 strip and a _____ strip instead of a 16 strip: $10 + \_\_\_\_ = 16$.

---

6 6

---

14 The missing strip below has a length of _____ inches,

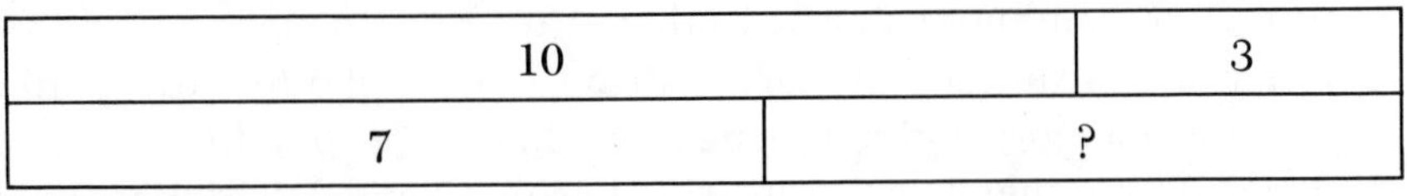

and the four addition and subtraction facts we can deduce are

_____ + _____ = 13 _____ + _____ = 13

_____ − _____ = _____ _____ − _____ = _____

---

6 $6 + 7$ $7 + 6$ $13 - 7 = 6$ $13 - 6 = 7$

---

15 Use four strips of cardboard, including a 10, to complete the subtraction fact $14 - 8 = \_\_\_\_$. You used strips _____, _____, _____, and _____.

---

6 10 4 8 6 Your strips should have looked like this:

---

16 Use your cardboard strips to complete the following subtraction facts:

a. 15 − 9 − _____ *b.* 13 − 4 = _____

c. 11 − 8 = _____ *d.* 17 − 9 = _____

e. 12 − 5 = _____ *f.* 16 − 7 = _____

g. 14 − 9 = _____ *h.* 17 − 12 = _____

i. 13 − 6 = _____ *j.* 16 − 8 = _____

---

a. 6 *b.* 9 *c.* 3 *d.* 8 *e.* 7 *f.* 9 *g.* 5 *h.* 5
i. 7 *j.* 8

---

17 In order to do more complicated subtractions, you will need to memorize all the subtraction facts for numbers under 20. The easiest way to do this is the following:

1. Pick out any two of your cardboard strips.
2. See if you can subtract the shorter strip from the larger strip by using your memory and not using any other strips to help.
3. See if you can subtract the larger strip from the shorter strip + 10 by using your memory and not using any other strips to help.
4. If you cannot remember the answers to steps 2 and 3, find them out by using some of your other strips as we did on page 8.
5. Memorize these answers if you didn't already know them.
6. Repeat steps 1 to 5 until you have memorized the answers to all possible combinations and no longer need the help of the strips.
7. When you have finished all the above steps, answer the questions below from memory:

a. $13 - 9 =$ _____ *b.* $9 - 1 =$ _____

c. $11 - 5 =$ _____ *d.* $17 - 8 =$ _____

e. $12 - 7 =$ _____ *f.* $15 - 8 =$ _____

g. $14 - 6 =$ _____ *h.* $18 - 9 =$ _____

i. $11 - 7 =$ _____ *j.* $16 - 9 =$ _____

---

a. 4 *b.* 8 *c.* 6 *d.* 9 *e.* 5 *f.* 7 *g.* 8 *h.* 9 *i.* 4 *j.* 7 If you made any mistakes, return to the beginning of Frame 17 and repeat the procedure.

---

18 What other subtraction fact can you deduce from $17 - 8 = 9$?

_____ − _____ = _____

And which two addition facts can you deduce?

_____ + _____ = _____

_____ + _____ = _____

$17 - 9 = 8$ $8 + 9 = 17$ $9 + 8 = 17$

---

19 *a.* If I have a pile of 6 oranges and I take away 0 oranges, how many do I end up with?

$6 - 0 =$ ______

b. If I have a pile of 6 oranges and I take away 6 oranges, how many do I end up with?

$6 - 6 =$ ______

---

a. 6; 6 *b.* 0; 0

---

20 *a.* $8 - 0 =$ ______ *b.* $9 - 9 =$ ______

c. $17 - 0 =$ ______ *d.* $13 - 13 =$ ______

---

a. 8 *b.* 0 *c.* 17 *d.* 0

---

21 *a.* If I have a pile of 7 oranges, can I take away 8 oranges?

b. Remembering that we are only concerned with the numbers of counting, that is, 0, 1, 2, 3, 4, . . . , can we answer the question $7 - 8 = ?$

---

a. no *b.* no

---

22 The technical term for the numbers 0, 1, 2, 3, . . . we use in counting is *natural numbers.* And for natural numbers subtraction of a larger number from a smaller number is not defined. Thus $9 - 5 =$ ______, but $5 - 9$ is not defined for ______ numbers.

4; natural

---

23 Complete the following subtraction facts with natural numbers where possible. Where it is not possible, explain why.

a. $17 - 9 =$ ______ *b.* $8 - 13 =$ ______

c. $16 - 16 =$ ______ *d.* $0 - 13 =$ ______

e. $11 - 0 =$ ______

---

a. 8 *b.* impossible for natural numbers because it is not defined
c. 0 *d.* impossible for natural numbers because it is not defined
e. 11

---

24 Two-digit numbers ending in 0 can be subtracted like one-digit numbers. Thus

$7 - 3 = 4$ and $70 - 30 = 40$

Similarly

$80 - 20 =$ _____ $70 - 50 =$ _____ $90 - 10 =$ _____

---

60 20 80

---

25 *a.* $55 =$ _____ $+ 5$
$24 =$ _____ $+ 4$

b. $50 - 20 =$ _____
$5 - 4 =$ _____

c. $55 =$ _____ $+ 5$
$24 =$ _____ $+ 4$
$55 - 24 =$ _____ $+ 1 =$ _____

---

a. 50; 20 *b.* 30; 1 *c.* 50; 20; 30; 31

---

26 *a.* $67 =$ _____ $+ 7$
$32 =$ _____ $+ 2$
$67 - 32 =$ _____ $+ 5 =$ ___

b. $98 =$ _____ $+ 8$
$33 =$ _____ $+ 3$
$98 - 33 =$ _____ $+$ _ $=$ _____

---

a. 60; 30; 30; 35 *b.* 90; 30; $60 + 5 = 65$

---

27 Use the same method to complete the following subtractions:

a. $87 - 36 =$ _____ *b.* $34 - 13 =$ _____

c. $79 - 63 =$ _____ *d.* $57 - 20 =$ _____

---

a. $87 = 80 + 7$
$36 = 30 + 6$
$87 - 36 = 50 + 1 = 51$

b. $34 = 30 + 4$
$13 = 10 + 3$
$34 - 13 = 20 + 1 = 21$

c. $79 = 70 + 9$
$63 = 60 + 3$
$79 - 63 = 10 + 6 = 16$

d. $57 = 50 + 7$
$20 = 20 + 0$
$57 - 20 = 30 + 7 = 37$

---

28 Subtracting 37 from 65 is more difficult:

$65 = \_\_\_\_\_ + 5$
$37 = \_\_\_\_\_ + 7$

but we cannot subtract 7 from 5, so we use 15 instead.

$$\begin{aligned} 65 &= \_\_\_\_\_ + 15 \\ 37 &= \_\_\_\_\_ + \ \ 7 \\ \hline 65 - 37 &= \_\_\_\_\_ + \_\_ = \_\_\_\_\_ \end{aligned}$$

---

60 30 50 30 $20 + 8 = 28$

29

$$\begin{aligned} 42 &= \_\_\_\_\_ + 12 \\ 27 &= \_\_\_\_\_ + \ \ 7 \\ \hline 42 - 27 &= \_\_\_\_\_ + \_\_ = \_\_\_\_\_ \end{aligned}$$

---

30 20 $10 + 5 = 15$

---

30 Use this method to complete the following subtractions:

a. $93 - 47 = \_\_\_\_\_$ *b.* $34 - 17 = \_\_\_\_\_$

c. $88 - 69 = \_\_\_\_\_$ *d.* $70 - 37 = \_\_\_\_\_$

---

a.
$$\begin{aligned} 93 &= 80 + 13 \\ 47 &= 40 + \ \ 7 \\ \hline 93 - 47 &= 40 + \ \ 6 = 46 \end{aligned}$$

b.
$$\begin{aligned} 34 &= 20 + 14 \\ 17 &= 10 + \ \ 7 \\ \hline 34 - 17 &= 10 + \ \ 7 = 17 \end{aligned}$$

c.
$$\begin{aligned} 88 &= 70 + 18 \\ 69 &= 60 + \ \ 9 \\ \hline 88 - 69 &= 10 + \ \ 9 = 19 \end{aligned}$$

d.
$$\begin{aligned} 70 &= 60 + 10 \\ 37 &= 30 + \ \ 7 \\ \hline 70 - 37 &= 30 + \ \ 3 = 33 \end{aligned}$$

---

31 Complete the following subtractions:

a. $88 - 16 = \_\_\_\_\_$ *b.* $88 - 39 = \_\_\_\_\_$

c. $60 - 24 = \_\_\_\_\_$ *d.* $62 - 27 = \_\_\_\_\_$

---

a. 72 *b.* 49 *c.* 36 *d.* 35

You have finished the program for Sec. 1-2. Now take the posttest to see how much you have learned.

POSTTEST 1-2

1. $13 - 8 =$ ______
2. $6 - 0 =$ ______
3. $37 - 23 =$ ______
4. $9 - 9 =$ ______
5. $65 - 28 =$ ______
6. $80 - 31 =$ ______
7. $65 - 25 =$ ______
8. $67 - 38 =$ ______

When you have finished the posttest, correct it yourself using the answers on page 46. Then check all the questions you got wrong on the list below:

1. ______	4. ______	7. ______
2. ______	5. ______	8. ______
3. ______	6. ______	

If you compare your performance with that on Pretest 1-2, you will see how much you have improved. If you made more than one mistake on this posttest, review the material and take the pretest again. Don't move on to the next section until you can get all the questions right.

Now proceed to the next section whose number you circled on page 2. If no other sections were circled, take Final Test 1 (page 44).

Section 1-3
Multiplication of Numbers up to 10

Do this section if you made a mistake in Questions 9 to 15 of Diagnostic Test 1. Begin by taking the pretest below, which will show you how much of Sec. 1-3 you already know and what you will be able to do by the time you have finished it.

PRETEST 1-3

1. $8 \times 7 =$ ______
2. $9 \times 6 =$ ______
3. $7 \times 1 =$ ______
4. $8 \times 8 =$ ______
5. $6 \times 5 =$ ______
6. $7 \times 6 =$ ______
7. $8 \times 9 =$ ______
8. $9 \times 3 =$ ______
9. $5 \times 4 =$ ______
10. $10 \times 3 =$ ______

When you have finished the pretest, correct it yourself using the answers on page 46. Then check all the questions you got wrong on the list below:

1.	______	5.	______	8.	______
2.	______	6.	______	9.	______
3.	______	7.	______	10.	______
4.	______				

If you made any mistakes (even if it was only *one* mistake), read the introduction and go through the program.

If you made no mistakes at all, skip Sec. 1-3, because you already know its material, and go on to Sec. 1-4.

Introduction

In this section you use cardboard strips to discover how to multiply the small natural numbers from 1 to 10. You have to memorize the answers, since they constitute the basic facts of multiplication, often referred to as *multiplication tables.*

Equipment

You will need a set of cardboard strips; instructions for making them are on pages 3 and 4.

FRAMES FOR SECTION 1-3

1 Multiplication by natural numbers is a form of *multiple addition,* and we use it to save time on addition sums. Thus instead of writing $2+2+2+2+2$, we can write $2 \times$ ______.

5

---

2 As with addition, there are a number of basic multiplication facts. We know that if we have 3 piles with 2 objects in each pile, we shall have ______ objects altogether and that this fact can be written in the form $3 \times 2 =$ ______.

---

6 6

---

3 Multiplication facts are usually referred to as multiplication tables, and you will need to memorize some of them. But first let us find out what the facts are. This is something that you can do for yourself.

$7 \times 1 =$ _____

$7 \times 2 = 7 + 7 =$ _____

The result of 7×2 can be obtained by using some of your cardboard strips.

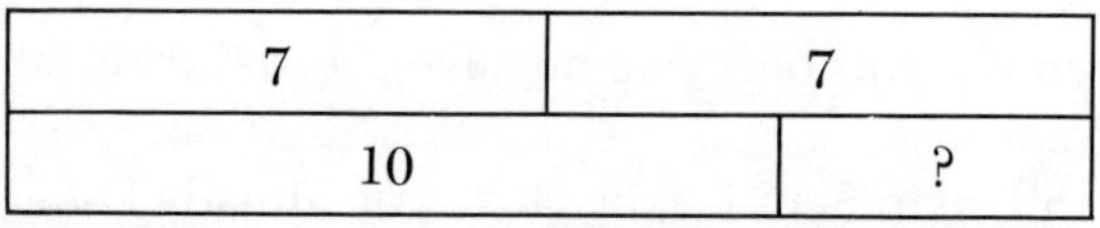

Two 7 strips can be matched by a 10 strip and a _____ strip, so $10 +$ _____ $= 7 \times 2$.

---

7 14 4 4

---

4 To find 7×3, add a third 7 strip:

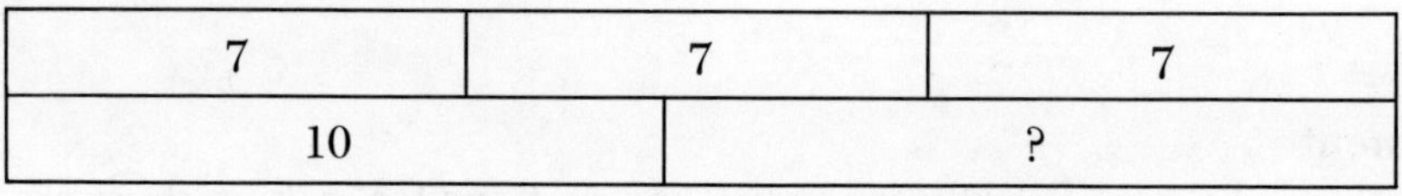

Can you match 7×3 by a single 10 strip and one other strip? Explain your answer. (*Do not* try this question without handling the strips.)

---

no It would need a strip longer than 10, and you do not have one.

---

5 *a.* Can you match 7×3 using two strips and one other strip?

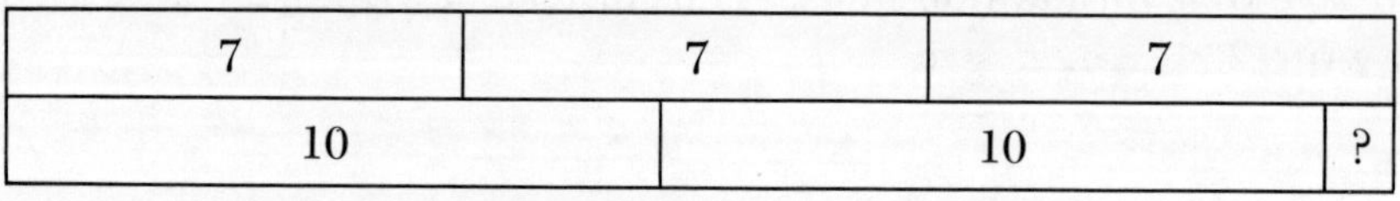

b. If so, what is the other strip?

---

a. yes *b.* a 1 strip

---

6 So $7 \times 3 = 10 + 10 + 1 =$ _____.

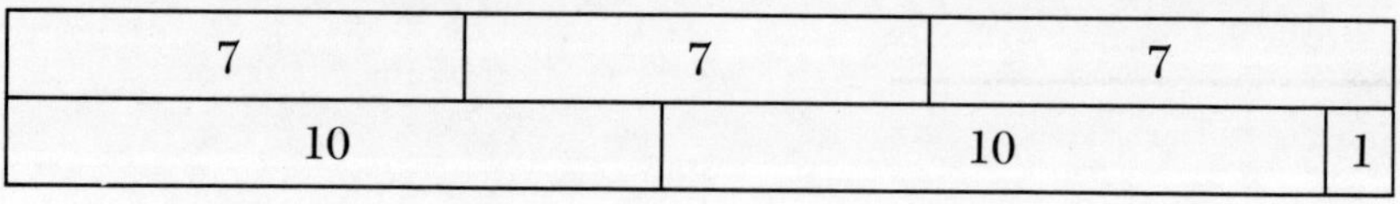

Use the same method to find $7 \times 4 =$ _____.

21 28 Your strips should look like this:

7	7	7	7
10	10	8	

7 Use the same method to find $7 \times 5 =$ ______. (Remember to use as many 10s as possible.)

35

7	7	7	7	7
10	10	10	5	

8 Use the same method to find $7 \times 7 =$ ______.

49

7	7	7	7	7	7	7
10	10	10	10	9		

9 Use the same method to find $7 \times 9 =$ ______.

63

7	7	7	7	7	7	7	7	7
10	10	10	10	10	10	3		

10 We can use this same method with other numbers besides 7. Suppose we want to find 8×4. We take four of the 8 strips and then see how many 10 strips are needed and what other strip is needed.

8	8	8	8
10	10	10	?

$8 \times 4 =$ ______.

32

11 In case you are wondering whether the order of the numbers is important in multiplication, let's see if it affects the answer.

6×8 means take eight of the 6 strips

but

8×6 means take six of the 8 strips

Use strips to find:

a. $6 \times 8 =$ _____ *b.* $8 \times 6 =$ _____

---

a. 48 *b.* 48

6	6	6	6	6	6	6	6

10	10	10	10	8

8	8	8	8	8	8

---

12 Use strips to fill in the following multiplication tables:

a.	*b.*	*c.*
$2 \times 1 =$ _____	$3 \times 1 =$ _____	$4 \times 1 =$ _____
$2 \times 2 =$ _____	$3 \times 2 =$ _____	$4 \times 2 =$ _____
$2 \times 3 =$ _____	$3 \times 3 =$ _____	$4 \times 3 =$ _____
$2 \times 4 =$ _____	$3 \times 4 =$ _____	$4 \times 4 =$ _____
$2 \times 5 =$ _____	$3 \times 5 =$ _____	$4 \times 5 =$ _____
$2 \times 6 =$ _____	$3 \times 6 =$ _____	$4 \times 6 =$ _____
$2 \times 7 =$ _____	$3 \times 7 =$ _____	$4 \times 7 =$ _____
$2 \times 8 =$ _____	$3 \times 8 =$ _____	$4 \times 8 =$ _____
$2 \times 9 =$ _____	$3 \times 9 =$ _____	$4 \times 9 =$ _____
$2 \times 10 =$ _____	$3 \times 10 =$ _____	$4 \times 10 =$ _____

---

a.	*b.*	*c.*
2	3	4
4	6	8
6	9	12
8	12	16
10	15	20
12	18	24
14	21	28
16	24	32
18	27	36
20	30	40

---

13 Now use your strips to fill in the following multiplication tables:

a.	*b.*	*c.*
$5 \times 1 =$ ____	$6 \times 1 =$ ____	$7 \times 1 =$ ____
$5 \times 2 =$ ____	$6 \times 2 =$ ____	$7 \times 2 =$ ____
$5 \times 3 =$ ____	$6 \times 3 =$ ____	$7 \times 3 =$ ____
$5 \times 4 =$ ____	$6 \times 4 =$ ____	$7 \times 4 =$ ____
$5 \times 5 =$ ____	$6 \times 5 =$ ____	$7 \times 5 =$ ____
$5 \times 6 =$ ____	$6 \times 6 =$ ____	$7 \times 6 =$ ____
$5 \times 7 =$ ____	$6 \times 7 =$ ____	$7 \times 7 =$ ____
$5 \times 8 =$ ____	$6 \times 8 =$ ____	$7 \times 8 =$ ____
$5 \times 9 =$ ____	$6 \times 9 =$ ____	$7 \times 9 =$ ____
$5 \times 10 =$ ____	$6 \times 10 =$ ____	$7 \times 10 =$ ____

---

a.	*b.*	*c.*
5	6	7
10	12	14
15	18	21
20	24	28
25	30	35
30	36	42
35	42	49
40	48	56
45	54	63
50	60	70

---

14 Use strips to fill in the following tables:

a.	*b.*	*c.*
$8 \times 1 =$ ____	$9 \times 1 =$ ____	$10 \times 1 =$ ____
$8 \times 2 =$ ____	$9 \times 2 =$ ____	$10 \times 2 =$ ____
$8 \times 3 =$ ____	$9 \times 3 =$ ____	$10 \times 3 =$ ____
$8 \times 4 =$ ____	$9 \times 4 =$ ____	$10 \times 4 =$ ____
$8 \times 5 =$ ____	$9 \times 5 =$ ____	$10 \times 5 =$ ____
$8 \times 6 =$ ____	$9 \times 6 =$ ____	$10 \times 6 =$ ____
$8 \times 7 =$ ____	$9 \times 7 =$ ____	$10 \times 7 =$ ____
$8 \times 8 =$ ____	$9 \times 8 =$ ____	$10 \times 8 =$ ____
$8 \times 9 =$ ____	$9 \times 9 =$ ____	$10 \times 9 =$ ____
$8 \times 10 =$ ____	$9 \times 10 =$ ____	$10 \times 10 =$ ____

a.	*b.*	*c.*
8	9	10
16	18	20
24	27	30
32	36	40
40	45	50
48	54	60
56	63	70
64	72	80
72	81	90
80	90	100

15 You will now need to memorize all these multiplication tables. This is not as difficult as it sounds. The following procedure is a good one.

1. Pick out any two cardboard strips from your pile.
2. See if you can multiply the two numbers on them by using your memory and without using other strips to help.
3. If you cannot remember the answer, find it by using some of your other strips as we did on page 8.
4. Memorize the answer if you didn't already know it.
5. Repeat steps 1 to 4 until you have memorized the answers to all possible combinations and no longer need the help of your strips.
6. When you have finished all the above steps, answer the questions below from memory:

a. $9 \times 8 =$ _____ *b.* $4 \times 7 =$ _____

c. $6 \times 9 =$ _____ *d.* $3 \times 8 =$ _____

e. $9 \times 7 =$ _____ *f.* $6 \times 5 =$ _____

g. $9 \times 9 =$ _____ *h.* $7 \times 8 =$ _____

a. 72 *b.* 28 *c.* 54 *d.* 24 *e.* 63 *f.* 30 *g.* 81 *h.* 56 If you made no mistakes, skip to the posttest. If you made any mistakes at all, continue with Frame 16.

16 It seems that your memorizing isn't quite perfect yet! Go back and repeat the procedure outlined in Frame 15, and when you are quite sure that you have memorized all the answers, do the questions below:

a. $8 \times 8 =$ _____ *b.* $9 \times 5 =$ _____

c. $7 \times 6 =$ _____ *d.* $8 \times 9 =$ _____

e. $7 \times 7 =$ _____ *f.* $9 \times 6 =$ _____

g. $10 \times 7 =$ _____ *h.* $6 \times 4 =$ _____

---

a. 64 *b.* 45 *c.* 42 *d.* 72 *e.* 49 *f.* 54 *g.* 70 *h.* 24

You have finished the program for Sec. 1-3. Now take the posttest to see how much you have learned.

POSTTEST 1-3

1. $7 \times 9 =$ _____ 2. $6 \times 8 =$ _____

3. $1 \times 6 =$ _____ 4. $9 \times 9 =$ _____

5. $8 \times 5 =$ _____ 6. $6 \times 6 =$ _____

7. $6 \times 9 =$ _____ 8. $7 \times 8 =$ _____

9. $3 \times 8 =$ _____ 10. $10 \times 4 =$ _____

When you have finished the posttest, correct it yourself using the answers on page 46. Then check all the questions you got wrong on the list below:

1. 5. _____ 8. _____

2. _____ 6. _____ 9. _____

3. _____ 7. _____ 10. _____

4. _____

If you compare your performance with that on Pretest 1-3, you will see how much you have improved. If you made more than one mistake on this posttest, review the material and take the pretest again. Don't move on to the next section until you can get all the questions right.

Now proceed to Sec. 1-4.

Section 1-4
Simple Division

Do this section if you made a mistake in Questions 9 to 15 of Diagnostic Test 1. Begin by taking the pretest below, which will show you how much of Sec. 1-4 you already know and what you will be able to do by the time you have finished it.

PRETEST 1-4

1. $0 \times 6 =$ _____
2. $7 \div 1 =$ _____
3. $30 \div 5 =$ _____
4. $8 \div 8 =$ _____
5. $63 \div 9 =$ _____
6. $0 \div 4 =$ _____
7. $42 \div 6 =$ _____
8. $70 \div 10 =$ _____

When you have finished the pretest, correct it yourself using the answers on page 46. Then check all the questions you got wrong on the list below:

1. _____
2. _____
3. _____
4. _____
5. _____
6. _____
7. _____
8. _____

If you made any mistakes (even if it was only *one* mistake), read the introduction and go through the program.

If you made no mistakes at all, skip Sec. 1-4, because you already know its material, and go on to the next section whose number you circled on page 2. If no further numbers were circled, take Final Test 1 (page 44).

Introduction

In this section you will learn how to derive division facts from multiplication facts. You also learn about multiplications and divisions involving 0.

FRAMES FOR SECTION 1-4

1 *a.* If we have 7 boxes each containing 8 apples, how many apples do we have altogether?

b. If we have 8 boxes each containing 7 apples, how many apples do we have altogether?

---

a. 56 *b.* 56

---

2 *a.* If we have 5 boxes each containing 4 apples, how many apples do we have altogether?

b. If we have 4 boxes each containing 5 apples, how many apples do we have altogether?

---

a. 20 *b.* 20

---

3 These results are related to our discovery in Sec. 1-3 that the order of the two numbers does not affect their product, i.e., that $8 \times 7 = 7 \times$ ______ and $4 \times 5 =$ ______ $\times 4$.

---

8 5

---

4 *a.* If we have 3 boxes each containing 8 apples, how many apples do we have altogether?

b. If we have 3 boxes each containing 5 apples, how many apples do we have altogether?

c. If we have 3 boxes each containing 0 apples, how many apples do we have altogether?

---

a. 24 *b.* 15 *c.* 0

---

5 $8 \times 3 =$ ______ $5 \times 3 =$ ______ $0 \times 3 =$ ______

---

24 15 0

---

6 *a.* If we have 9 boxes each containing 0 apples, how many apples do we have altogether?

b. $0 \times 9 =$ ______

---

a. 0 *b.* 0

---

7 *a.* If we have 63 boxes each containing 0 apples, how many apples do we have altogether?

b. $0 \times 63 =$ ______

---

a. 0 *b.* 0

---

8 In general, if n is a natural number, $0 \times n =$ ______.

0

---

9 Now since we have shown that the order of two numbers does not affect their product,

$9 \times 6 = 6 \times$ ______ and $9 \times 0 = 0 \times$ ______

Therefore $9 \times 0 =$ ______.

---

9 9 0

---

10 Similarly $5 \times 0 =$ ______ and $7 \times 0 =$ ______, and if n is a natural number, $n \times 0 =$ ______.

---

0 0 0

---

11 *a.* Suppose we have 20 apples and we want to divide them equally into 5 boxes; how many apples should we put into each box?

b. Suppose we have 20 apples and we want to divide them equally into 4 boxes; how many apples should we put into each box? (Refer back to Frame 2 if you have any difficulty.)

---

a. 4 *b.* 5

---

12 Notice how these two division facts

$20 \div 5 =$ ______ and $20 \div 4 =$ ______

are derived from the multiplication facts

$5 \times 4 =$ ______ and $4 \times 5 =$ ______

---

4 5 20 20

---

13 What other multiplication fact and what two division facts can be derived from $3 \times 7 = 21$?

$7 \times 3 = 21$ $21 \div 7 = 3$ $21 \div 3 = 7$

---

14 Division and multiplication facts are so closely related that remembering multiplication facts alone is sufficient. You only have to remember that $9 \times$ ______ $= 36$ in order to answer $36 \div 9 =$ ______; and remembering that $6 \times$ ______ $= 42$ will answer $42 \div 6 =$ ______.

---

4 4 7 7

---

15 Use your knowledge of multiplication facts to complete the following:

a. $72 \div 8 =$ ______ *b.* $45 \div 9 =$ ______

c. $63 \div 7 =$ ______ *d.* $60 \div 10 =$ ______

e. $54 \div 6 =$ ______ *f.* $35 \div 5 =$ ______

g. $32 \div 4 =$ ______ *h.* $56 \div 7 =$ ______

---

a. 9 *b.* 5 *c.* 9 *d.* 6 *e.* 9 *f.* 7 *g.* 8 *h.* 8
Check your answers very carefully. If you made no mistakes, skip to Frame 17. If you made any mistakes at all, continue with Frame 16.

---

16 Examine all your mistakes in Frame 15 carefully and see why the printed answers are correct. Review your multiplication tables if you think you have forgotten any of them. You can find the tables in the answers to Frames 12 to 14 of Sec. 1-3. Then answer the following questions:

a. $81 \div 9 =$ ______ *b.* $42 \div 6 =$ ______

c. $56 \div 8 =$ ______ *d.* $40 \div 4 =$ ______

e. $63 \div 9 =$ ______ *f.* $49 \div 7 =$ ______

g. $45 \div 5 =$ ______ *h.* $64 \div 8 =$ ______

---

a. 9 *b.* 7 *c.* 7 *d.* 10 *e.* 7 *f.* 7 *g.* 9 *h.* 8

---

17 *a.* $5 \times 6 = 30$, so $30 \div 6 =$ ______.

b. $4 \times 7 = 28$, so $28 \div 7 =$ ______.

c. $0 \times 3 = 0$, so $0 \div 3 =$ ______.

---

a. 5 *b.* 4 *c.* 0

---

18 Since $0 \times 5 = 0$, $0 \div 5 =$ ______, and since $0 \times 9 = 0$, $0 \div 9 =$ ______; similarly $0 \div 6 =$ ______, and $0 \div 10 =$ ______.

---

0 0 0 0

---

19 But can we find an answer to $24 \div 0$? To find $24 \div 3$, we recall the multiplication fact $3 \times$ _____ $= 24$; and to find $24 \div 6$, we recall the multiplication fact $6 \times$ _____ $= 24$.

a. But can you find a number to fit the multiplication $0 \times$ _____ $= 24$?

b. If not, why not?

---

8 4 *a.* no *b.* Because any number multiplied by 0 gives 0, and it is impossible to find a number which when multiplied by 0 gives 24.

---

20 So $24 \div 0$ is *not defined* because there is no multiplication fact from which it would be possible to derive a value for it. Complete the following division facts, where possible. Where it is not possible, explain why.

a. $36 \div 0 =$ _____ *b.* $10 \div 2 =$ _____

c. $0 \div 7 =$ _____ *d.* $30 \div 10 =$ _____

e. $4 \div 0 =$ _____

---

a. impossible, because division by 0 is not defined *b.* 5 *c.* 0 *d.* 3 *e.* impossible, because division by 0 is not defined

---

21 $7 \times 2 = 14$ so $14 \div 2 =$ _____ and $14 \div 7 =$ _____.

$7 \times 1 = 7$ so $7 \div 1 =$ _____ and $7 \div 7 =$ _____.

Similarly

$8 \div 1 =$ _____ and $8 \div 8 =$ _____

$3 \div 1 =$ _____ and $3 \div 3 =$ _____

---

7 2 7 1 8 1 3 1

---

22 Complete the following divisions where possible. Where it is not possible, explain why.

a. $0 \div 6 =$ _____ *b.* $9 \div 1 =$ _____

c. $12 \div 0 =$ _____ *d.* $10 \div 10 =$ _____

---

a. 0 *b.* 9 *c.* impossible because division by 0 is not defined *d.* 1

You have finished the program for Sec. 1-4. Now take the posttest to see how much you have learned.

POSTTEST 1-4

1. $0 \times 5 =$ ______
2. $6 \div 1 =$ ______
3. $40 \div 5 =$ ______
4. $9 \div 9 =$ ______
5. $36 \div 6 =$ ______
6. $0 \div 7 =$ ______
7. $54 \div 9 =$ ______
8. $80 \div 10 =$ ______

When you have finished the posttest, correct it yourself using the answers on page 46. Then check all the questions you got wrong on the list below:

1. ______	4. ______	7. ______
2. ______	5. ______	8. ______
3. ______	6. ______	

If you compare your performance with that on Pretest 1-4, you will see how much you have improved. If you made more than one mistake on this posttest, review the material and take the pretest again. Don't move on to the next section until you can get all the questions right.

Now proceed to the next section whose number you circled on page 2. If no other sections were circled, take Final Test 1 (page 44).

Section 1-5 Some Operations with Three-digit Numbers

Do this section if you made a mistake in Questions 16 to 20 of Diagnostic Test 1. Begin by taking the pretest below, which will show you how much of Sec. 1-5 you already know and what you will be able to do by the time you have finished it.

PRETEST 1-5

1. $371 + 283 =$ ______
2. $538 + 395 =$ ______
3. $481 - 256 =$ ______
4. $916 - 764 =$ ______
5. $373 - 193 =$ ______
6. $90 \times 9 =$ ______
7. $35 \times 10 =$ ______
8. $360 \div 4 =$ ______

When you have finished the pretest, correct it yourself using the answers on page 47. Then check all the questions you got wrong on the list below:

1. _____ 4. _____ 7. _____

2. _____ 5. _____ 8. _____

3. _____ 6. _____

If you made any mistakes (even if it was only *one* mistake), read the introduction and go through the program.

If you made no mistakes at all, skip Sec. 1-5, because you already know its material, and go on to take Final Test 1 (page 44).

Introduction

In this section you will learn how to add and subtract three-digit numbers and how to multiply and divide two- and three-digit numbers by 10. You will also learn how to multiply two-digit numbers ending in 0 by one-digit numbers and how to perform the corresponding divisions.

FRAMES FOR SECTION 1-5

1 In Sec. 1-1 we showed that two-digit numbers ending in 0 could be added like one-digit numbers. Thus $70 + 20 =$ _____ because $7 + 2 =$ _____, and $30 + 40 =$ _____ because $3 + 4 =$ _____.

---

90 9 70 7

---

2 The same principle applies to addition of three-digit numbers ending in 0. Since $12 + 33 =$ _____, $120 + 330 =$ _____, and since $27 + 55 =$ _____, $270 + 550 =$ _____.

---

45 450 82 820

---

3 Add the following numbers:

a. $230 + 640 =$ _____ *b.* $520 + 390 =$ _____

c. $340 + 460 =$ _____

---

a. 870 *b.* 910 *c.* 800

---

4 By adapting the addition process we used for two-digit numbers in Sec. 1-1, we can add three-digit numbers ending in 0 directly. Thus

a.
$$\begin{array}{r} 620 = 600 + 20 \\ 270 = \_\_\_ + 70 \\ \hline 620 + 270 = \_\_\_ + 90 = \_\_\_\_\_ \end{array}$$

b.
$$\begin{array}{r} 540 = \_\_\_ + 40 \\ 180 = \_\_\_ + 80 \\ \hline 540 + 180 = \_\_\_ + 120 = \_\_\_\_\_ \end{array}$$

---

a. 200; 800; 890 *b.* 500; 100; 600; 720

---

5 Use the same method to add the following:

a. 510 + 380 *b.* 260 + 470 *c.* 350 + 550

---

a.
$$\begin{array}{r} 510 = 500 + 10 \\ 380 = 300 + 80 \\ \hline 510 + 380 = 800 + 90 = 890 \end{array}$$

b.
$$\begin{array}{r} 260 = 200 + 60 \\ 470 = 400 + 70 \\ \hline 260 + 470 = 600 + 130 = 730 \end{array}$$

c.
$$\begin{array}{r} 350 = 300 + 50 \\ 550 = 500 + 50 \\ \hline 350 + 550 = 800 + 100 = 900 \end{array}$$

---

6 This method can be extended to three-digit numbers that do not end in 0 by separating all three digits, adding, and recombining. Thus

a.
$$\begin{array}{r} 564 = 500 + 60 + 4 \\ 232 = 200 + \_\_ + 2 \\ \hline 564 + 232 = 700 + \_\_ + 6 = \_\_\_\_\_ \end{array}$$

b.
$$\begin{array}{r} 493 = 400 + \_\_\_ + 3 \\ 288 = 200 + \_\_\_ + 8 \\ \hline 493 + 288 = 600 + \qquad + 11 = 600 + \_\_\_ = \_\_\_\_\_ \end{array}$$

---

a. 30; 90; 796 *b.*
$$\begin{array}{r} 400 + 90 + 3 \\ 200 + 80 + 8 \\ \hline 600 + 170 + 11 = 600 + 181 = 781 \end{array}$$

---

7 *a.*
$$\begin{array}{r} 586 = 500 + 80 + \_\_ \\ 297 = \_\_\_ + 90 + 7 \\ \hline 586 + 297 = \_\_\_ + 170 + \_\_ = \_\_\_\_\_ \end{array}$$

b. Use this method to add 278 and 396.

---

a. 6; 200; 700 + 170 + 13 = 883

b.
$$\begin{array}{r} 278 = 200 + 70 + 8 \\ 396 = 300 + 90 + 6 \\ \hline 278 + 396 = 500 + 160 + 14 = 674 \end{array}$$

---

8 Use this method to add the following:

a. $431 + 479 =$ _____ *b.* $218 + 437 =$ _____

c. $592 + 195 =$ _____

---

a.
$$\begin{aligned} 431 &= 400 + 30 + 1 \\ 479 &= 400 + 70 + 9 \\ \hline 431 + 479 &= 800 + 100 + 10 = 910 \end{aligned}$$

b.
$$\begin{aligned} 218 &= 200 + 10 + 8 \\ 437 &= 400 + 30 + 7 \\ \hline 218 + 437 &= 600 + 40 + 15 = 655 \end{aligned}$$

c.
$$\begin{aligned} 592 &= 500 + 90 + 2 \\ 195 &= 100 + 90 + 5 \\ \hline 592 + 195 &= 600 + 180 + 7 = 787 \end{aligned}$$

---

9 In Sec. 1-2 we showed that two-digit numbers ending in 0 could be subtracted like one-digit numbers. Thus $50 - 10 =$ _____ because $5 - 1 =$ _____, and $70 - 20 =$ _____ because $7 - 2 =$ _____.

---

40 4 50 5

---

10 This same principle applies to three-digit numbers ending in 0 being subtracted like two-digit numbers; and we can modify the methods of Sec. 1-2 accordingly. Thus

a.
$$\begin{aligned} 550 &= 500 + 50 \\ 410 &= \_\_\_ + 10 \\ \hline 550 - 410 &= \_\_\_ + 40 = \_\_\_\_\_ \end{aligned}$$

b.
$$\begin{aligned} 690 &= \_\_\_ + 90 \\ 260 &= \_\_\_ + 60 \\ \hline 690 - 260 &= \_\_\_ + 30 = \_\_\_\_\_ \end{aligned}$$

---

a. 400; 100; 140 *b.* 600; 200; 400; 430

---

11 Use the same method to subtract the following:

a. $850 - 540 =$ _____ *b.* $780 - 350 =$ _____

---

a.
$$\begin{aligned} 850 &= 800 + 50 \\ 540 &= 500 + 40 \\ \hline 850 - 540 &= 300 + 10 = 310 \end{aligned}$$

b.
$$\begin{aligned} 780 &= 700 + 80 \\ 350 &= 300 + 50 \\ \hline 780 - 350 &= 400 + 30 = 430 \end{aligned}$$

---

12 Extending the method to other three-digit numbers, we have:

a.

$$
\begin{aligned}
864 &= 800 + 60 + 4\\
712 &= 700 + \_\_ + 2\\
\hline
864 - 712 &= 100 + \_\_ + 2 = \_\_\_
\end{aligned}
$$

b.

$$
\begin{aligned}
495 &= 400 + \_\_ + 5\\
182 &= 100 + \_\_ + 2\\
\hline
495 - 182 &= 300 + \_\_ + 3 = \_\_\_
\end{aligned}
$$

---

a. 10; 50; 152 *b.* 90; 80; 10; 313

---

13 Use this method to subtract the following:

a. 976 − 755 *b.* 328 − 127

---

a.

$$
\begin{aligned}
976 &= 900 + 70 + 6\\
755 &= 700 + 50 + 5\\
\hline
976 - 755 &= 200 + 20 + 1 = 221
\end{aligned}
$$

b.

$$
\begin{aligned}
328 &= 300 + 20 + 8\\
127 &= 100 + 20 + 7\\
\hline
328 - 127 &= 200 + 0 + 1 = 201
\end{aligned}
$$

---

14 Use this method to subtract the following:

a. 856 − 805 *b.* 824 − 220

---

a.

$$
\begin{aligned}
856 &= 800 + 50 + 6\\
805 &= 800 + 0 + 5\\
\hline
856 - 805 &= 0 + 50 + 1 = 51
\end{aligned}
$$

b.

$$
\begin{aligned}
824 &= 800 + 20 + 4\\
220 &= 200 + 20 + 0\\
\hline
824 - 220 &= 600 + 0 + 4 = 604
\end{aligned}
$$

---

15 When there are larger digits in the second number (the number we are subtracting), we can rewrite the first number and make use of subtraction facts, such as 17 − 9 = _____ and 170 − 90 = _____. So to find 874 − 493 we write

$$
\begin{aligned}
874 &= 700 + 170 + 4\\
493 &= 400 + \_\_\_ + 3\\
\hline
874 - 493 &= 300 + \_\_\_ + 1 = \_\_\_
\end{aligned}
$$

---

8 80 90 80 381

---

16 *a.* Similarly we can use the subtraction fact that $160 - 70 =$ _____ to find $765 - 374$.

$$\begin{array}{r} 765 = 600 + \_\_\_ + 5 \\ 374 = 300 + \_\_\_ + 4 \\ \hline 765 - 374 = 300 + \_\_\_ + 1 = \_\_\_ \end{array}$$

b. We can use the subtraction fact that $14 - 8 =$ _____ to find $434 - 218$.

$$\begin{array}{r} 434 = 400 + 20 + \_ \\ 218 = 200 + 10 + \_ \\ \hline 434 - 218 = 200 + 10 + \_ = \_\_\_ \end{array}$$

---

a. 90; 160; 70; 90; 391 *b.* 6; 14; 8; 6; 216

---

17 Use these methods to find the following:

a. $843 - 819$ *b.* $517 - 345$

---

a.

$$\begin{array}{r} 843 = 800 + 30 + 13 \\ 819 = 800 + 10 + 9 \\ \hline 843 - 819 = 0 + 20 + 4 = 24 \end{array}$$

b.

$$\begin{array}{r} 517 = 400 + 110 + 7 \\ 345 = 300 + 40 + 5 \\ \hline 517 - 345 = 100 + 70 + 2 = 172 \end{array}$$

---

18 Use these methods to find the following:

a. $805 - 394$ *b.* $690 - 273$

---

a.

$$\begin{array}{r} 805 = 700 + 100 + 5 \\ 394 = 300 + 90 + 4 \\ \hline 805 - 394 = 400 + 10 + 1 = 411 \end{array}$$

b.

$$\begin{array}{r} 690 = 600 + 80 + 10 \\ 273 = 200 + 70 + 3 \\ \hline 690 - 273 = 400 + 10 + 7 = 417 \end{array}$$

---

19 So far we have noted that addition and subtraction of two- and three-digit numbers ending in 0 is analogous to that of the corresponding one- and two-digit numbers. Thus

a. $2 + 7 =$ _____, $20 + 70 =$ _____, and $200 + 700 =$ _____.

b. $9 - 3 =$ _____, $90 - 30 =$ _____, and $900 - 300 =$ _____.

a. 9; 90; 900   *b.* 6; 60; 600

20 Multiplication, which is multiple addition, must therefore be similar. Thus

a. $2 \times 2 =$ ______, $20 \times 2 =$ ______, and $200 \times 2 =$ ______.

b. $5 \times 7 =$ ______, and $50 \times 7 =$ ______.

a. 4; 40; 400   *b.* 35; 350

21 *a.* $80 \times 6 =$ ______   *b.* $50 \times 8 =$ ______

c. $30 \times 4 =$ ______

a. 480   *b.* 400   *c.* 120

22 Division follows from multiplication. Since $30 \times 4 = 120$, we know that $120 \div 4 =$ ______, and since $50 \times 8 = 400$, we know that $400 \div 8 =$ ______. Similarly $480 \div 6 =$ ______.

30   50   80

23 *a.* $540 \div 9 =$ ______   *b.* $630 \div 7 =$ ______   *c.* $320 \div 8 =$ ______

a. 60   *b.* 90   *c.* 40

24 *a.* $2 \times 10 =$ ______   *b.* $40 \div 10 =$ ______

$5 \times 10 =$ ______ $60 \div 10 =$ ______

$7 \times 10 =$ ______ $80 \div 10 =$ ______

a. 20; 50; 70   *b.* 4; 6; 8

25 Multiplying a number by 10 is equivalent to putting a ______ on the end, and dividing a number by 10 is equivalent to taking a ______ off the end. So

$16 \times 10 =$ ______   and $490 \div 10 =$ ______

$23 \times 10 =$ ______   and $530 \div 10 =$ ______

0   0   160   49   230   53

26 *a.* $25 \times 10 =$ _____ *b.* $50 \times 9 =$ _____

c. $70 \times 3 =$ _____ *d.* $190 \div 10 =$ _____

e. $280 \div 7 =$ _____ *f.* $560 \div 8 =$ _____

---

a. 250 *b.* 450 *c.* 210 *d.* 19 *e.* 40 *f.* 70

You have finished the program for Sec. 1-5. Now take the posttest to see how much you have learned.

POSTTEST 1-5

1. $567 + 392 =$ _____
2. $285 + 497 =$ _____
3. $763 - 548 =$ _____
4. $527 - 283 =$ _____
5. $664 - 394 =$ _____
6. $80 \times 7 =$ _____
7. $59 \times 10 =$ _____
8. $640 \div 8 =$ _____

When you have finished the posttest, correct it yourself using the answers on page 47. Then check all the questions you got wrong on the list below:

1. _____ 4. _____ 7. _____

2. _____ 5. _____ 8. _____

3. _____ 6. _____

If you compare your performance with that on Pretest 1-5, you will see how much you have improved. If you made more than one mistake on this posttest, review the material and take the pretest again.

Now proceed to Final Test 1.

FINAL TEST 1

---

1. $26 + 67 =$ _____
2. $15 + 55 =$ _____
3. $0 + 32 =$ _____
4. $23 + 39 =$ _____
5. $85 - 27 =$ _____
6. $70 - 33 =$ _____
7. $91 - 52 =$ _____
8. $74 - 74 =$ _____
9. $9 \times 8 =$ _____
10. $8 \times 4 =$ _____
11. $7 \times 5 =$ _____
12. $63 \div 7 =$ _____
13. $6 \div 6 =$ _____
14. $9 \times 0 =$ _____
15. $0 \div 4 =$ _____
16. $256 + 587 =$ _____

17. $739 - 279 =$ ______ 18. $942 - 638 =$ ______

19. $50 \times 6 =$ ______ 20. $720 \div 9 =$ ______

The answers to this test are on page 47. Check the questions you got wrong on the list below and compare your results with those on Diagnostic Test 1 to see how much you have improved.

1.	______	5.	______	9.	______	16.	______
2.	______	6.	______	10.	______	17.	______
3.	______	7.	______	11.	______	18.	______
4.	______	8.	______	12.	______	19.	______
				13.	______	20.	______
				14.	______		
				15.	______		

If you made more than one mistake, review the material and then take the diagnostic test again. Make sure that you can get all the questions right before you move on to the next chapter.

Now proceed to Chap. 2, Multiple Operations with Natural Numbers.

ANSWERS TO TESTS FOR CHAPTER 1

DIAGNOSTIC TEST 1

1. 83	2. 80	3. 18	4. 94
5. 36	6. 39	7. 26	8. 0
9. 42	10. 81	11. 45	12. 7
13. 1	14. 0	15. 0	16. 732
17. 320	18. 304	19. 400	20. 90

PRETEST 1-1

1. 15
2. 4
3. 48
4. 84
5. 92
6. 23
7. 87
8. 80

POSTTEST 1-1

1. 15
2. 5
3. 66
4. 83
5. 93
6. 58
7. 76
8. 90

PRETEST 1-2

1. 8
2. 8
3. 11
4. 0
5. 24
6. 48
7. 60
8. 19

POSTTEST 1-2

1. 5
2. 6
3. 14
4. 0
5. 37
6. 49
7. 40
8. 29

PRETEST 1-3

1. 56
2. 54
3. 7
4. 64
5. 30
6. 42
7. 72
8. 27
9. 20
10. 30

POSTTEST 1-3

1. 63
2. 48
3. 6
4. 81
5. 40
6. 36
7. 54
8. 56
9. 24
10. 40

PRETEST 1-4

1. 0
2. 7
3. 6
4. 1
5. 7
6. 0
7. 7
8. 7

POSTTEST 1-4

1. 0
2. 6
3. 8
4. 1
5. 6
6. 0
7. 6
8. 8

PRETEST 1-5

1. 654
2. 933
3. 225
4. 152
5. 180
6. 810
7. 350
8. 90

POSTTEST 1-5

1. 959
2. 782
3. 215
4. 244
5. 270
6. 560
7. 590
8. 80

FINAL TEST 1

1. 93	2. 70	3. 32	4. 62
5. 58	6. 37	7. 39	8. 0
9. 72	10. 32	11. 35	12. 9
13. 1	14. 0	15. 0	16. 843
17. 460	18. 304	19. 300	20. 80

Chapter 2
Multiple Operations with Natural Numbers

We have already defined single arithmetic operations for small natural numbers. In this chapter we extend our definitions to cover combinations of operations. We have to do this before we can learn to carry out arithmetic operations on large natural numbers because our methods of adding, subtracting, multiplying, and dividing large numbers all depend on combinations of simple operations on small numbers.

Section 2-1 introduces the commutative principle, and Sec. 2-2 introduces the associative principle. As you will see, these two principles are extremely important throughout algebra. Section 2-2 also introduces the use of parentheses in defining multiple operations. Sections 2-3 and 2-4 introduce the distributive laws, which are the basis of the multiplication and division of large numbers and are also important in algebra.

DIAGNOSTIC TEST 2

1. The commutative principle applied to multiplication means that $A \times B =$ ______.
2. Which of the following operations are commutative?

   a. Addition *b.* Subtraction *c.* Division
3. The associative principle applied to addition means that $(A + B) + C =$ ______.
4. Which of the following operations are associative?

   a. Subtraction *b.* Multiplication *c.* Division
5. $18 + (6 \div 3) =$ ______
6. $(11 - 5) \times 2 =$ ______
7. $4 \cdot 9 - 3 \cdot 5 =$ ______
8. The distributive law applied to division over subtraction states that $(A - B) \div C =$ __________.
9. Which of the following statements are true?

   a. $(A - B) \times C = (A \times C) - (B \times C)$

   b. $(A \div B) \times C = (A \times C) \div (B \times C)$

   c. $(A \div B) - C = (A - C) \div (B - C)$

   d. $(A \times B) + C = (A + C) \times (B + C)$

Use the distributive law to find:

10. $29 \times 7 =$ ______
11. $90 \div 6 =$ ______
12. $7(8 + 9) =$ ______

When you have finished the diagnostic test, correct it yourself using the answers on page 78. Then check all the questions you got wrong on the list below:

1. ______	3. ______	8. ______
2. ______	4. ______	9. ______
	5. ______	10. ______
	6. ______	11. ______
	7. ______	12. ______

If you made no mistakes at all, you need not do any work in this chapter, so go straight on to Chap. 3. But if you made any mistakes (even if it was only *one* mistake), follow these instructions:

	section	page
If you made any mistakes in Question 1 or 2, circle Sec. 2-1.	2-1	50
If you made any mistakes in Questions 3 to 7, circle Sec. 2-2.	2-2	55
If you made any mistakes in Questions 8 to 12, circle Sec. 2-3.	2-3	62

You should now work through the sections whose numbers you have circled, taking them in order. The sections begin on the pages indicated.

Section 2-1
The Commutative Principle

Do this section if you made a mistake in Question 1 or 2 of Diagnostic Test 2. Begin by taking the pretest below, which will show you how much of Sec. 2-1 you already know and what you will be able to do by the time you have finished it.

PRETEST 2-1

Which of the following situations are commutative?

1. Being the son of
2. Being at the same school as
3. Being far away from
4. Being north of

Which of the following operations are commutative?

5. Subtraction
6. Multiplication
7. Division

8. The commutative principle applied to addition means that $A + B =$ ______.

Use the commutative principle to find:

9. $5 \times 70 =$ ______
10. $9 \times 50 =$ ______

When you have finished the pretest, correct it yourself using the answers on page 78. Then check all the questions you got wrong on the list below:

1.	______	5.	______	8.	______
2.	______	6.	______	9.	______
3.	______	7.	______	10.	______
4.	______				

If you made any mistakes (even if it was only *one* mistake), read the introduction and go through the program.

If you made no mistakes at all, skip Sec. 2-1, because you already know its material, and go on to the next section whose number you circled on page 49. If no further numbers were circled, take Final Test 2 (page 76).

Introduction

In this section you will learn about the commutative principle. You also learn how to determine whether relationships are commutative and what arithmetical operations are commutative.

FRAMES FOR SECTION 2-1

1 A number of general principles apply to some arithmetical operations with whole numbers. One of these is the *commutative principle*, or principle of interchangeability. Let us start with some everyday examples to illustrate what we mean by commutative. If *X* sits next to *Y*, does *Y* sit next to *X*?

---

yes

---

2 Do statements *a* and *b* give the same information?

a. *X* sits next to *Y*. *b.* *Y* sits next to *X*.

---

yes

---

3 If *X* sits on the right of *Y*, does *Y* sit on the right of *X*?

---

no

---

4 Do statements *a* and *b* give the same information?

a. *X* sits on the right of *Y*. *b.* *Y* sits on the right of *X*.

---

no

---

5 A *relationship* between *two* objects is said to be *commutative* if it is unaffected by *order*, i.e., if the statements *X* . . . *Y* and *Y* . . . *X* are interchangeable because they give the same information. Thus "sitting next to" (is/is not) ________ commutative; and "sitting on the right of" (is/is not) ________ commutative.

---

is; is not

---

6 *a.* Are statements 1 and 2 interchangeable?

1. X comes after Y.
2. Y comes after X.

b. Is "coming after" commutative?

c. Do statements 3 and 4 give the same information?

3. The earth is 93 million miles away from the sun.
4. The sun is 93 million miles away from the earth.

d. Is "being 93 million miles away from" commutative?

---

a. no *b.* no *c.* yes *d.* yes

---

7 Which of the following are commutative?

a. Being close to

b. Being the father of

c. Having the same father as

d. Being redder than

---

a and *c*

---

8 Is "being equal to" commutative?

---

yes

---

9 An arithmetic *operation* involving *two* numbers is said to be *commutative* if the *order* of the two numbers does not affect the result. Thus addition is commutative if $X + Y$ is always equal to $Y + X$, and subtraction is commutative if $X - Y$ is always equal to ______.

$Y - X$

---

10 *a.* Now $6 + 8 =$ ______, and $8 + 6 =$ ______.

b. Is the addition of 6 and 8 commutative?

c. Are all additions commutative? (Try some other examples if you have any doubts.)

---

a. 14; 14 *b.* yes *c.* yes

---

11 These two addition facts, $6 + 8 = 14$ and $8 + 6 = 14$, give rise to two subtraction facts. What are they?

$14 - 8 = 6$ $14 - 6 = 8$

---

12 If subtraction were commutative, the corresponding statements $8 - 14 = 6$ and $6 - 14 = 8$ would also have to be true. Are they in fact true?

---

no

---

13 $8 - 14$ and $6 - 14$ are not defined for natural numbers, that is, the numbers 0, 1, 2, 3, . . . , because if you start with a pile of 8 apples you cannot take away more than ______ and if you start with a pile of 6 apples you cannot take away more than ______.

---

8 6

---

14 *a.* Are $12 - 7$ and $7 - 12$ equivalent operations?

b. Is subtraction commutative?

---

a. no *b.* no

---

15 Is multiplication commutative? (Invent your own examples to test this out.)

---

yes; one example is the multiplication of 5 and 3: $3 \times 5 = 15$ and $5 \times 3 = 15$, so 3×5 and 5×3 are the same.

---

16 Is division commutative?

---

no; consider 10 and 2, for example: $10 \div 2 = 5$, but $2 \div 10$ is not defined for natural numbers since we cannot divide 2 apples among 10 people without chopping up the apples.

---

17 Which of the following situations are commutative?

a. Being greater than

b. Being not equal to

c. Being 1 inch away from

---

b and *c*

---

18 Which of the following operations are commutative?

a. Addition
b. Subtraction
c. Multiplication
d. Division

---

a and *c*

---

19 We can use the commutative principle to multiply one-digit numbers by two-digit numbers ending in 0. In Chap. 1 we showed that $60 \times 7 =$ _____ and $30 \times 9 =$ _____. The commutative principle then implies that $7 \times 60 =$ _____ and $9 \times 30 =$ _____.

---

420 270 420 270

---

20 Use the commutative principle to find:

a. 6×50 *b.* 4×80 *c.* 7×90

---

a. $6 \times 50 = 50 \times 6 = 300$
b. $4 \times 80 = 80 \times 4 = 320$
c. $7 \times 90 = 90 \times 7 = 630$

You have finished the program for Sec. 2-1. Now take the posttest to see how much you have learned.

POSTTEST 2-1

Which of the following situations are commutative?

1. Being married to
2. Being taller than
3. Being on the same side as
4. Being closer than

Which of the following operations are commutative?

5. Addition
6. Subtraction
7. Division

8. The commutative principle applied to multiplication means that $A \times B =$ _____.

Use the commutative principle to find:

9. $7 \times 80 =$ _____
10. $6 \times 90 =$ _____

When you have finished the posttest, correct it yourself using the answers on page 78. Then check all the questions you got wrong on the list below:

1. ______ 5. ______ 8. ______

2. ______ 6. ______ 9. ______

3. ______ 7. ______ 10. ______

4. ______

If you compare your performance with that on Pretest 2-1, you will see how much you have improved. If you made more than one mistake on this posttest, review the material and take the pretest again. Don't move on to the next section until you can get all the questions right.

Now proceed to the next section whose number you circled on page 49. If no other sections were circled, take Final Test 2 (page 76).

Section 2-2 The Associative Principle and the Use of Parentheses

Do this section if you made a mistake in Questions 3 to 7 of Diagnostic Test 2. Begin by taking the pretest below, which will show you how much of Sec. 2-2 you already know and what you will be able to do by the time you have finished it.

PRETEST 2-2

1. Which of the following are associative?

   a. Addition *b.* Subtraction

   c. Multiplication *d.* Division

2. $31 + 15 + 19 =$ ______ 3. $32 - (23 - 15) =$ ______

4. $(6 + 15) \div 3 =$ ______

Use the associative principle to find:

5. $5 \times 14 =$ ______ 6. $4 \times 18 =$ ______

7. The associative principle applied to multiplication means that $(a \times b) \times c =$ ______.

8. $7 \cdot 9 - 32 =$ ______ 9. $(13 + 24) - (11 + 8) =$ ______

10. $8 \cdot 6 - 5 \cdot 3 =$ ______

When you have finished the pretest, correct it yourself using the answers on page 78. Then check all the questions you got wrong on the list below:

1. ______ 5. ______ 8. ______
2. ______ 6. ______ 9. ______
3. ______ 7. ______ 10. ______
4. ______

If you made any mistakes (even if it was only *one* mistake), read the introduction and go through the program.

If you made no mistakes at all, skip Sec. 2-2, because you already know its material, and go on to the next section whose number you circled on page 49. If no further numbers were circled, take Final Test 2 (page 76).

Introduction

In this section you will learn how to use parentheses to define the order in which operations are to be performed. You will also learn about the associative principle and which arithmetical operations are associative. The dot notation for multiplication is then introduced.

FRAMES FOR SECTION 2-2

1 Hitherto we have dealt with expressions like $12 \div 2$, which involve only one operation. But we also need to be able to tackle expressions like $12 \div 2 + 4$, which involve more than one operation. The problem is how to group the numbers. If, for example, we group the 2 with the 4, our steps are:

a. Add 2 and 4 to get ______.

b. Divide 12 by this answer to get ______.

But if we group the 2 with the 12, our steps are:

c. Divide 12 by 2 to get ______.

d. Add 4 to this answer to get ______.

e. If we begin by grouping the 2 with the 4, we get a final answer of ______. But if we begin by grouping the 2 with the 12 we get a final answer of ______.

---

a. 6 *b.* 2 *c.* 6 *d.* 10 *e.* 2; 10

---

2 This could clearly lead to confusion unless we define our expression more carefully. We do this by using parentheses and by agreeing that operations within parentheses come first. So if we write $(12 \div 2) + 4$, this is equal to ______ $+ 4 =$ ______. But if we write $12 \div (2 + 4)$, this is equal to $12 \div$ ______ $=$ ______. We avoid confusion by starting with the operation (inside/outside) ______ the parentheses.

---

6 10 6 2 inside

---

3 *a.* $(5 \times 4) + 3 =$ ______ $+ 3 =$ ______

b. $5 \times (4 + 3) = 5 \times$ ______ $=$ ______

c. Does the position of the parentheses affect the final result?

---

a. 20; 23 *b.* 7; 35 *c.* yes

---

4 *a.* $(4 \times 7) + 3 =$ ______ *b.* $4 \times (7 + 3) =$ ______

c. Does the position of the parentheses affect the final result?

---

a. 31 *b.* 40 *c.* yes

---

5 *a.* $(7 \times 5) - 4 =$ ______ *b.* $7 \times (5 - 4) =$ ______

---

a. 31 *b.* 7

---

6 *a.* $(30 - 6) \div 3 =$ ______ *b.* $30 - (6 \div 3) =$ ______

---

a. 8 *b.* 28

---

7 Even when both operations are the same, the position of the parentheses can be important.

a. $(35 - 23) - 11 =$ ______ *b.* $35 - (23 - 11) =$ ______

---

a. 1 *b.* 23

---

8 But with two additions

a. $(5 + 11) + 13 =$ ______ *b.* $5 + (11 + 13) =$ ______

it is clear that the position of the parentheses (does/does not) ______ affect the final result.

---

a. 29 *b.* 29; does not

---

9 *a.* $(6+7)+8=$ ______ *b.* $6+(7+8)=$ ______

c. In general $(a+b)+c$ (does/does not) ______ equal $a+(b+c)$.

---

a. 21 *b.* 21 *c.* does

---

10 We call the fact that $(a+b)+c=a+(b+c)$ the *associative principle* of addition because the value of the expression $a+b+c$ does not depend on whether b is associated with a or with c. We also say that the operation of addition is *associative.*

a. But does $(a-b)-c$ equal $a-(b-c)$? [Try $(15-6)-4$ and $15-(6-4)$ or look back to Frame 7.]

b. Is subtraction associative?

---

a. no [$(15-6)-4=5$, but $15-(6-4)=13$. Frame 7 also showed they were not equal.] *b.* no

---

11 *a.* $(16 \div 4) \div 2=$ ______ *b.* $16 \div (4 \div 2)=$ ______

c. Is division associative?

---

a. 2 *b.* 8 *c.* no

---

12 *a.* $(3 \times 2) \times 4=$ ______ *b.* $(5 \times 2) \times 3=$ ______

$3 \times (2 \times 4)=$ ______ $5 \times (2 \times 3)=$ ______

c. Is multiplication associative?

---

a. 24; 24 *b.* 30; 30 *c.* yes

---

13 Which of the following are equal?

a. $(a+b)+c$ and $a+(b+c)$

b. $(a-b)-c$ and $a-(b-c)$

c. $(a \times b) \times c$ and $a \times (b \times c)$

d. $(a \div b) \div c$ and $a \div (b \div c)$

---

a and *c*

---

14 Which of the following operations are associative?

a. Addition *b.* Subtraction

c. Multiplication *d.* Division

---

a and *c*

---

15 State the associative principle for multiplication in symbolic form: $(a \times b) \times c =$ __________.

$a \times (b \times c)$

---

16 Since addition and multiplication are associative operations, we can discard parentheses for multiple additions and multiple multiplications. $11 + 12 + 8 =$ _____ because both $(11 + 12) + 8$ and $11 + (12 + 8)$ are equal to _____, and $5 \times 2 \times 4 =$ _____ because both $(5 \times 2) \times 4$ and $5 \times (2 \times 4)$ are equal to _____.

---

31 31 40 40

---

17 *a.* $13 + 22 + 11 =$ ____ *b.* $1 \times 7 \times 6 =$ ____

---

a. 46 *b.* 42

---

18 We can use this property to multiply numbers which were not included in our multiplication tables. Thus

$3 \times 27 = 3 \times (3 \times 9) = (3 \times 3) \times 9 =$ _____ $\times 9 =$ _____

and

$4 \times 16 = 4 \times (2 \times 8) = (4 \times 2) \times 8 =$ _____ $\times 8 =$ _____

---

9 81 8 64

---

19 Use this method to find:

a. 3×24 *b.* 5×18

---

a. 72 *b.* 90

---

20 *a.* $17 + 23 + 12 =$ _____ *b.* $19 - (11 - 2) =$ _____

c. $(6 + 8) \div 2 =$ _____ *d.* $5 \times 8 \times 2 =$ _____

---

a. 52 *b.* 10 *c.* 7 *d.* 80

---

21 Evaluate the following expressions:

a. $15 + 9 + 8 + 7 =$ ______

b. $(15 + 9) - (8 + 7) =$ ___________

c. $15 + (9 - 8) + 7 =$ ___________

---

a. 39 *b.* $24 - 15 = 9$ *c.* $15 + 1 + 7 = 23$

---

22 Sometimes a $\cdot$ is used instead of a $\times$ to denote the operation of multiplication. Thus

$3 \cdot 4 = 3 \times 4 = 12$ and $5 \cdot 6 =$ ______

30

---

23 In multiple operations the $\cdot$ is always taken first. So $5 \cdot 6 + 7 = (5 \times 6) + 7 =$ ______, and $5 + 6 \cdot 7 = 5 + (6 \times 7) =$ ______.

---

37 47

---

24 Evaluate the following expressions:

a. $6 \cdot 9 - 5$ *b.* $23 - 3 \cdot 4$ *c.* $7 \cdot 8 + 6$

---

a. $54 - 5 = 49$ *b.* $23 - 12 = 11$ *c.* $56 + 6 = 62$

---

25 Evaluate the following expressions:

a. $(7 + 5) - (3 + 4)$ *b.* $7 + (5 - 3) + 4$

c. $7 \cdot 5 - 3 \cdot 4$ *d.* $7 \times (5 - 3) \times 4$

---

a. $12 - 7 = 5$ *b.* $7 + 2 + 4 = 13$ *c.* $35 - 12 = 23$
d. $7 \times 2 \times 4 = 7 \times 8 = 56$

You have finished the program for Sec. 2-2. Now take the posttest to see how much you have learned.

POSTTEST 2-2

1. Which of the following operations are associative?

   a. Addition *b.* Subtraction

   c. Multiplication *d.* Division

2. $5 \times 6 \times 3 =$ ______

3. $40 - (11 + 23) =$ ______

4. $(20 - 12) \div 4 =$ ______

5. $5 \times 16 =$ ______

6. $4 \times 14 =$ ______

7. The associative principle applied to addition means that $(a + b) + c =$ ______.

8. $6 \cdot 5 - 11 =$ ______

9. $(4 + 25) - (13 + 6) =$ ______

10. $7 \cdot 8 - 3 \cdot 4 =$ ______

When you have finished the posttest, correct it yourself using the answers on page 78. Then check all the questions you got wrong on the list below:

1. ______	5. ______	8. ______
2. ______	6. ______	9. ______
3. ______	7. ______	10. ______
4. ______		

If you compare your performance with that on Pretest 2-2, you will see how much you have improved. If you made more than one mistake on this posttest, review the material and take the pretest again. Don't move on to the next section until you can get all the questions right.

Now proceed to the next section whose number you circled on page 49. If no other sections were circled, take Final Test 2 (page 76).

Section 2-3
The Distributive Principle Applied to Multiplication

Do this section if you made a mistake in Questions 8 to 12 of Diagnostic Test 2. Begin by taking the pretest below, which will show you how much of Sec. 2-3 you already know and what you will be able to do by the time you have finished it.

PRETEST 2-3

1. The distributive law of multiplication over addition states that $(A + B) \times C =$ ______.

Use the distributive law of multiplication over addition to find:

2. $52 \times 9 =$ ______
3. $34 \times 7 =$ ______

Use the distributive law of multiplication over subtraction to find:

4. $49 \times 8 =$ ______
5. $78 \times 6 =$ ______
6. $(A - 6) \times 8 =$ ______ $- 48$

When you have finished the pretest, correct it yourself using the answers on page 78. Then check all the questions you got wrong on the list below:

1. ______ 4. ______

2. ______ 5. ______

3. ______ 6. ______

If you made no mistakes at all, skip Sec. 2-3 and go on to Sec. 2-4.

Introduction

In this section you will learn about the distributive law as applied to multiplication over addition and multiplication over subtraction. You will also learn how it can be used to simplify the multiplication of large numbers.

FRAMES FOR SECTION 2-3

1 *a.* Suppose we have 5 strips of cardboard each 3 inches long and we put them in one long row:

3	3	3	3	3

What is the total length?

b. Suppose we have 5 strips of cardboard each 6 inches long and we put them in one long row:

6	6	6	6	6

What is the total length?

---

a. 15 inches *b.* 30 inches

---

2 *a.* Suppose we have 5 strips of cardboard each *A* inches long and we put them in one long row:

A	A	A	A	A

The total length is $A \times$ ______.

b. Suppose we have 5 strips of cardboard each *B* inches long and we put them in one long row:

B	B	B	B	B

The total length is ______ $\times$ ______.

---

a. 5 *b.* $B \times 5$

---

3 If we put these two rows end to end:

A	A	A	A	A	B	B	B	B	B

the total length will be (______ $\times$ ______) + (______ $\times$ ______).

$(A \times 5) + (B \times 5)$

---

4 If we now change the order of the strips from the order in *a* below to the order in *b*, will the total length be altered?

a.

A	A	A	A	A	B	B	B	B	B

b.

A	B	A	B	A	B	A	B	A	B

In *b* if we join each *A* strip with the *B* strip on its right and put the whole group in the form of *c*, will the total length be altered?

c.

A + B	A + B	A + B	A + B	A + B

---

no; no

---

5 Since $4c$ now contains 5 strips of length $A + B$, we know that its total length must be $(A + B) \times \underline{\qquad}$. We also know from Frame 3 that the total length of a is $(A \times 5) + (B \times 5)$ and from Frame 4 that length of $a =$ length of c. Therefore we deduce that

$$(A \times \underline{\qquad}) + (B \times \underline{\qquad}) = (A + B) \times \underline{\qquad}.$$

---

5 5 5 5

---

6 With 8 strips of each type instead of 5, we can produce the same kinds of arrangement:

a.	A	A	A	A	A	A	A	A	B	B	B	B	B	B	B	B

b.	A	B	A	B	A	B	A	B	A	B	A	B	A	B	A	B

c.	A + B	A + B	A + B	A + B	A + B	A + B	A + B	A + B

Now length of $a = (A \times \underline{\qquad}) + (B \times \underline{\qquad})$, and length of $c = (A + B) \times \underline{\qquad}$. Since the length hasn't changed, we deduce that

$$(A + B) \times \underline{\qquad} = (A \times \underline{\qquad}) + (B \times \underline{\qquad})$$

---

8 8 8 8 8 8

---

7 With 6 strips of each type:

a.	A	A	A	A	A	A	B	B	B	B	B	B

b.	A	B	A	B	A	B	A	B	A	B	A	B

c.	A + B	A + B	A + B	A + B	A + B	A + B

we deduce that

$$(A + B) \times \underline{\qquad} = (A \times \underline{\qquad}) + (B \times \underline{\qquad})$$

---

6 6 6

---

8 If we had n strips of cardboard of each type, we could deduce that

$$(A + B) \times \underline{\qquad} = (A \times \underline{\qquad}) + (B \times \underline{\qquad})$$

This is known as the *distributive principle* (or law) of multiplication over addition.

$(A + B) \times n = (A \times n) + (B \times n)$

---

9 This distributive law saves you from having to remember too many multiplication tables. If, for example, you want to multiply 13 by 6, you could consider this as the problem of finding the length of 6 strips of length 13 inches:

13	13	13	13	13	13

We can chop each 13-inch strip into a 10-inch strip and a 3-inch strip:

10	3	10	3	10	3	10	3	10	3	10	3

and rearrange the order:

10	10	10	10	10	10	3	3	3	3	3	3

According to the distributive law,

$$(10 + 3) \times 6 = (\_\_\_\_ \times \_\_\_\_) + (\_\_\_\_ \times \_\_\_\_)$$

so

$$13 \times 6 = \_\_\_\_ + \_\_\_\_ = \_\_\_\_$$

$(10 \times 6) + (3 \times 6)$ $\quad 60 + 18 = 78$

---

10 Similarly in order to multiply 14 by 7 we can replace 14 by 10 + 4 and use the distributive law $(10 + 4) \times 7 = (\_\_\_\_ \times \_\_\_\_) + (\_\_\_\_ \times \_\_\_\_)$, so $14 \times 7 = \_\_\_\_ + \_\_\_\_ = \_\_\_\_$.

$(10 \times 7) + (4 \times 7)$ $\quad 70 + 28 = 98$

---

11 Use the distributive law to find 17×7.

According to the distributive law, $(10 + 7) \times 7 = (10 \times 7) + (7 \times 7)$, so $17 \times 7 = 70 + 49 = 119$.

---

12 Use the distributive law to find

a. 18×8 *b.* 16×9

---

a. $(10 + 8) \times 8 = (10 \times 8) + (8 \times 8)$, so $18 \times 8 = 80 + 64 = 144$.
b. $(10 + 6) \times 9 = (10 \times 9) + (6 \times 9)$, so $16 \times 9 = 90 + 54 = 144$.

---

13 *a.* We can also use the distributive law to find 34×8:

$(30 + 4) \times 8 = (30 \times 8) + (4 \times 8) =$ ______ + ______ = ______

b. Similarly

$72 \times 6 = (70 + 2) \times 6 = (70 \times$ ______$) + ($______ $\times$ ______$)$

= ______ + ______ = ______

---

a. 240; 32; 272 *b.* 6; (2×6); $420 + 12 = 432$

---

14 *a.* Use the distributive law to find 37×5:

$37 \times 5 = (30 \times$ ______$) + ($______ $\times$ ______$) =$ ______ + ______ = ______

b. Use the distributive law to find 66×4.

---

a. 5; (7×5); $150 + 35 = 185$

b. $66 \times 4 = (60 \times 4) + (6 \times 4) = 240 + 24 = 264$

---

15 *a.* Use the distributive law to find 32×9.

b. Use the distributive law to find 53×7.

---

a. $32 \times 9 = (30 \times 9) + (2 \times 9) = 270 + 18 = 288$

b. $53 \times 7 = (50 \times 7) + (3 \times 7) = 350 + 21 = 371$

---

16 *a.* Suppose we start with 4 strips of cardboard, each of which is Y inches long and place them end to end:

Y	Y	Y	Y

Total length $= Y \times$ ______ inches

b. If we now chop each strip into two parts, one of which is 3 inches long, the other part will be ______ inches long.

---

a. 4 *b.* $Y - 3$

---

17 The total length of *a* is ______ $\times$ ______ inches, and after chopping each strip into two parts to get *b*, the total length is ______ inches.

a.

Y	Y	Y	Y

b.

3	$Y-3$	3	$Y-3$	3	$Y-3$	3	$Y-3$

c.

3	3	3	3	$Y-3$	$Y-3$	$Y-3$	$Y-3$

When these strips are rearranged to give c, the total length remains ______ inches.

$Y \times 4$ $Y \times 4$ $Y \times 4$

18

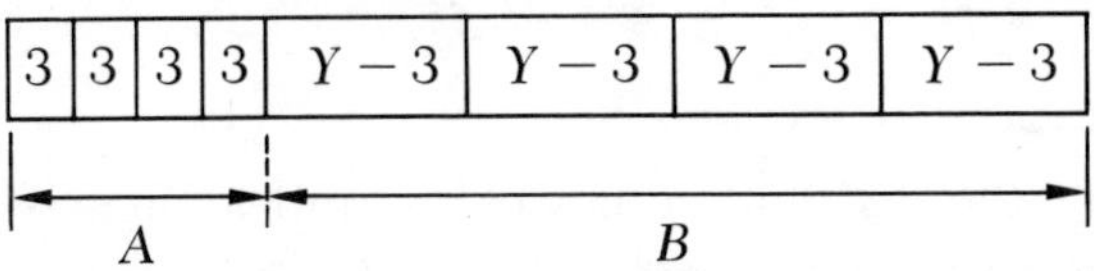

If we now divide c from Frame 17 at the dotted line, we can see that A, the length of the left part, is ______ $\times$ 4 and B, the length of the right part, is ______ $\times$ 4.

---

3 $(Y - 3)$

---

19 So we know from Frame 18 that

Length $A = 3 \times 4$ and length $B = (Y - 3) \times 4$

and we know from Frame 17 that

Total length $= Y \times 4$

Now since length B = total length − length A, we can deduce that

$(Y - 3) \times 4 =$ (______ $\times$ ______) − (______ $\times$ ______)

This is an example of the distributive law of multiplication over subtraction.

$(Y \times 4) - (3 \times 4)$

---

20 To take a more general example:

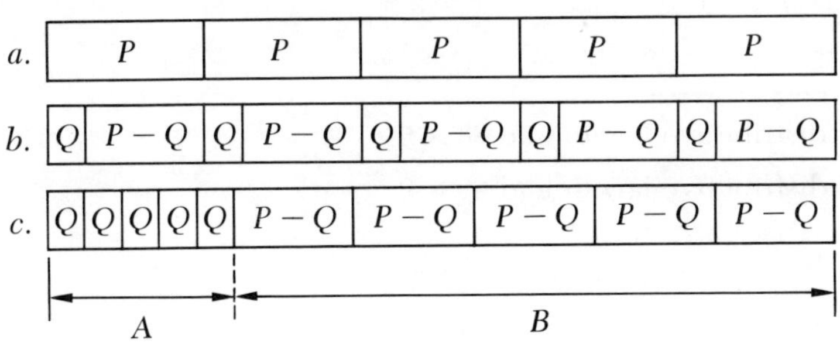

Total length of c = total length of $a = P \times$ ______. Length $A = Q \times$ ______, and length B = ______ $\times$ ______.

---

5 5 $(P - Q)$ 5

---

21 In Frame 20 since length $B =$ total length $-$ length A, we can deduce that

$(P - Q) \times 5 = (\_\_\_\_ \times \_\_\_\_) - (\_\_\_\_ \times \_\_\_\_)$

which is another example of the distributive law of multiplication over subtraction.

$(P \times 5) - (Q \times 5)$

---

22 In its most general form, the distributive law of multiplication over subtraction states that

$(A - B) \times n = (\_\_\_\_ \times \_\_\_\_) - (\_\_\_\_ \times \_\_\_\_)$

$(A \times n) - (B \times n)$

---

23 *a.* We can now use the distributive law of multiplication over subtraction to find 39×5:

$(40 - 1) \times 5 = (40 \times 5) - (1 \times 5)$

so $39 \times 5 = \_\_\_\_ - \_\_\_\_ = \_\_\_\_$

b. Similarly $58 \times 7 = (60 - 2) \times 7 = (60 \times \_\_\_\_) - (\_\_\_\_ \times \_\_\_\_)$

$= \_\_\_\_ - \_\_\_\_ = \_\_\_\_$

---

a. $200 - 5 = 195$ *b.* $7; (2 \times 7); 420 - 14 = 406$

---

24 *a.* Use the distributive law to find 28×6:

$28 \times 6 = (30 \times \_\_\_\_) - (\_\_\_\_ \times \_\_\_\_) = \_\_\_\_ - \_\_\_\_ = \_\_\_\_$

b. Use the distributive law to find 89×8.

---

a. $6; (2 \times 6); 180 - 12 = 168$

b. $89 \times 8 = (90 \times 8) - (1 \times 8) = 720 - 8 = 712$

---

25 *a.* Use the distributive law to find 88×5.

b. Use the distributive law to find 69×9.

---

a. $88 \times 5 = (90 \times 5) - (2 \times 5) = 450 - 10 = 440$

b. $69 \times 9 = (70 \times 9) - (1 \times 9) = 630 - 9 = 621$

---

26 Complete the following statements of the distributive law of multiplication over addition and the distributive law of multiplication over subtraction:

$(A + B) \times n = \_\_\_\_$ $(A - B) \times n = \_\_\_\_$

$(A \times n) + (B \times n)$ $(A \times n) - (B \times n)$

---

27 Use one of these distributive laws to calculate the following:

a. $88 \times 6 =$ ______ *b.* $31 \times 9 =$ ______

---

a. $88 \times 6 = (90 \times 6) - (2 \times 6) = 540 - 12 = 528$
b. $31 \times 9 = (30 \times 9) + (1 \times 9) = 270 + 9 = 279$

You have finished the program for Sec. 2-3. Now take the posttest to see how much you have learned.

POSTTEST 2-3

1. The distributive law of multiplication over subtraction states that $(A - B) \times C =$ ______.

Use the distributive law of multiplication over addition to find:

2. $43 \times 5 =$ ______
3. $65 \times 8 =$ ______

Use the distributive law of multiplication over subtraction to find:

4. $69 \times 7 =$ ______
5. $38 \times 9 =$ ______
6. $(B + 4) \times 6 =$ ______ $+ 24$

When you have finished the posttest, correct it yourself using the answers on page 78. Then check all the questions you got wrong on the list below:

1. ______ 3. ______ 5. ______
2. ______ 4. ______ 6. ______

If you compare your performance with that on Pretest 2-3, you will see how much you have improved. If you made more than one mistake on this posttest, review the material and take the pretest again. Don't move on to the next section until you can get all the questions right.

Now proceed to Sec. 2-4.

Section 2-4
The Distributive Principle Applied to Division

Do this section if you made a mistake in Questions 8 to 12 of Diagnostic Test 2. Begin by taking the pretest below, which will show you how much of Sec. 2-4 you already know and what you will be able to do by the time you have finished it.

PRETEST 2-4

1. Which of the following are true?

   a. $(A + B) - c = (A - c) + (B - c)$ *b.* $(A + B) \div c = (A \div c) + (B \div c)$

   c. $(A \div B) + c = (A + c) \div (B + c)$ *d.* $(A \times B) - c = (A - c) \times (B - c)$

Use the distributive law of division over addition to find:

2. $91 \div 7 =$ _____ 3. $96 \div 6 =$ _____

Use the distributive law of division over subtraction to find:

4. $87 \div 3 =$ _____ 5. $72 \div 4 =$ _____

6. Use the distributive law to find $9(8 + 6) =$ _____.

When you have finished the pretest, correct it yourself using the answers on page 79. Then check all the questions you got wrong on the list below:

1. _____ 3. _____ 5. _____

2. _____ 4. _____ 6. _____

If you made any mistakes (even if it was only *one* mistake), read the introduction and go through the program.

If you made no mistakes at all, skip Sec. 2-4, because you already know its material, and go on to take Final Test 2 (page 76).

Introduction

In this section you will learn that the distributive law also applies to division over addition and to division over subtraction but does not apply to any other pairs of operations apart from these two and the two from Sec. 2-3. You are shown how this law can be used to simplify the division of large numbers and are introduced to the convention which omits multiplication signs before parentheses.

FRAMES FOR SECTION 2-4

1 In Sec. 2-3 you discovered that the distributive law applied to multiplication over addition, i.e., that

$(a+b) \times n = (a \times \_\_\_\_) + (\_\_\_\_ \times \_\_\_\_)$

and that the distributive law applied to multiplication over subtraction, i.e., that

$(a-b) \times n = (\_\_\_\_) - (\_\_\_\_)$

---

n $\quad b \times n \quad$ $(a \times n) - (b \times n)$

---

2 If the distributive law also applied to *division* over addition, $(a+b) \div n$ would equal $(a \div n) + (b \div n)$. Let us try an example to see if this is true.

$(20+8) \div 4 = \_\_\_\_ \div 4 = \_\_\_\_$

and

$(20 \div 4) + (8 \div 4) = \_\_\_\_ + \_\_\_\_ = \_\_\_\_$

So for this particular example the distributive law (does/does not) ____ apply.

---

28 $\quad$ 7 $\quad$ $5+2=7$ $\quad$ does

---

3 In general it can be shown that the distributive law applies to division over addition, i.e., that

$(a+b) \div n = (a \div \_\_\_\_) + (\_\_\_\_ \div \_\_\_\_)$

---

n $\quad (b \div n)$

---

4 If the distributive law also applies to division over subtraction, then $(a-b) \div n$ would equal $(\_\_\_\_) - (\_\_\_\_)$.

$(a \div n) - (b \div n)$

---

5 Let us try an example:

$(30-6) \div 3 = \_\_\_\_ \div 3 = \_\_\_\_$

and

$(30 \div 3) - (6 \div 3) = \_\_\_\_ - \_\_\_\_ = \_\_\_\_$

Does the distributive law apply to this particular example of division over subtraction?

---

24 8 $10-2=8$ yes

---

6 The distributive law can also be shown to be true for the general use of division over subtraction, so we now have four valid applications of the distributive law.

$(a+b)\times n=$ _____ $(a-b)\times n=$ _____

$(a+b)\div n=$ _____ $(a-b)\div n=$ _____

$(a\times n)+(b\times n)$ $(a\times n)-(b\times n)$ $(a\div n)+(b\div n)$
$(a\div n)-(b\div n)$

---

7 *a.* We can now use the distributive law of division over addition to find $88\div 4$:

$(80+8)\div 4=(80\div 4)+(8\div 4)=$ _____ $+$ _____ $=$ _____

b. Similarly $69\div 3=(60+9)\div 3=(60\div$ _____$)+($_____ $\div$ _____$)$
$=$ _____ $+$ _____ $=$ _____

---

a. 20; 2; 22 *b.* 3; $(9\div 3)$; $20+3=23$

---

8 *a.* Use the distributive law to find $55\div 5$:

$55\div 5=(50+5)\div 5=($_____ $\div$ _____$)+($_____ $\div$ _____$)$
$=$ _____ $+$ _____ $=$ _____

b. Use the distributive law to find $48\div 2$.

---

a. $(50\div 5)+(5\div 5)=10+1=11$
b. $(40+8)\div 2=(40\div 2)+(8\div 2)=20+4=24$

---

9 Use the distributive law to find:

a. $96\div 3=$ _____ *b.* $48\div 4=$ _____

---

a. $(90+6)\div 3=(90\div 3)+(6\div 3)=30+2=32$
b. $(40+8)\div 4=(40\div 4)+(8\div 4)=10+2=12$

---

10 *a.* The distributive law of division over addition can also be used to find $91\div 7$ because

$91\div 7=(70+21)\div 7=($_____ $\div$ _____$)+($_____ $\div$ _____$)$
$=$ _____ $+$ _____ $=$ _____

b. Use the distributive law of division over addition to find $96\div 8$.

---

a. $(70 \div 7) + (21 \div 7)$; $10 + 3 = 13$
b. $96 \div 8 = (80 + 16) \div 8 = (80 \div 8) + (16 \div 8) = 10 + 2 = 12$

---

11 Use the distributive law of division over addition to find:

a. $75 \div 5 =$ ______ *b.* $96 \div 6 =$ ______

---

a. $(50 + 25) \div 5 = (50 \div 5) + (25 \div 5) = 10 + 5 = 15$
b. $(60 + 36) \div 6 = (60 \div 6) + (36 \div 6) = 10 + 6 = 16$

---

12 *a.* The distributive law of division over subtraction can be used to find $57 \div 3$ because

$(60 - 3) \div 3 = ($______ $\div$ ______$) - ($______ $\div$ ______$)$

$=$ ______ $-$ ______ $=$ ______

b. Use the distributive law of division over subtraction to find $84 \div 3$.

---

a. $(60 \div 3) - (3 \div 3) = 20 - 1 = 19$
b. $84 \div 3 = (90 - 6) \div 3 = (90 \div 3) - (6 \div 3) = 30 - 2 = 28$

---

13 To find $81 \div 3$ we can use the distributive law of division over subtraction:

$(90 - 9) \div 3 = ($______ $\div$ ______$) - ($______ $\div$ ______$)$

$81 \div 3 =$ ______ $-$ ______ $=$ ______

or we can use the distributive law of division over addition

$(60 + 21) \div 3 = ($______ $\div$ ______$)$ $($______ $\div$ ______$)$

$81 \div 3 =$ ______ $+$ ______ $=$ ______

$(90 \div 3) - (9 \div 3)$ $30 - 3 = 27$ $(60 \div 3) + (21 \div 3)$ $20 + 7 = 27$

---

14 *a.* Use the distributive law of division over subtraction to find $76 \div 4$.

b. Use the distributive law of division over addition to find $56 \div 4$.

---

a. $(80 - 4) \div 4 = (80 \div 4) - (4 \div 4) = 20 - 1 = 19$
b. $(40 + 16) \div 4 = (40 \div 4) + (16 \div 4) = 10 + 4 = 14$

---

15 The distributive law of division over addition can be summarized for the general case as

$(a + b) \div c = (\_\_\_\_\_) + (\_\_\_\_\_)$

and the distributive law of division over subtraction can be summarized as

$(a - b) \div c = \_\_\_\_\_\_\_\_$

$(a \div c) + (b \div c)$ $\quad$ $(a \div c) - (b \div c)$

---

16 State the distributive law for multiplication over addition and multiplication over subtraction, using a, b, and c.

$(a + b) \times c = (a \times c) + (b \times c)$ $\quad$ $(a - b) \times c = (a \times c) - (b \times c)$

---

17 We have now stated four distributive laws, and we want to find out if there are any others. Let's try some examples:

a. Consider $(a \times b) \div c$ and $(a \div c) \times (b \div c)$:

$(4 \times 8) \div 2 = \_\_\_\_\_$ and $(4 \div 2) \times (8 \div 2) = \_\_\_\_\_$

Is there a distributive law of division over multiplication?

b. Consider $(a \div b) + c$ and $(a + c) \div (b + c)$:

$(12 \div 2) + 3 = \_\_\_\_\_$ and $(12 + 3) \div (2 + 3) = \_\_\_\_\_$

Is there a distributive law of addition over division?

c. Consider $(a \times b) - c$ and $(a - c) \times (b - c)$:

$(5 \times 4) - 2$ and $(5 - 2) \times (4 - 2) = \_\_\_\_\_$

Is there a distributive law of subtraction over multiplication?

---

a. 16; 8; no $\quad$ *b.* 9; 3; no $\quad$ *c.* 18; 6; no

---

18 In a similar manner it can be shown that there are no distributive laws other than the four which we have already discovered. These four are:

a. Multiplication over ________

b. Multiplication over ________

c. ________ over ________

d. ________ over ________

---

a. addition $\quad$ *b.* subtraction $\quad$ *c.* division over addition
d. division over subtraction

---

19 *a.* $(a + b) + c = a + (b + c)$ is a statement of the (commutative/associative/distributive) ______ principle for ______.

b. $a \times b = b \times a$ is a statement of the ______ principle for ______.

c. $(a + b) \div c = (a \div c) + (b \div c)$ is a statement of the ______ principle for ______.

---

a. associative; addition *b.* commutative; multiplication
c. distributive; division over addition

---

20 State the four distributive laws in symbolic form.

$(a + b) \times c = (a \times c) + (b \times c)$ $(a - b) \times c = (a \times c) - (b \times c)$
$(a + b) \div c = (a \div c) + (b \div c)$ $(a - b) \div c = (a \div c) - (b \div c)$

---

21 It is quite common to omit the × sign when it comes immediately before parentheses, so $7(8 + 3)$ is a shorter version of $7 \times (8 + 3)$, and $6(5 - 4)$ is a shorter version of ______.

$6 \times (5 - 4)$

---

22 Evaluate the following expressions:

a. $6(3 + 5)$ *b.* $9(8 - 3)$

---

a. $6 \times 8 = 48$ *b.* $9 \times 5 = 45$

---

23 Since multiplication is commutative, the distributive law $(a + b) \times c = (a \times c) + (b \times c)$ can also be written as $c \times (a + b) = (c \times$ ______$) + (c \times$ ______$)$ or in an even shorter version as $c(a + b) = c \cdot$ ______ $+ c \cdot$ ______.

---

a *b* *a* *b*

---

24 Use this version of the distributive law to evaluate

a. $7(8 + 9)$ *b.* $6(7 + 8)$

---

a. $7(8 + 9) = 7 \cdot 8 + 7 \cdot 9 = 56 + 63 = 119$
b. $6(7 + 8) = 6 \cdot 7 + 6 \cdot 8 = 42 + 48 = 90$

---

25 Evaluate the following expressions:

a. $6(13-4)$ *b.* $9(8+7)$

---

a. $6 \times 9 = 54$ *b.* $9 \cdot 8 + 9 \cdot 7 = 72 + 63 = 135$

You have finished the program for Sec. 2-4. Now take Posttest 2-4 to see how much you have learned.

POSTTEST 2-4

1. Which of the following are true?

   a. $(A \times B) \div c = (A \div c) \times (B \div c)$ *b.* $(A \times B) + c = (A + c) \times (B + c)$

   c. $(A - B) \div c = (A \div c) - (B \div c)$ *d.* $(A \div B) - c = (A - c) \div (B - c)$

Use the distributive law of division over addition to find

2. $78 \div 6 =$ _____ 3. $92 \div 4 =$ _____

Use the distributive law of division over subtraction to find

4. $58 \div 2 =$ _____ 5. $54 \div 3 =$ _____

6. Use the distributive law to find $8(9 + 7) =$ _____.

When you have finished the posttest, correct it yourself using the answers on page 79. Then check all the questions you got wrong on the list below:

1. _____ 3. _____ 5. _____

2. _____ 4. _____ 6. _____

If you compare your performance with that on Pretest 2-4, you will see how much you have improved. If you made more than one mistake on this posttest, review the material and take the pretest again.

Now proceed to Final Test 2.

FINAL TEST 2

1. The commutative principle applied to addition means that $A + B =$ _____.
2. Which of the following operations are commutative?

   a. Subtraction *b.* Multiplication

   c. Division
3. The associative principle applied to multiplication means that $(A \times B) \times C =$ ________.

4. Which of the following operations are associative?

   a. Addition *b.* Subtraction

   c. Division

5. $7 + (4 \times 3) =$ ______

6. $(10 - 6) \div 2 =$ ______

7. $7 \cdot 8 - 6 \cdot 4 =$ ______

8. The distributive law applied to division over addition states that $(A + B) \div C =$ ____________.

9. Which of the following statements are true?

   a. $(A + B) \times C = (A \times C) + (B \times C)$

   b. $(A - B) + C = (A + C) - (B + C)$

   c. $(A \div B) + C = (A + C) \div (B + C)$

   d. $(A \times B) - C = (A - C) \times (B - C)$

Use the distributive law to find:

10. $59 \times 9 =$ ______

11. $98 \div 7 =$ ______

12. $8(9 + 8) =$ ______

The answers to this test are on page 79. Check the questions you got wrong on the list below and compare your results with those on Diagnostic Test 2 to see how much you have improved.

1. ______	3. ______	8. ______
2. ______	4. ______	9. ______
	5. ______	10. ______
	6. ______	11. ______
	7. ______	12. ______

If you made more than one mistake, review the material and then take the diagnostic test again. Make sure that you can get all the questions right before you move on to the next chapter.

Now proceed to Chap. 3, Operations with Large Natural Numbers.

ANSWERS TO TESTS FOR CHAPTER 2

DIAGNOSTIC TEST 1

1. $B \times A$
2. a
3. $A + (B + C)$
4. b
5. 20
6. 12
7. 21
8. $(A \div C) - (B \div C)$
9. only a
10. 203
11. 15
12. 119

PRETEST 2-1

2, 3, and 6 are commutative
1, 4, 5, and 7 are not

8. $B + A$
9. 350
10. 450

POSTTEST 2-1

1, 3, and 5 are commutative
2, 4, 6, and 7 are not

8. $B \times A$
9. 560
10. 540

PRETEST 2-2

1. a and c
2. 65
3. 24
4. 7
5. 70
6. 72
7. $a \times (b \times c)$
8. 31
9. 18
10. 33

POSTTEST 2-2

1. a and c
2. 90
3. 6
4. 2
5. 80
6. 56
7. $a + (b + c)$
8. 19
9. 10
10. 44

PRETEST 2-3

1. $(A \times C) + (B \times C)$
2. 468
3. 238
4. 392
5. 468
6. $A \times 8$

POSTTEST 2-3

1. $(A \times C) - (B \times C)$
2. 215
3. 520
4. 483
5. 342
6. $B \times 6$

PRETEST 2-4

1. only *b*
2. 13
3. 16
4. 29
5. 18
6. 126

POSTTEST 2-4

1. only *c*
2. 13
3. 23
4. 29
5. 18
6. 128

FINAL TEST 2

1. $B + A$
2. *b*
3. $A \times (B \times C)$
4. *a*
5. 19
6. 2
7. 32
8. $(A \div C) + (B \div C)$
9. only *a*
10. 531
11. 14
12. 136

Chapter 3
Operations with Large Natural Numbers

We have already defined arithmetic operations with small natural numbers (Chap. 1) and investigated the commutative, associative, and distributive principles (Chap. 2). These were needed before we could carry out arithmetic operations on large natural numbers by the methods described in this chapter.

Section 3-1 covers addition, and Sec. 3-2 covers subtraction. Then Sec. 3-3 introduces the multiplication of large numbers by one-digit numbers, and Sec. 3-4 extends this to multiplication by large numbers, using the method of long multiplication. Sections 3-5 to 3-7 show you how to divide numbers of increasing complexity by the method of long division.

DIAGNOSTIC TEST 3

1. Add 5,435, 2,672, and 1,988.
2. Add 4,080, 9, 356, and 5,295.
3. Subtract 3,859 from 4,806.
4. Subtract 4,932 from 8,000.
5. Multiply 3,050 by 920.
6. Multiply 4,297 by 647.
7. Multiply 3,084 by 508.
8. Divide 32,061 by 78.
9. Divide 70,432 by 37.
10. Divide 96,584 by 236.

When you have finished the diagnostic test, correct it yourself using the answers on page 159. Then check all the questions you got wrong on the list below:

1. ______	3. ______	5. ______	8. ______
2. ______	4. ______	6. ______	9. ______
		7. ______	10. ______

If you made no mistakes at all, you need not do any work in this chapter, so go straight on to Chap. 4. But if you made any mistakes (even if it was only *one* mistake), follow these instructions:

	section	page
If you made any mistakes in Question 1 or 2, circle Sec. 3-1.	3-1	82
If you made any mistakes in Question 3 or 4, circle Sec. 3-2.	3-2	94
If you made any mistakes in Questions 5 to 7, circle Secs. 3-3 and 3-4.	3-3	106
	3-4	117
If you made any mistakes in Questions 8 to 10, circle Secs. 3-5, 3-6, and 3-7.	3-5	124
	3-6	131
	3-7	145

You should now work through the sections whose numbers you have circled, taking them in numerical order. The sections begin on the pages indicated.

Section 3-1
Addition of Large Numbers

Do this section if you made a mistake in Question 1 or 2 of Diagnostic Test 3. Begin by taking the pretest below, which will show you how much of Sec. 3-1 you already know and what you will be able to do by the time you have finished it.

PRETEST 3-1

Add the following numbers:

1. 321, 745, and 44
2. 1,153, 625, and 1,222
3. 3,230, 4,256, 7,682, and 5,576
4. 2,654, 3,329, 7,615, and 406
5. 1,040, 7, 295, and 2,163

When you have finished the pretest, correct it yourself using the answers on page 160. Then check all the questions you got wrong on the list below:

1. ______ 3. ______ 5. ______

2. ______ 4. ______

If you made any mistakes (even if it was only *one* mistake), read the introduction and go through the program.

If you made no mistakes at all, skip Sec. 3-1, because you already know its material, and go on to the next section whose number you circled on page 81. If no further numbers were circled, take Final Test 3 (page 159).

Introduction

In this section you will learn how to add large natural numbers accurately and rapidly. The methods will be carefully explained so you can understand why they work.

FRAMES FOR SECTION 3-1

1 In Chap. 1 we showed that we can add numbers between 10 and 100 by adding the units and the tens separately and then combining. Thus:

a.

	Tens		Units	
38 =	30	+	8	
11 =	10	+	1	
38 + 11 =	__	+	9	= __

b.

	Tens		Units	
26 =	20	+	6	
42 =	__	+	2	
26 + 42 =	__	+	8	= __

a. 40; 49 *b.* 40; 60; 68

---

2 *a.*

$$\begin{array}{r} 35 = \_\_ + 5 \\ 13 = \_\_ + 3 \\ \hline 35 + 13 = \_\_ + \_ = \_\_ \end{array}$$

b.

$$\begin{array}{r} 43 = \_\_ + \_\_ \\ 25 = \_\_ + \_\_ \\ \hline 43 + 25 = \_\_ + \_\_ = \_\_\_ \end{array}$$

a. 30; 10; 40 + 8 = 48 *b.* 40 + 3; 20 + 5; 60 + 8 = 68

---

3 This method can be written in much shorter form by crossing out the three 0s and three plus signs, so

$$\begin{array}{r} 35 = 3\not{0} \not{+} 5 \\ 13 = 1\not{0} \not{+} 3 \\ \hline 35 + 13 = 4\not{0} \not{+} 8 \end{array} \quad \text{becomes} \quad \begin{array}{r} 35 \\ +13 \\ \hline 48 \end{array}$$

and

$$\begin{array}{r} 43 = 4\not{0} \not{+} 3 \\ 25 = 2\not{0} \not{+} 5 \\ \hline 43 + 25 = 6\not{0} \not{+} 8 \end{array} \quad \text{becomes} \quad \begin{array}{r} -- \\ +-- \\ \hline -- \end{array}$$

$$\begin{array}{r} 43 \\ +25 \\ \hline 68 \end{array}$$

---

4 *a.* Similarly

$$\begin{array}{r} 34 = 3\not{0} \not{+} \_\_ \\ 45 = 4\not{0} \not{+} \_\_ \\ \hline 34 + 45 = 7\not{0} \not{+} \_\_ \end{array} \quad \text{becomes} \quad \begin{array}{r} 34 \\ +45 \\ \hline -- \end{array}$$

b. Use the shorter form alone to add 23 and 52:

$$\begin{array}{r} 23 \\ +52 \\ \hline -- \end{array}$$

a. 4; 5; 9; 79 *b.* 75

---

5 Use this shorter form for the following additions:

a. $\begin{array}{r} 31 \\ +56 \\ \hline -- \end{array}$ *b.* $\begin{array}{r} 27 \\ +32 \\ \hline -- \end{array}$ *c.* $\begin{array}{r} 20 \\ 67 \\ +10 \\ \hline -- \end{array}$

a. 87 *b.* 59 *c.* 97

---

6 This simple method breaks down when one column adds up to more than 9.

a. With $55 + 28$, for example, $5 + 8 = 13$, and we cannot write 13 in a single column. Se we rewrite the sum as follows:

$$\begin{array}{r} 55 \\ +\underline{28} \end{array} \Rightarrow \begin{array}{r} {\scriptstyle 1\,3} \\ 50 \\ +\underline{20} \\ -\,- \end{array}$$ which can be added as before.

b. Similarly with $66 + 29$, we find that $6 + 9 =$ ______, so we write the sum

$$\begin{array}{r} 66 \\ +\underline{29} \end{array} \Rightarrow \begin{array}{r} -\,- \\ 60 \\ +\underline{20} \\ -\,- \end{array}$$

---

a. 83 *b.* 15; 95

---

7 *a.* Since $5 + 7 =$ ______,

$$\begin{array}{r} 25 \\ +\underline{57} \end{array} \Rightarrow \begin{array}{r} -\,- \\ 20 \\ +\underline{50} \\ -\,- \end{array}$$

b. Similarly,

$$\begin{array}{r} 23 \\ +\underline{38} \end{array} \Rightarrow \begin{array}{r} -\,- \\ 20 \\ +\underline{30} \\ -\,- \end{array}$$

---

a. 12; 12; 82 *b.* 11; 61

---

8 *a.*

$$\begin{array}{r} 37 \\ +\underline{46} \end{array} \Rightarrow \begin{array}{r} -\,- \\ 30 \\ +\underline{40} \\ -\,- \end{array}$$

b.

$$\begin{array}{r} 68 \\ +\underline{28} \end{array} \Rightarrow \begin{array}{r} -\,- \\ 60 \\ +\underline{20} \\ -\,- \end{array}$$

---

a. 13; 83 *b.* 16; 96

---

9 This procedure is a simple two-stage process. With $65 + 28$, for example,

a. Add the units column, record the result, and record the units digit of the answer:

$$\begin{array}{r} 65 \\ +\underline{28} \end{array} \quad \begin{array}{r} -\,- \\ 6\,0 \\ +\underline{2\,0} \end{array} \quad \begin{array}{r} -\,{\scriptstyle 3} \\ 6\,0 \\ +\underline{2\,0} \\ - \end{array}$$

b. Add the tens column:

$$\begin{array}{r} -\,{\scriptstyle 3} \\ 6\,0 \\ +\underline{2\,0} \\ -\,- \end{array}$$

---

a.
```
 1 3
 60
+20
---
  3
```
b.
```
 1 3
 60
+20
---
 93
```

---

10 Similarly with 24 + 38:

a. Add the units column, record the result, and record the units digit of the answer:

```
 24          — —
+38    ⇒     20
---         +30
            ---
```

b. Add the tens column:

```
 — —
 20
+30
---
```

---

a.
```
 1 2
 20
+30
---
  2
```
b.
```
 1 2
 20
+30
---
 62
```

---

11 Similarly with 17 + 59:

a. Add the units column, record the result, and record the units digit of the answer:

```
 17          — —
+59    ⇒     10
---         +50
            ---
```

b. Add the tens column:

```
 — —
 10
+50
---
```

---

a.
```
 1 6
 10
+50
---
  6
```
b.
```
 1 6
 10
+50
---
 76
```

---

12 But this simple two-stage process can be applied to the original sum without rewriting it. Thus with 17 + 59:

a. Add the units column, record the units digit of the answer, and transfer 1 ten to the tens column:

```
  1
 17
+59
---
```

b. Add the revised tens column:

```
  1
 17
+59
---
```

---

a.

```
   1
  17
 +59
 ---
   6
```

b.

```
   1
  17
 +59
 ---
  76
```

Compare this answer with the answer to Frame 11.

---

13 Similarly with 24 + 38:

a. Add the units column, record the units digit of the answer, and transfer 1 ten to the tens column:

```
  24
 +38
 ---
```

b. Add the revised tens column:

```
  24
 +38
 ---
```

---

a.

```
   1
  24
 +38
 ---
   2
```

b.

```
   1
  24
 +38
 ---
  62
```

Compare this answer with the answer to Frame 10.

---

14 The same method can be used for adding more than two numbers. Thus to add 32, 17, and 24:

a. Add the units column, record the units digit of the answer, and transfer 1 ten to the tens column:

```
  32
  17
 +24
 ---
```

b. Add the revised tens column:

```
  32
  17
 +24
 ---
```

---

a.

```
   1
  32
  17
 +24
 ---
   3
```

b.

```
   1
  32
  17
 +24
 ---
  73
```

---

15 Sometimes more than 1 ten has to be transferred, as with 48 + 29 + 16.

a. Add the units column, record the units digit of the answer, and transfer ___ tens to the tens column:

```
 _
 48
 29
+16
 ——
  _
```

b. Add the revised tens column:

```
 _
 48
 29
+16
 ——
 __
```

---

a. 2;

```
 2
 48
 29
+16
 ——
  3
```

b.

```
 2
 48
 29
+16
 ——
 93
```

---

16 *a.* Use the same method to add 37, 28, and 19. Add the units column, record the units digit of the answer, and transfer tens. Then add the revised tens column:

```
 _
 37
 28
+19
 ——
 __
```

b. Use the same method to add 25, 33, and 12:

```
 _
 25
 33
+12
 ——
 __
```

---

a.

```
 2
 37
 28
+19
 ——
 84
```

b.

```
 1
 25
 33
+12
 ——
 70
```

---

17 Use the same method to add:

a. 41, 33, and 16:

```
 _
 41
 33
+16
 --
```

b. 39, 25, and 27:

```
 _
 39
 25
+27
 --
```

---

a.

```
  1
 41
 33
+16
 90
```

b.

```
  2
 39
 25
+27
 91
```

---

18 Use the same method to add the following numbers:

a. 27, 13, 34 *b.* 30, 28, 23 *c.* 48, 18, 24

---

a.

```
  1
 27
 13
+34
 74
```

b.

```
  1
 30
 28
+23
 81
```

c.

```
  2
 48
 18
+24
 90
```

---

19 With three-digit numbers the procedure is essentially the same. Thus to add 349, 287, and 158:

a. Add the units column, record the units digit of the answer, and transfer tens:

```
  2
 349
 287
+158
   4
```

b. Add the revised tens column, record the tens digit of the answer, and transfer hundreds:

```
 _2
 349
 287
+158
  _4
```

c. Add the revised hundreds column:

```
 _2
 349
 287
+158
 __4
```

```
b.   1 2        c.   1 2
     349             349
     287             287
    +158            +158
      94             794
```

20 Similarly to add 158, 278, and 391:

a. Add the units column, record the units digit of the answer, and transfer tens:

```
   _
  158
  278
 +391
    _
```

b. Add the revised tens column, record the tens digit of the answer, and transfer hundreds:

```
  _ _
  158
  278
 +391
   _ _
```

c. Add the revised hundreds column:

```
  _ _
  158
  278
 +391
  _ _ _
```

```
a.    1      b.   2 1      c.   2 1
     158          158           158
     278          278           278
    +391         +391          +391
       7           27           827
```

21 *a.* Use this same method to add 472, 287, and 194. Add the units column, record the units digit of the answer, and transfer tens. Add the revised tens column, record the tens digit of the answer, and transfer hundreds. Add the revised hundreds column:

```
 --
 472
 287
+194
 ---
```

b. Use this same method to add 315, 492, and 133:

```
 --
 315
 492
+133
 ---
```

---

a.
```
 21
 472
 287
+194
 953
```

b.
```
 11
 315
 492
+133
 940
```

---

22 Use this same method to add:

a. 238, 129, and 356

```
 --
 238
 129
+356
 ---
```

b. 122, 500, and 274

```
 122
 500
+274
 ---
```

---

a.
```
 12
 238
 129
+356
 723
```

b.
```
 122
 500
+274
 896
```

---

23 Use this method to add the following numbers:

a. 316, 249, 127 *b.* 402, 311, 240 *c.* 561, 198, 187

---

a. 2
316
249
+127
692

b. 402
311
+240
953

c. 21
561
198
+187
946

---

24 Sometimes the number transferred is the only member of a new column. Thus

a. _ _
63
+79
_ _ _

b. _ _
453
291
+724
_ _ _ _

---

a. 11
63
+79
142

b. 11
453
291
+724
1,468

---

25 Add the following numbers:

a. 64, 79, and 88

b. 325, 406, and 737

---

a. 22
64
79
+88
231

b. 1 1
325
406
+737
1,468

---

26 The same methods can be extended to even larger numbers.

a. _ _ _
1,432
2,967
+3,845
_ _ _ _

b. _ _ _ _
6,780
3,254
+2,987
_ _ _ _ _

---

a. 2 11
1,432
2,967
+3,845
8,244

b. 12 21
6,780
3,254
+ 2,987
13,021

---

27 Add the following numbers:

a. 1,234, 5,678, and 5,432

b. 2,697, 4,235, and 2,038

c. 4,325, 5,296, and 3,879

---

a.
```
  11 11
  1,234
  5,678
 +5,432
 12,344
```

b.
```
    12
  2,697
  4,235
 +2,038
  8,970
```

c.
```
  11 22
  4,325
  5,296
 +3,879
 13,500
```

---

28 A large number of adding mistakes arise from getting the columns muddled up. This can be avoided by writing down the numbers to be added very carefully so that it is quite clear which column each digit is in. It should be possible to draw a vertical line between each column.

a. To add 33, 202, and 1,079 we write

```
   -|-|
    |3|3
   2|0|2
+1,|0|7|9
 -|-|-|-
```

b. Add 4, 560, and 1,008.

---

a.
```
    1 1
     3 3
   2 0 2
+1,0 7 9
 1,3 1 4
```

b.
```
      1
        4
    5 6 0
+1,0 0 8
 1,5 7 2
```

---

29 *a.* Add 109, 35, 3,061, and 7.

b. Add 3,050, 23, 419, and 9,899.

---

a.
```
    12
    109
     35
  3,061
+     7
  3,212
```

b.
```
  11 12
  3,050
     23
    419
 +9,899
 13,391
```

---

30 *a.* Add 5,069, 478, 4,239, and 8.

b. Add 4,000, 299, 31, and 873.

a.

$$\begin{array}{r} {\scriptstyle 1\,3} \\ 5{,}069 \\ 478 \\ 4{,}239 \\ +\quad 8 \\ \hline 9{,}794 \end{array}$$

b.

$$\begin{array}{r} {\scriptstyle 1\;2\,1} \\ 4{,}000 \\ 299 \\ 31 \\ +\ 873 \\ \hline 5{,}203 \end{array}$$

Note: Once you are used to addition, it is normal to remember the transfer of tens, hundreds, and thousands and not to use "reminder digits." Thus the answers to Frame 30 are normally written as

$$\begin{array}{r} 5{,}069 \\ 478 \\ 4{,}239 \\ +\quad 8 \\ \hline 9{,}794 \end{array} \qquad \text{and} \qquad \begin{array}{r} 4{,}000 \\ 299 \\ 31 \\ +\ 873 \\ \hline 5{,}203 \end{array}$$

No reminder digits will be used for addition in the rest of this book.

You have finished the program for Sec. 3-1. Now take the posttest to see how much you have learned.

POSTTEST 3-1

Add the following numbers:

1. 432, 643, and 35
2. 2,136, 422, and 1,442
3. 5,643, 9,724, 4,830, and 9,976
4. 5,729, 2,304, 2,456, and 513
5. 2,008, 9, 325, and 5,260

When you have finished the posttest, correct it yourself using the answers on page 160. Then check all the questions you got wrong on the list below:

1. ______ 3. ______ 5. ______

2. ______ 4. ______

If you compare your performance with that on Pretest 3-1, you will see how much you have improved. If you made more than one mistake on this posttest, review the material and take the pretest again. Don't move on to the next section until you can get all the questions right.

Now proceed to the next section whose number you circled on page 81. If no other sections were circled, take Final Test 3 (page 159).

Section 3-2
Subtraction of Large Numbers

Do this section if you made a mistake in Question 3 or 4 of Diagnostic Test 3. Begin by taking the pretest below, which will show you how much of Sec. 3-2 you already know and what you will be able to do by the time you have finished it.

PRETEST 3-2

Subtract the following numbers:

1. 255 from 325
2. 491 from 1,484
3. 746 from 4,253
4. 1,925 from 2,608
5. 2,653 from 5,000

When you have finished the pretest, correct it yourself using the answers on page 160. Then check all the questions you got wrong on the list below:

1. _____ 3. _____ 5. _____

2. _____ 4. _____

If you made any mistakes (even if it was only *one* mistake), read the introduction and go through the program.

If you made no mistakes at all, skip Sec. 3-2, because you already know its material, and go on to the next section whose number you circled on page 81. If no further numbers were circled, take Final Test 3 (page 159).

Introduction

In this section you will learn how to subtract large natural numbers accurately and rapidly. The methods will be carefully explained so you can understand why they work.

FRAMES FOR SECTION 3-2

1 In Chap. 1 we showed that we can subtract numbers between 10 and 100 by subtracting the units and the tens separately and then combining. Thus

		Tens		Units				Tens		Units	
a.	48 =	40	+	8		*b.*	56 =	__	+	6	
	27 =	20	+	7			22 =	__	+	2	
	48 − 27 =	__	+	1 =	_____		56 − 22 =	__	+	4 =	_____

a. 20; 21 *b.* 50; 20; 30; 34

2 *a.*

$$\begin{array}{r} 65 = \_ + 5 \\ 13 = \_ + 3 \\ \hline 65 - 13 = \_ + \_ = \_\_\_\_ \end{array}$$

b.

$$\begin{array}{r} 78 = \_ + 8 \\ 51 = \_ + 1 \\ \hline 78 - 51 = \_ + \_ = \_\_\_\_ \end{array}$$

---

a. 60; 10; 50 + 2 = 52 *b.* 70; 50; 20 + 7 = 27

---

3 This method can be written in much shorter form by crossing out the three 0s and three plus signs, so

$$\begin{array}{r} 65 = 6\not{0} \not{+} 5 \\ 13 = 1\not{0} \not{+} 3 \\ \hline 65 - 13 = 5\not{0} \not{+} 2 \end{array} \quad \text{becomes} \quad \begin{array}{r} 65 \\ -13 \\ \hline 52 \end{array}$$

and

$$\begin{array}{r} 78 = 7\not{0} \not{+} 8 \\ 51 = 5\not{0} \not{+} 1 \\ \hline 78 - 51 = 2\not{0} \not{+} 7 \end{array} \quad \text{becomes} \quad \begin{array}{r} \_\,\_ \\ -\_\,\_ \\ \hline \_\,\_ \end{array}$$

$$\begin{array}{r} 78 \\ -51 \\ \hline 27 \end{array}$$

---

4 *a.*

$$\begin{array}{r} 86 = 8\not{0} \not{+} \_ \\ 32 = 3\not{0} \not{+} \_ \\ \hline 86 - 32 = 5\not{0} \not{+} \_ \end{array} \quad \text{becomes} \quad \begin{array}{r} 8\;6 \\ -3\;2 \\ \hline \_\,\_ \end{array}$$

b. The shorter form alone can be used to subtract 33 from 76:

$$\begin{array}{r} 7\;6 \\ -3\;3 \\ \hline \_\,\_ \end{array}$$

---

a.

$$\begin{array}{r} 86 = 8\not{0} \not{+} 6 \\ 32 = 3\not{0} \not{+} 2 \\ \hline 86 - 32 = 5\not{0} \not{+} 4 \end{array} \quad \text{becomes} \quad \begin{array}{r} 86 \\ -32 \\ \hline 54 \end{array}$$

b.

$$\begin{array}{r} 76 \\ -33 \\ \hline 43 \end{array}$$

---

5 Use this shorter form for the following subtractions:

a. $\begin{array}{r} 5\;9 \\ -1\;4 \\ \hline \_\,\_ \end{array}$ *b.* $\begin{array}{r} 6\;3 \\ -1\;3 \\ \hline \_\,\_ \end{array}$ *c.* $\begin{array}{r} 7\;3 \\ -3\;0 \\ \hline \_\,\_ \end{array}$

---

a. 45 *b.* 50 *c.* 43

---

6 This simple method breaks down for an example like $63-27$ because $3-7$ is undefined. The longer method we used in Chap. 1 solved this problem by replacing $3-7$ by $13-7$ in the units column.

$$\begin{array}{r} 63 = \_\_ + 13 \\ 27 = \_\_ + \ \ 7 \\ \hline 63 - 27 = \_\_ + \ \ 6 = \_\_\_\_ \end{array}$$

This meant replacing 60 in the tens column by ____.

$$\begin{array}{r} 63 = 50 + 13 \\ 27 = 20 + \ \ 7 \\ \hline 63 - 27 = 30 + \ \ 6 = 36 \end{array} \qquad 50$$

---

7 Now instead of replacing 60 in the tens column by 50, we can replace it by $60-10$ and write

$$\begin{array}{r} 63 = (60 - 10) + 13 \\ 27 = \ 20 \qquad\quad + \ \ 7 \\ \hline 63 - 27 = \ \_\_ \qquad\quad + \ \ 6 = \_\_\_\_ \end{array} \qquad \text{because } (60-10)-20 = \_\_\_\_$$

$$\begin{array}{r} 63 = (60 - 10) + 13 \\ 27 = \ 20 \qquad\quad + \ \ 7 \\ \hline 63 - 27 = \ 30 \qquad\quad + \ \ 6 = 36 \end{array} \qquad 30$$

---

8 We can use this same method to find $76-39$:

$$\begin{array}{r} 76 = (70 - 10) + \_\_ \\ 39 = \ 30 \qquad\quad + \_\_ \\ \hline 76 - 39 = \ \_\_ \qquad\quad + \_\_ = \_\_\_\_ \end{array} \qquad \text{because } (70-10)-30 = \_\_\_\_$$

$$\begin{array}{r} 76 = (70 - 10) + 16 \\ 39 = \ 30 \qquad\quad + \ \ 9 \\ \hline 76 - 39 = \ 30 \qquad\quad + \ \ 7 = 37 \end{array} \qquad 30$$

---

9 *a.* Using the same method to find $55-17$, we can write

$$\begin{array}{r} 55 \\ \underline{-17} \end{array} \quad \Rightarrow \quad \begin{array}{r} (50 - 10) + \_\_ \\ 10 + \_\_ \\ \hline \_\_ + \_\_ = \_\_\_\_ \end{array}$$

b. Or if we put the -10 at the bottom, we can write

$$\begin{array}{r} 50 + \_\_ \\ -(10 + \_\_) \\ -10 \qquad \\ \hline \_\_ + \_\_ = \_\_\_\_ \end{array}$$

$$a.\quad \begin{array}{lr} (50-10) & +\ 15 \\ \underline{10} & \underline{+\ \ 7} \\ 30 & +\ \ 8 = 38 \end{array} \qquad b.\quad \begin{array}{r} 50 + 15 \\ -(10 + \ \ 7) \\ \underline{-10} \\ 30 + \ \ 8 = 38 \end{array}$$

---

10 Similarly to find 83 − 36 we can write

$$a.\quad \begin{array}{r} 83 \\ \underline{-36} \end{array} \Rightarrow \begin{array}{lr} (80-10) & +\ \_\_\_ \\ \underline{30} & \underline{+\ \_\_\_} \\ \_\_ & +\ \_\_\_ = \_\_\_\_ \end{array}$$

or

$$b.\quad \begin{array}{r} 80 + \_\_\_ \\ -(30 + \_\_\_) \\ \underline{-10} \\ \_\_ + \_\_\_ = \_\_\_\_ \end{array}$$

$$a.\quad \begin{array}{lr} (80-10) & +\ 13 \\ \underline{30} & \underline{+\ \ 6} \\ 40 & +\ \ 7 = 47 \end{array} \qquad b.\quad \begin{array}{r} 80 + 13 \\ -(30 + \ \ 6) \\ \underline{-10} \\ 40 + \ \ 7 = 47 \end{array}$$

---

11 The method of Frames 9*b* and 10*b* can be shortened by crossing out the 0s and plus signs, so

$$\begin{array}{r} 5\not{0} \not{+} 15 \\ -(1\not{0} \not{+}\ \ 7) \\ \underline{-1\not{0}\phantom{\not{+}\ \ 7)}} \\ 3\not{0} \not{+}\ \ 8 \end{array} \Rightarrow \begin{array}{r} {}^{1} \\ 5\ 5 \\ -1\ 7 \\ \underline{{}^{-1}} \\ 3\ 8 \end{array}$$

and

$$\begin{array}{r} 8\not{0} \not{+} 13 \\ -3\not{0} \not{+}\ \ 6 \\ \underline{-1\not{0}\phantom{\not{+}\ \ 6}} \\ 4\not{0} \not{+}\ \ 7 \end{array} \Rightarrow \begin{array}{r} \_ \\ \_\ \_ \\ -\_\ \_ \\ \underline{-\_} \\ \_\ \_ \end{array}$$

$$\begin{array}{r} {}^{1} \\ 8\ 3 \\ -3\ 6 \\ \underline{{}^{-1}} \\ 4\ 7 \end{array}$$

---

12 In subtracting 25 from 71,

a.

```
   70 + _              1
 −(20 + _)    ⇒      7 _
 −10                −2 _
 ______             − 1
   40 + _           ____
                     _ _
```

b.[*] The shorter form alone can be used to subtract 38 from 76:

```
  1
 7 6
−3 8
− 1
____
 _ _
```

---

a.

```
   70 + 11             1
 −(20 +  5)   ⇒      7 1
 −10                −2 5
 ______             − 1
   40 +  6          ____
                     4 6
```

b. 38

---

13 *a.* Subtract 29 from 70:

```
  1
 7 0
−2 9
− 1
____
 _ _
```

b. Subtract 34 from 82:

```
  1
 8 2
−3 4
− 1
____
 _ _
```

---

a. 41 *b.* 48

---

14 This subtraction method has two stages for 43 − 18.

a. Subtract the units column, adding a ten if necessary: 13 − 8 = ______

Then record the units digit of the answer and put −1 in the tens column if 1 ten was transferred.

```
  43
 −18
 − 1
 ___
   _
```

Notice that the 1 above the top now has been omitted.

b. Subtract the revised tens column:

```
  4 3
 −1 8
 − 1
 ____
  _ _
```

---

a. 5; 5 *b.* 25

---

15 Using this same method for 86 − 49,

a. Subtract the units column, adding a ten if necessary; then record the unit digit of the answer and put −1 in the tens column if 1 ten was transferred:

```
 86
−49
−
 __
  _
```

b. Subtract the revised tens column:

```
 8 6
−4 9
−
 ___
 _ _
```

---

a. 1; 7 *b.* 1; 37

---

16 *a.* Use the same method to subtract 41 from 80. Subtract the units column, transferring 1 ten if necessary; then record the units digit of the answer. Subtract the revised tens column:

```
 8 0
−4 1
−
 ___
 _ _
```

b. Use the same method to subtract 19 from 42:

```
 4 2
−1 9
−
 ___
 _ _
```

---

a. 1; 39 *b.* 1; 23

---

17 Use the same method to complete the following:

```
a.   7 3     b.   5 6     c.   9 4
    −3 8         −2 5         −6 5
    −            ___          −
     ___         _ _           ___
     _ _                       _ _
```

---

a. 1; 35 *b.* 31 *c.* 1; 29

---

18 Subtract the following:

a. 28 from 65 *b.* 39 from 89 *c.* 46 from 90

---

a.
```
  65
 −28
 − 1
 ---
  37
```
b.
```
  89
 −39
 ---
  50
```
c.
```
  90
 −46
 − 1
 ---
  44
```

---

19 With three-digit numbers the procedure is essentially the same. Thus to subtract 227 from 813:

a. Subtract the units column, transferring 1 ten if necessary; then record the units digit of the answer:

```
  813
 −227
 −  1
 ----
    6
```

b. Subtract the revised tens column, transferring 1 hundred if necessary; then record the tens digit of the answer:

```
  813
 −227
 − 11
 ----
   _6
```

c. Subtract the revised hundreds column:

```
  8 1 3
 −2 2 7
 −1 1
 ------
  _ _ 6
```

---

b.
```
  813
 −227
 − 11
 ----
   86
```
c.
```
  813
 −227
 − 11
 ----
  586
```

---

20 Use the same method to subtract 448 from 826.

a. Subtract the units column, transferring 1 ten if necessary; then record the units digit of the answer:

```
  826
 -448
 - 1
 ----
    8
```

b. Subtract the revised tens column, transferring 1 hundred if necessary; then record the tens digit of the answer:

```
  826
 -448
 -_1
 ----
   _8
```

c. Subtract the revised hundreds column:

```
  8 2 6
 -4 4 8
 -_ 1
 ------
  _ _ 8
```

---

b.
```
  826
 -448
 -11
 ----
   78
```

c.
```
  826
 -448
 -11
 ----
  378
```

21 Use the same method to subtract 386 from 912.

a. Subtract the units column, transferring 1 ten if necessary; then record the units digit of the answer:

```
 912
-386
- _
  _
```

b. Subtract the revised tens column, transferring 1 hundred if necessary; then record the tens digit of the answer:

```
 9 1 2
-3 8 6
- _ _
  _ _
```

c. Subtract the revised hundreds column:

```
 9 1 2
-3 8 6
- _ _
 _ _ _
```

```
a.   912     b.   912     c.   912
    -386         -386         -386
    -  1         - 11         - 11
       6           26          526
```

---

22 *a.* Use the same method to subtract 735 from 913. Subtract the units column, transferring 1 ten if necessary; then record the units digit of the answer. Subtract the revised tens column, transferring 1 hundred if necessary; then record the tens digit of the answer. Subtract the revised hundreds column:

```
 9 1 3
-7 3 5
- _ _
 _ _ _
```

b. Use the same method to subtract 294 from 743:

```
 7 4 3
-2 9 4
- _ _
 _ _ _
```

---

a. 11; 178 *b.* 11; 449

---

23 Use this same method to subtract:

a. 346 from 723

```
 7 2 3
–3 4 6
–_ _
 _ _ _
```

b. 492 from 865

```
 8 6 5
–4 9 2
–_
 _ _ _
```

---

a. 11; 377 *b.* 1; 373

---

24 Use the same method to subtract:

a. 178 from 468

```
 4 6 8
–1 7 8
–_
 _ _ _
```

b. 587 from 640

```
 6 4 0
–5 8 7
–_ _
 _ _ _
```

---

a. 1; 290 *b.* 11; 53

---

25 Use this same method to subtract:

a. 467 from 963

```
 9 6 3
–4 6 7
–_ _
 _ _ _
```

b. 314 from 567

```
 5 6 7
–3 1 4
 _ _ _
```

---

a. 11; 496 *b.* 253

---

26 Subtract the following numbers:

a. 276 from 845 *b.* 357 from 762 *c.* 874 from 929

---

a.
```
 845
–276
–11
 569
```

b.
```
 762
–357
– 1
 405
```

c.
```
 929
–874
–1
  55
```

---

27 Subtract the following numbers:

a. 193 from 570 *b.* 368 from 762 *c.* 487 from 728

```
a.   570     b.   762     c.   728
    −193         −368         −487
    −11          −11          −1
     377          394          241
```

28 Extending this method to four-digit numbers we get:

```
a.   1,2 0 1              b.   9,8 4 3
    −  9 2 7                  −3,8 7 5
    −_ _ _                    −_ _ _
       _ _ _                   _ _ _ _
```

```
a.   1,201     b.   9,843
    −  927         −3,875
    −1 11          −1 11
       274          5,968
```

29 Subtract the following numbers:

a. 5,830 from 6,308   *b.* 2,935 from 6,431   *c.* 789 from 8,827

```
a.   6,308     b.   6,431     c.   8,827
    −5,830         −2,935         −  789
    −1 1           −1 11          −   11
       478          3,496          8,038
```

30 Subtract the following numbers:

a. 5,193 from 7,230   *b.* 999 from 3,256   *c.* 1,876 from 2,038

```
a.   7,230     b.   3,256     c.   2,038
    −5,193         −  999         −1,876
    −   11         −1 11          −1 1
     2,037          2,257            162
```

Note: Once you are used to subtraction, it is normal to remember the transfer of tens, hundreds, and thousands and not to use reminder digits. Thus the answers to Frame 30 are normally written as

```
  7,230      3,256      2,038
 −5,193     −  999     −1,876
  2,037      2,257        162
```

No reminder digits will be used for subtraction in the rest of the book.

You have finished the program for Sec. 3-2. Now take the posttest to see how much you have learned.

POSTTEST 3-2

Subtract the following numbers:

1. 466 from 546
2. 733 from 1,731
3. 678 from 5,182
4. 2,914 from 3,409
5. 5,287 from 7,000

When you have finished the posttest, correct it yourself using the answers on page 160. Then check all the questions you got wrong on the list below:

1. _____ 3. _____ 5. _____

2. _____ 4. _____

If you compare your performance with that on Pretest 3-2, you will see how much you have improved. If you made more than one mistake on this posttest, review the material and take the pretest again. Don't move on to the next section until you can get all the questions right.

Now proceed to the next section whose number you circled on page 81. If no other sections were circled, take Final Test 3 (page 159).

Section 3-3
Multiplication of Large Numbers by One-digit Numbers

Do this section if you made a mistake in Questions 5 to 7 of Diagnostic Test 3. Begin by taking the pretest below, which will show you how much of Sec. 3-3 you already know and what you will be able to do by the time you have finished it.

PRETEST 3-3

Multiply the following numbers:

1. 34 and 6
2. 789 and 8
3. 670 and 7
4. 756 and 9
5. 206 and 5

When you have finished the pretest, correct it yourself using the answers on page 160. Then check all the questions you got wrong on the list below:

1. _____ 3. _____ 5. _____
2. _____ 4. _____

If you made any mistakes (even if it was only *one* mistake), read the introduction and go through the program.

If you made no mistakes at all, skip Sec. 3-3, because you already know its material, and go on to Sec. 3-4, which begins on page 117.

Introduction

In this section you will learn how to multiply two- and three-digit numbers by one-digit numbers. You see how these multiplications are based upon the multiplication of one-digit numbers (Chap. 1) and the distributive law (Chap. 2).

FRAMES FOR SECTION 3-3

1 The distributive law of multiplication over addition states that $(a+b) \times n =$ __________.

$(a \times n) + (b \times n)$

---

2 In Chap. 2 we showed how the distributive law can be used in the multiplication of large numbers. Thus

a. $28 \times 6 = (20 + 8) \times 6 = (20 \times 6) + (8 \times 6)$
$= \_\_\_\_ + \_\_\_\_ = \_\_\_\_\_$

and

b. $37 \times 5 = (30 + 7) \times 5 = 150 + \_\_\_\_ = \_\_\_\_\_$

---

a. $120 + 48 = 168$ *b.* $150 + 35 = 185$

---

3 *a.* $54 \times 3 = (50 + \_\_) \times 3 = 150 + \_\_\_\_ = \_\_\_\_\_$
b. $29 \times 7 = (20 + \_\_) \times 7 = \_\_\_\_ + \_\_\_\_ = \_\_\_\_\_$

---

a. $(50 + 4) \times 3 = 150 + 12 = 162$ *b.* $(20 + 9) \times 7 = 140 + 63 = 203$

---

4 Instead of writing in Frame 3*a* that $54 \times 3 = (50 + 4) \times 3 = 150 + 12 = 162$, one could write

```
  54
  ×3
  12    4 × 3
 150   50 × 3
 162   54 × 3
```

and Frame 3*b*, which states that $29 \times 7 = (20 + 9) \times 7 = 140 + 63 = 203$, could be written as

```
  29
  ×7
  __    9 × 7
 ___   20 × 7
 ___   29 × 7
```

```
  29
  ×7
  63
 140
 203
```

---

5 *a.* Use this method to multiply 35×9:

```
 35
 ×9
 --    5 × 9
---   30 × 9
---   35 × 9
```

b. and to multiply 43 by 8:

```
 43
 ×8
 --
---
---
```

---

a.
```
 35
 ×9
 45
270
315
```

b.
```
 43
 ×8
 24
320
344
```

---

6 *a.* Use this same method to multiply 28 by 7:

```
 28
 ×7
 --
---
---
```

b. Use the same method to multiply 57 by 6.

---

a.
```
 28
 ×7
 56
140
196
```

b.
```
 57
 ×6
 42
300
342
```

---

7 Use the same method to multiply:

b. 75 by 9

---

a.
```
 68
 ×4
 32
240
272
```

b.
```
 75
 ×9
 45
630
675
```

8 This method can be shortened to a simple two-stage operation as follows. We shall use 68 × 4 again as an example.

a. Multiply units column:

$8 \times 4 = 32$

Record units digit of answer, and transfer 3 tens:

$$\begin{array}{r} {}^{3} \\ 68 \\ \times 4 \\ \hline 2 \end{array}$$

b. Multiply tens column and add transferred tens:

$(6 \times 4) + 3 = \_\_\_ + 3 = \_\_\_$

Then record first two digits of answer:

$$\begin{array}{r} {}^{3} \\ 6\,8 \\ \times 4 \\ \hline \_\,\_\,2 \end{array}$$

$24 + 3 = 27$ $\begin{array}{r} {}^{3} \\ 68 \\ \times 4 \\ \hline 272 \end{array}$ Compare this with the answer to Frame 7*a*.

---

9 Using 75 × 9 as another example,

a. Multiply units column:

$5 \times 9 = \_\_\_\_$

Record units digit of answer, and transfer 4 tens:

$$\begin{array}{r} {}^{4} \\ 75 \\ \times 9 \\ \hline \_ \end{array}$$

b. Multiply tens column and add transferred tens:

$(7 \times 9) + \_\_\_ = \_\_\_ + \_\_\_ = \_\_\_\_$

Then record first two digits of answer:

$$\begin{array}{r} {}^{4} \\ 7\,5 \\ \times 9 \\ \hline \_\,\_\,\_ \end{array}$$

---

a. 45; 5 *b.* 4; 63 + 4 = 67; 675 Compare this with the answer to Frame 7*b*.

---

10 Using 28 × 7 as another example,

a. Multiply units column, record units digit of answer, and transfer _____ tens:

$$\begin{array}{r} \_ \\ 28 \\ \times 7 \\ \hline \_ \end{array}$$

b. Multiply tens column and add transferred tens:

(2 × 7) + ____ = _____

Then record first two digits of answer:

$$\begin{array}{r} \_ \\ 28 \\ \times 7 \\ \hline \_\,\_\,\_ \end{array}$$

---

a. 5; $\begin{array}{r} {\scriptstyle 5} \\ 28 \\ \times 7 \\ \hline 6 \end{array}$ *b.* 5; 19; $\begin{array}{r} {\scriptstyle 5} \\ 28 \\ \times 7 \\ \hline 196 \end{array}$

Compare this with your answer to Frame 6*a*.

---

11 *a.* Use the same method to multiply 32 by 9. Multiply units column, record units digit of answer, and transfer tens; multiply tens column, add transferred tens, and record first two digits of answer:

$$\begin{array}{r} \_ \\ 32 \\ \times 9 \\ \hline \_\,\_\,\_ \end{array}$$

b. Use the same method to multiply 39 by 5.

$$\begin{array}{r} \_ \\ 39 \\ \times 5 \\ \hline \_\,\_\,\_ \end{array}$$

---

a.
```
  1
 32
 ×9
288
```
b.
```
  4
 39
 ×5
195
```

---

12 Use the same method to multiply:

a. 46 by 7

```
 _
 4 6
 × 7
 ---
 _ _ _
```

b. 39 by 8

```
 _
 3 9
 × 8
 ---
 _ _ _
```

---

a.
```
  4
 46
 ×7
322
```
b.
```
  7
 39
 ×8
312
```

---

13 *a.* Multiply 73 by 4:

```
 _
 7 3
 × 4
 ---
 _ _ _
```

b. Multiply 67 by 6.

---

a.
```
  1
 73
 ×4
292
```
b.
```
  4
 67
 ×6
402
```

---

14 *a.* Multiply 53 by 9.

b. Multiply 46 by 8.

---

a.
```
  2
 53
 ×9
477
```
b.
```
  4
 46
 ×8
368
```

---

15 *a.* Multiply 38 by 4.

b. Multiply 83 by 7.

---

a.
```
  3
 38
 ×4
152
```
b.
```
  2
 83
 ×7
581
```

---

16 Multiplying three-digit numbers is essentially similar. Let us use the longer method first so you can see what's going on. According to the distributive law,

$$235 \times 7 = (200 \times 7) + (30 \times 7) + (\_ \times 7)$$

We write this as

```
   2 3 5
     × 7
  -------
     _ _      _ × 7
   _ _ _     30 × 7
 _ _ _ _    200 × 7
 -------
 _ _ _ _    235 × 7
```

5;

```
   235
    ×7
  ----
    35     5 × 7
   210    30 × 7
  1400   200 × 7
  ----
  1645
```

---

17 The shorter method is a three-stage process for 235×7.

a. Multiply units column, record units digit of answer, and transfer tens:

```
  3
235
 ×7
---
  5
```

b. Multiply tens column, add transferred tens, record tens digit of answer, and transfer hundreds:

```
23
235
 ×7
---
 45
```

c. Multiply hundreds column, add transferred hundreds, and record first two digits of answer:

```
  2 3
 2 3 5
   × 7
 -----
 _ _ 4 5
```

1,645 Compare this with the answer to Frame 16.

---

18 Using the shorter method for 483 × 6,

a. Multiply units column, record units digit of answer, and transfer tens:

```
  1
483
 ×6
----
  8
```

b. Multiply tens column, add transferred tens, record tens digit of answer, and transfer hundreds:

```
_1
483
 ×6
----
 _8
```

c. Multiply hundreds column, add transferred hundreds, and record first two digits of answer:

```
 _ 1
 4 8 3
   × 6
-------
_ _ _ 8
```

---

b.

```
 41
483
 ×6
----
 98
```

c.

```
 41
483
 ×6
----
2898
```

---

19 Using the same method for 386 × 9,

a. Multiply units column, record units digit of answer, and transfer tens:

```
 _
386
 ×9
----
  _
```

b. Multiply tens column, add transferred tens, record tens digit of answer, and transfer hundreds:

```
_ _
3 8 6
  × 9
------
  _ _
```

c. Multiply hundreds column, add transferred hundreds, and record first two digits of answer:

```
_ _
3 8 6
  × 9
-------
_ _ _ _
```

a.	*b.*	*c.*
5	75	75
386	386	386
×9	×9	×9
4	74	3474

20 *a.* Use the same method to multiply 535 by 8. Multiply units column, record units digit of answer, and transfer tens. Multiply tens column, add transferred tens, record tens digit of answer, and transfer hundreds. Multiply hundreds column, add transferred hundreds, and record first two digits of answer:

```
 _ _
 5 3 5
   × 8
 -----
_ _ _ _
```

b. Use the same method to multiply 386 by 3.

```
 _ _
 3 8 6
   × 3
 -----
_ _ _ _
```

a.	*b.*
24	21
535	386
×8	×3
4280	1158

21 Use the same method to multiply:

a. 854 by 6

```
 _ _
 8 5 4
   × 6
 -----
_ _ _ _
```

b. 412 by 5

```
   _
 4 1 2
   × 5
 -----
_ _ _ _
```

a.	*b.*
32	1
854	412
×6	×5
5124	2060

22 Use the same method to multiply:

a. 325 by 4

```
 _ _
 3 2 5
   × 4
 -------
 _ _ _ _
```

b. 607 by 7

```
   _
 6 0 7
   × 7
 -------
 _ _ _ _
```

---

a.
```
  12
 325
  ×4
1300
```

b.
```
   4
 607
  ×7
4249
```

---

23 *a.* Multiply 384 by 2:

```
 _
 3 8 4
   × 2
 -----
 _ _ _
```

b. Multiply 268 by 5.

---

a.
```
  1
 384
  ×2
 768
```

b.
```
  34
 268
  ×5
1340
```

---

24 *a.* Multiply 218 by 4.

b. Multiply 403 by 9.

---

a.
```
  3
 218
  ×4
 872
```

b.
```
  2
 403
  ×9
3627
```

---

25 *a.* Multiply 429 by 5.

b. Multiply 560 by 3.

---

a.
```
  14
 429
  ×5
2145
```

b.
```
  1
 560
  ×3
1680
```

26 *a.* Multiply 725 by 8. *b.* Multiply 538 by 9.

---

a.

$$\begin{array}{r} {}^{2\,4} \\ 725 \\ \times 8 \\ \hline 5800 \end{array}$$

b.

$$\begin{array}{r} {}^{3\,7} \\ 538 \\ \times 9 \\ \hline 4842 \end{array}$$

Note: Once you are used to multiplication, it is normal to stop using reminders of the transfer of tens and hundreds and to remember these extra digits. Thus the answers to Frame 26 are normally written as

$$\begin{array}{r} 725 \\ \times 8 \\ \hline 5800 \end{array} \quad \text{and} \quad \begin{array}{r} 538 \\ \times 9 \\ \hline 4842 \end{array}$$

Since there is little room for reminder digits when multiplying by larger numbers, they will not be used for multiplication in the rest of the book.

You have finished the program for Sec. 3-3. Now take the posttest to see how much you have learned.

POSTTEST 3-3

Multiply the following numbers:

1. 43 and 7
2. 958 and 9
3. 590 and 6
4. 376 and 8
5. 408 and 5

When you have finished the posttest, correct it yourself using the answers on page 160. Then check all the questions you got wrong on the list below:

1. ______ 3. ______ 5. ______

2. ______ 4. ______

If you compare your performance with that on Pretest 3-3, you will see how much you have improved. If you made more than one mistake on this posttest, review the material and take the pretest again. Don't move on to the next section until you can get all the questions right.

Now proceed to Sec. 3-4.

Section 3-4
Long Multiplication

Do this section if you made a mistake in Questions 5 to 7 of Diagnostic Test 3. Begin by taking the pretest below, which will show you how much of Sec. 3-4 you already know and what you will be able to do by the time you have finished it.

PRETEST 3-4

Multiply the following numbers:

1. 799 and 713
2. 468 and 559
3. 4,329 and 586
4. 2,008 and 405
5. 6,050 and 840

When you have finished the pretest, correct it yourself using the answers on page 160. Then check all the questions you got wrong on the list below:

1. ______ 3. ______ 5. ______

2. ______ 4. ______

If you made any mistakes (even if it was only *one* mistake), read the introduction and go through the program.

If you made no mistakes at all, skip Sec. 3-4, because you already know its material, and go on to the next section whose number you circled on page 81. If no further numbers were circled, take Final Test 3 (page 159).

Introduction

In this section you will learn how to multiply large natural numbers by large natural numbers. It is assumed that you remember the commutative, associative, and distributive properties from Chap. 2. The section uses them to construct "long" multiplications from the "short" multiplications by one-digit numbers which were covered in Sec. 3-3.

FRAMES FOR SECTION 3-4

1 In Chap. 1 we showed that multiplication by 10 is equivalent to adding a _____ digit on the end of a number. Thus

$5 \times 10 =$ _____

$13 \times 10 =$ _____

$302 \times 10 =$ _____

$430 \times 10 =$ _____

---

0 50 130 3,020 4,300

---

2 We can then use the associative principle to multiply by two-digit numbers ending in 0. Thus:

a. $412 \times 20 = 412 \times (2 \times 10) = (412 \times 2) \times 10 =$ _____ $\times 10 =$ _____

and

b. $623 \times 40 = (623 \times 4) \times 10 =$ _____ $\times 10 =$ _____

---

a. 824; 8,240 *b.* 2,492; 24,920

---

3 *a.* $535 \times 60 = (535 \times 6) \times 10 =$ _____ $\times 10 =$ _____

b. $716 \times 70 =$ _____

---

a. 3,210; 32,100 *b.* 50,120

---

4 *a.* $481 \times 30 =$ _____

b. $704 \times 20 =$ _____

c. $590 \times 90 =$ _____

---

a. 14,430 *b.* 14,080 *c.* 53,100

---

5 *a.* $2 \times 100 = 2 \times (10 \times 10) = (2 \times 10) \times 10 =$ _____ $\times 10 =$ _____

b. $7 \times 100 =$ _____ $\times 10 =$ _____

c. $16 \times 100 =$ _____ $\times 10 =$ _____

d. $231 \times 100 =$ _____

e. Multiplying by 100 is the same as multiplying by 10 _____ times, so all you need do is add _____ 0 digits on the end.

---

a. 20; 200 *b.* 70; 700 *c.* 160; 1,600 *d.* 23,100
e. two; two

---

6 *a.* $33 \times 400 = (33 \times 4) \times 100 =$ ______

b. $189 \times 600 = (189 \times 6) \times 100 =$ ______

---

a. 13,200 *b.* 113,400

---

7 *a.* $718 \times 500 =$ ______

b. $586 \times 700 =$ ______

c. $950 \times 900 =$ ______

---

a. 359,000 *b.* 410,200 *c.* 855,000

---

8 So far we have multiplied large numbers only by one-digit numbers or by multiples of 10 and one-digit numbers. Thus we know that $421 \times 30 =$ ______ and $421 \times 8 =$ ______. Now the distributive law tells us that

$(30 + 8) \times 421 = ($______ $\times 421) + ($______ $\times 421)$

and since multiplication is commutative, we can rewrite each multiplication the other way round. So

$421 \times (30 + 8) = (421 \times$ ______$) + (421 \times$ ______$)$

and

$421 \times 38 =$ ______ + ______ = ______

12,630 3,368 30 8 30 8 12,630 + 3,368 = 15,998

---

9 Similarly

$584 \times 26 = (584 \times$ ______$) + (584 \times 6) =$ ______ + ______ = ______

---

20 11,680 + 3,504 = 15,184

---

10 The multiplications of Frames 8 and 9 could have been written in shorter form as follows:

a.
```
   421
   ×38
  ----
  3368   421 × 8
 12630   421 × 30
 -----   421 × 38
```

b.
```
   584
   ×26
  ----
  ----   584 × 6
 -----   584 × 20
 -----   584 × 26
```

---

a. 15,998 *b.* 3504
11680
15184

---

11 In this method, called *long multiplication,* we use the distributive law to reduce the problem to a series of short multiplications and a final addition:

a.
```
   459
   ×63
  ----
  ----   459 × 3
 -----   459 × 60
 -----   459 × 63
```

b.
```
   327
   ×57
  ----
  ----
 -----
 -----
```

---

a. 1377
27540
28917

b. 2289
16350
18639

---

12 *a.*
```
   492
   ×28
  ----
  ----
 -----
 -----
```

b.
```
   504
   ×71
  ----
  ----
 -----
 -----
```

---

a. 3936
9840
13776

b. 504
35280
35784

---

13 *a.* Multiply 340 by 23:

```
   3 4 0
  × 2 3
 -------
  - - - -
- - - - -
---------
- - - - -
```

b. Multiply 371 by 49.

```
a.  1020     b.    371
    6800           ×49
    ----          ----
    7820          3339
                 14840
                 -----
                 18179
```

---

14 *a.* Multiply 703 by 37.

b. Multiply 481 by 24.

```
a.    703     b.    481
      ×37           ×24
     ----          ----
     4921          1924
    21090          9620
    -----         -----
    26011         11544
```

---

15 The same principles apply to multiplications by numbers greater than 100, and the same basic method can be used.

```
a.       3 5 8
       × 4 3 6
       -------
       - - - -     358 × 6
     - - - - -     358 × 30
   - - - - - -     358 × 400
   -----------
   - - - - - -     358 × 436
```

b. Multiply 237 by 576.

```
a.    2148     b.      237
     10740            ×576
    143200            ----
    ------            1422
    156088           16590
                    118500
                    ------
                    136512
```

---

16 *a.* Multiply 603 by 148.

b. Multiply 1,238 by 983.

---

a.

```
   603
  ×148
  4824
 24120
 60300
 89244
```

b.

```
   1238
   ×983
   3714
  99040
1114200
1216954
```

---

17 In dealing with large numbers, it is a good idea to check that your answer is approximately correct. For example, in Frame 16*a*, 603 is roughly equal to _00, and 148 is between _00 and _00. So we can deduce that the answer to 603 × 148 will be between _0,000 and _ _0,000.

---

600 100 200 60,000 120,000

---

18 With Frame 16*b*, 1,238 × 983, we see that 1,238 is roughly equal to _,000 and 983 is roughly equal to _,000, so the answer will be roughly equal to ________. (To multiply by 1,000, one simply adds ____ 0s.)

1,000 1,000 1,000,000 three

---

19 *a.* Multiply 4,632 by 704. *b.* Multiply 5,210 by 329.

Check your answers by finding approximate answers.

---

a.

```
   4632
   ×704
  18528   4,632 × 4
3242400   4,632 × 700
3260928
```

Since 4,632 is roughly 5,000 and 704 is roughly 700, the answer should be about 3,500,000.

b.

```
   5210
   ×329
  46890   5,210 × 9
 104200   5,210 × 20, or 521 × 2 × 100
1563000   5,210 × 300, or 521 × 3 × 1,000
1714090
```

Since 5,210 is roughly 5,000 and 329 is roughly 300, the answer should be about 1,500,000.

---

20 *a.* Multiply 2,093 by 310. *b.* Multiply 4,320 by 509.

Check your answers by finding the approximate answers.

---

a.

```
   2093
  ×310
  20930   2,093 × 10
 627900   2,093 × 300
 648830
```

Since 2,093 is roughly 2,000 and 310 is roughly 300, the answer should be about 600,000.

b.

```
    4320
   ×509
   38880   4,320 × 9
 2160000   4,320 × 500, or 432 × 5 × 1,000
 2198880
```

Since 4,320 is roughly 4,000 and 509 is roughly 500, the answer should be about 2,000,000.

---

21 *a.* Multiply 2,004 by 430. *b.* Multiply 6,040 by 207.

Check your answers by finding the approximate answers.

---

a.

```
   2004
  ×430
  60120   2,004 × 30
 801600   2,004 × 400
 861720
```

Answer should be about 2,000 × 400 = 800,000.

b.

```
    6040
   ×207
   42280   6,040 × 7
 1208000   6,040 × 200, or 604 × 2 × 1,000
 1250280
```

Answer should be about 6,000 × 200 = 1,200,000.

You have finished the program for Sec. 3-4. Now take the posttest to see how much you have learned.

POSTTEST 3-4

Multiply the following numbers:

1. 489 and 317
2. 482 and 558
3. 6,278 and 597
4. 4,006 and 805
5. 7,050 and 960

When you have finished the posttest, correct it yourself using the answers on page 160. Then check all the questions you got wrong on the list below:

1. ______ 3. ______ 5. ______

2. ______ 4. ______

If you compare your performance with that on Pretest 3-4, you will see how much you have improved. If you made more than one mistake on this posttest, review the material and take the pretest again. Don't move on to the next section until you can get all the questions right.

Now proceed to the next section whose number you circled on page 81. If no other numbers were circled, take Final Test 3 (page 159).

Section 3-5 Division of Two-digit Numbers by One-digit Numbers

Do this section if you made a mistake in Questions 8 to 10 of Diagnostic Test 3. Begin by taking the pretest below, which will show you how much of Sec. 3-5 you already know and what you will be able to do by the time you have finished it.

PRETEST 3-5

Divide the following numbers:

1. 71 by 8
2. 57 by 6
3. 39 by 5
4. 73 by 9
5. 44 by 7

When you have finished the pretest, correct it yourself using the answers on page 160. Then check all the questions you got wrong on the list below:

1. ______ 3. ______ 5. ______

2. ______ 4. ______

If you made any mistakes (even if it was only *one* mistake), read the introduction and go through the program.

If you made no mistakes at all, skip Sec. 3-5, because you already know its material, and go on to Sec. 3-6 which begins on page 131.

Introduction

In this section you will learn to divide two-digit numbers by one-digit numbers. Since many of these divisions do not have natural-number answers, the concept of a remainder is introduced. Because you are probably familiar with simple fractions, the remainder is often expressed in fractional form. You also get an idea of how the distributive law is applied to division problems, and you learn how to check your answers by multiplication.

FRAMES FOR SECTION 3-5

1 If you want to share 12 chocolates equally among 4 people, you give each person ______ chocolates because $12 \div 4 =$ ______.

---

3 3

---

2 But if you have 13 chocolates to share out equally and you are not allowed to cut any of them into pieces, you will have ______ chocolate(s) left over. We call this the *remainder,* and we write this kind of division as

$13 \div 4 = 3$ remainder 1

because each person gets ______ chocolates and ______ chocolate(s) *remain(s)* unshared.

---

1 3 1

---

3 If you share 15 chocolates among 4 people, each person gets ______ and ______ are left unshared.

$15 \div 4 =$ ______ remainder ______

---

3 3 3 3

---

4 *a.* If in Frame 2 we had been allowed to cut up the remaining chocolate, we would have cut it into ______ equal pieces and given 1 piece to each person. Each person would then have received $\frac{1}{4}$ of a chocolate extra and $3\frac{1}{4}$ chocolates altogether ($3\frac{1}{4}$ means $3 + \frac{1}{4}$).

b. In Frame 3 we have 3 chocolates left over to share among 4 people. We could have cut each chocolate into ______ equal pieces and given 1 piece of each chocolate to each person. Each person would then have received 3 extra pieces ($\frac{3}{4}$ of a chocolate) and ______ chocolates altogether.

---

a. 4 *b.* 4; $3\frac{3}{4}$

---

5 So we can also write

$13 \div 4 = 3\frac{1}{4}$ and $15 \div 4 =$ ______

$3\frac{3}{4}$

---

6 *a.* If 17 chocolates are to be shared equally among 4 people, there are two ways of answering the problem:

$17 \div 4 =$ ______ remainder ______

and

$17 \div 4 = \_\frac{1}{4}$

b. If 17 chocolates are to be shared equally among 3 people, we can write:

$17 \div 3 =$ ______ remainder ______

or

$17 \div 3 = \_\frac{-}{3}$

---

a. 4, 1; $4\frac{1}{4}$ *b.* 5; 2; $5\frac{2}{3}$

---

7 *a.* You may have noticed that we have been using the distributive law to obtain our answers:

$(16 + 1) \div 4 = (16 \div 4) + (1 \div 4)$

so

$17 \div 4 =$ ______ + ______ = ______

b. $19 = 15 +$ ____ so $19 \div 5 =$ ____

---

a. $4 + \frac{1}{4} = 4\frac{1}{4}$ *b.* $4; 3\frac{4}{5}$

---

8 Of course you could write

$$21 \div 4 = (12 + 9) \div 4 = 3\frac{9}{4}$$

or

$$21 \div 4 = (16 + 5) \div 4 = 4\frac{\_}{4}$$

or

$$21 \div 4 = (20 + 1) \div 4 = \_\_\_\_$$

but it is usual to write the answer with the largest whole number and the smallest remainder, that is, ____.

$4\frac{5}{4}$ $5\frac{1}{4}$ $5\frac{1}{4}$

---

9 $31 \div 4$ can be written as

$$(20 + 11) \div 4 = 5\frac{\_}{4}$$

or as $(24 + 7) \div 4 =$ ____

or as $(28 + 3) \div 4 =$ ____

but is normally written as ____.

$5\frac{11}{4}$ $6\frac{7}{4}$ $7\frac{3}{4}$ $7\frac{3}{4}$

---

10 Notice that if you are dividing by 4, the smallest possible remainder, if there is one, must always be less than ____. **Hint:** If you are sharing chocolates among 4 people, when does it become impossible to share out another round?

4

---

11 This means that if a number is not divisible by 4 and you are dividing by 4, you should look for the next lowest number divisible by 4. With 21, the next lowest number divisible by 4 is 20, so we write $(20 + 1) \div 4 = \_\frac{\_}{4}$ With 31, the next lowest number divisible by 4 is ____, so we write (____ + ____) $\div 4 =$ ____.

$5\frac{1}{4}$ 28 $(28+3)\div 4 = 7\frac{3}{4}$

---

12 With 37, the next lowest number divisible by 4 is ______, so we write $37 \div 4 =$ ______; and the next lowest number divisible by 5 is ______, so we write $37 \div 5 =$ ______.

---

36 $9\frac{1}{4}$ 35 $7\frac{2}{5}$

---

13 With 19 the next lowest number divisible by 3 is ______. The next lowest number divisible by 4 is ______, and the next lowest number divisible by 5 is ______.

$19 \div 3 =$ ______

$19 \div 4 =$ ______

$19 \div 5 =$ ______

---

18 16 15 $6\frac{1}{3}$ $4\frac{3}{4}$ $3\frac{4}{5}$

---

14 With 59, the next lowest number divisible by 6 is ______, and the next lowest number divisible by 7 is ______.

$59 \div 6 =$ ______

$59 \div 7 =$ ______

---

54 56 $9\frac{5}{6}$ $8\frac{3}{7}$

---

15 *a.* $65 \div 8 =$ ______ *b.* $65 \div 9 =$ ______

---

a. $8\frac{1}{8}$ *b.* $7\frac{2}{9}$

---

16 *a.* $35 \div 4 =$ ______

b. $43 \div 9 =$ ______

c. $27 \div 6 =$ ______

---

a. $8\frac{3}{4}$ *b.* $4\frac{7}{9}$ *c.* $4\frac{3}{6}$

---

17 *a.* $61 \div 8 =$ ______

b. $42 \div 7 =$ ______

c. $38 \div 5 =$ ______

a. $7\frac{5}{8}$ *b.* 6 *c.* $7\frac{3}{5}$

18 These divisions can all be checked by carrying out the corresponding multiplications. Thus if $48 \div 5 =$ ______, we know that ______ $\times 5 = 48$.

$9\frac{3}{5}$ $9\frac{3}{5}$

19 We can do this multiplication in two ways. One method ignores fractions. Thus $9\frac{3}{5}$ is 9 with remainder ______, so

$9\frac{3}{5} \times 5 = (9 \times 5) +$ ______ $= 45 +$ ______ $= 48$

3 3 3

20 The other method recognizes that $5 \times \frac{1}{5} =$ ______

$5 \times \frac{3}{5} =$ ______, etc.,

so

$9\frac{3}{5} \times 5 = (9 \times 5) + (\frac{3}{5} \times 5) = 45 +$ ______ $= 48$

1 3 3

21 In Frame 16*b* we found that $43 \div 9 = 4\frac{7}{9}$. Check this result by multiplication using either method.

Method 1	*Method 2*
$4\frac{7}{9} \times 9 = (4 \times 9) +$ remainder 7	$4\frac{7}{9} \times 9 = (4 \times 9) + (\frac{7}{9} \times 9)$
$= 36 + 7 = 43$	$= 36 + 7 = 43$

22 In Frame 17*a* we found that $61 \div 8 = 7\frac{5}{8}$. Check this result by multiplication.

Method 1	*Method 2*
$7\frac{5}{8} \times 8 = (7 \times 8) +$ remainder 5	$7\frac{5}{8} \times 8 = (7 \times 8) + (\frac{5}{8} \times 8)$
$= 56 + 5 = 61$	$= 56 + 5 = 61$

23 Find $68 \div 7$ and check your answer by multiplication.

$9\frac{5}{7}$

CHECK

Method 1

$9\frac{5}{7} \times 7 = (9 \times 7) + \text{remainder } 5$

$= 63 + 5 = 68$

Method 2

$(9 \times 7) + (\frac{5}{7} \times 7) = 63 + 5 = 68$

---

24 Find $75 \div 8$ and check your answer by multiplication.

$9\frac{3}{8}$

CHECK

Method 1

$9\frac{3}{8} \times 8 = (9 \times 8) + \text{remainder } 5$

$= 72 + 3 = 75$

Method 2

$9\frac{3}{8} \times 8 = (9 \times 8) + (\frac{3}{8} \times 8)$

$= 72 + 3 = 75$

You have finished the program for Sec. 3-5. Now take the posttest to see how much you have learned.

POSTTEST 3-5

Divide the following numbers:

1. 61 by 8   2. 53 by 6   3. 47 by 5

4. 88 by 9   5. 64 by 7

When you have finished the posttest, correct it yourself, using the answers on page 160. Then check all the questions you got wrong on the list below:

1. _____   3. _____   5. _____

2. _____   4. _____

If you compare your performance with that on Pretest 3-5, you will see how much you have improved. If you made more than one mistake on this posttest, review the material and take the pretest again. Don't move on to the next section until you can get all the questions right.

Now proceed to Sec. 3-6.

Section 3-6
Division of Large Numbers by One-digit Numbers

Do this section if you made a mistake in Questions 8 to 10 of Diagnostic Test 3. Begin by taking the pretest below, which will show you how much of Sec. 3-6 you already know and what you will be able to do by the time you have finished it.

PRETEST 3-6

Divide the following numbers:

1. 487 by 8 2. 740 by 6 3. 6,128 by 9
4. 7,034 by 5 5. 4,703 by 7

When you have finished the pretest, correct it yourself using the answers on page 161. Then check all the questions you got wrong on the list below:

1. ______ 3. ______ 5. ______
2. ______ 4.

If you made any mistakes (even if it was only *one* mistake), read the introduction and go through the program.

If you made no mistakes at all, skip Sec. 3-6, because you already know its material, and go on to Sec. 3-7, which begins on page 145.

Introduction

This section shows you how to divide large natural numbers by one-digit numbers. It assumes that you are now familiar with short division and introduces the process of long division. It also shows you how to check your answers by multiplication.

FRAMES FOR SECTION 3-6

1 The division of some large numbers by one-digit numbers can be carried out by separating each large number into divisible portions using the distributive law. Thus

a. $88 \div 4 = (80 + 8) \div 4 = (80 \div 4) + (8 \div 4) =$ ______ + ______ = ______

and

b. $96 \div 3 = (90 + 6) \div 3 =$ (____ ÷ ____) + (____ ÷ 3)

= ____ + ____ = ______

---

a. $20 + 2 = 22$ *b.* $90 \div 3$; 6; $30 + 2 = 32$

---

2 *a.* $91 \div 7 = (70 + 21) \div 7 = (\_\_\_\_ \div 7) + (\_\_\_\_ \div 7)$

$= \_\_\_\_ + \_\_\_\_ = \_\_\_\_\_$

b. $125 \div 5 = (100 + 25) \div 5 = (\_\_\_\_ \div 5) + (\_\_\_\_ \div 5)$

$= \_\_\_\_\_ + \_\_\_\_\_ = \_\_\_\_\_$

---

a. 70; 21; $10 + 3 = 13$ *b.* 100; 25; $20 + 5 = 25$

---

3 *a.* $336 \div 7 = (280 + \_\_\_\_) \div 7 = (\_\_\_\_ \div 7) + (\_\_\_\_ \div 7)$

$= \_\_\_\_\_ + \_\_\_\_\_ = \_\_\_\_\_$

b. $444 \div 6 = (420 + \_\_\_\_) \div 6 = (\_\_\_\_ \div 6) + (\_\_\_\_ \div 6)$

$= \_\_\_\_\_ + \_\_\_\_\_ = \_\_\_\_\_$

---

a. 56; 280; 56; $40 + 8 = 48$ *b.* 24; 420; 24; $70 + 4 = 74$

---

4 It is not always easy to spot how to split up a number, so we use a systematic method, called long division, which proceeds by a series of approximations. Suppose you want to divide 272 by 8.

a. $27 \div 8$ is more than 3 but less than 4, so $272 \div 8$ is more than _0 but less than _0.

b. The first digit of the answer to $227 \div 8$ must therefore be _____ and if A is the second digit of the answer,

$272 \div 8 = \_A = \_0 + A$

---

a. 30; 40 *b.* 3; 3A; 30

---

5 The answer is written on the top line, so we write in the first digit as shown:

$$\begin{array}{r} 3 \\ 8\overline{)272} \end{array}$$

Now since $272 = (30 + A) \times 8 = (30 \times 8) + (A \times 8)$, we can find $A \times 8$ by subtraction:

$$\begin{array}{rl} 3 & \\ 8\overline{)2\,7\,2} & \\ \underline{\_\,\_\,\_} & 30 \times 8 \\ \_\,\_ & A \times 8 \end{array}$$

and since $A \times 8 =$ ______, $A =$ ______. So we can complete the long division:

$$\begin{array}{rl} 3\,\_ & \text{Answer} \\ 8\overline{)2\,7\,2} & \\ \underline{\_\,\_\,\_} & \\ \_\,\_ & \end{array}$$

$$\begin{array}{r} 3 \\ 8\overline{)272} \\ \underline{240} \\ 32 \end{array} \qquad 32 \qquad 4 \qquad \begin{array}{r} 34 \\ 8\overline{)272} \\ \underline{240} \\ 32 \end{array}$$

---

6 Suppose you want to divide 477 by 9. Since $47 \div 9$ is between ○ and _, $477 \div 9$ is between ○0 and _0, and

$477 \div 9 = ○A = ○0 + A$

where A is the second digit of the answer. Now fill in the first digit of the answer:

$$\begin{array}{r} ○ \\ 9\overline{)477} \end{array}$$

Note: We use a circle instead of a dash to show that the same digit is involved each time.

---

5 6 50 60 $477 \div 9 = 5A = 50 + A$ $\begin{array}{r} 5 \\ 9\overline{)477} \end{array}$

---

7 $477 = (50 + A) \times 9 = (○0 \times 9) + (A \times 9)$, so we can find $A \times 9$ by subtraction. Complete the long division below:

$$\begin{array}{rl} ⑤\,\_ & \text{Answer} \\ 9\overline{)4\,7\,7} & \\ \underline{\_\,\_\,\_} & ○0 \times 9 \\ \_\,\_ & A \times 9 \end{array}$$

50×9

$$\begin{array}{r} 53 \\ 9\overline{)477} \\ \underline{450} \\ 27 \end{array} \quad 50 \times 9$$

---

8 Suppose you want to divide 468 by 6. Since $46 \div 6$ is between ○ and _, $468 = (○0 + A) \times 6$, where A is the second digit. Fill in the first digit of the answer:

$$\begin{array}{r} ○ \\ 6\overline{)468} \end{array}$$

Then complete the long division:

$$\begin{array}{rl} ○\_ & \text{Answer} \\ 6\overline{)4\,6\,8} & \\ \underline{\_\,\_\,\_} & ○0 \times 6 \\ \_\,\_ & A \times 6 \end{array}$$

---

7 8 $(70 + A) \times 6$

$$\begin{array}{r} 7 \\ 6\overline{)468} \end{array} \qquad \begin{array}{rl} 78 & \\ 6\overline{)468} & \\ \underline{420} & 70 \times 6 \\ 48 & \end{array}$$

---

9 Suppose you want to divide 395 by 5. Since $39 \div 5$ is between ○ and _, $395 = (○0 + A) \times 6$, where A is less than 10. Fill in the first digit of the answer and then complete the long division:

$$\begin{array}{rl} ○\_ & \text{Answer} \\ 5\overline{)3\,9\,5} & \\ \underline{\_\,\_\,\_} & ○0 \times 5 \\ \_\,\_ & A \times 5 \end{array}$$

---

7 8 70

$$\begin{array}{rl} 79 & \\ 5\overline{)395} & \\ \underline{350} & 70 \times 5 \\ 45 & \end{array}$$

---

10 Suppose you want to divide 595 by 7. $59 \div 7$ is between ○ and _. Fill in the first digit of the answer and then complete the long division:

$$\begin{array}{rl} ○\_ & \text{Answer} \\ 7\overline{)5\,9\,5} & \\ \underline{\_\,\_\,\_} & ○0 \times 7 \\ \_\,\_ & A \times 7 \end{array}$$

---

8 9

$$\begin{array}{r} 85 \\ 7\overline{)595} \\ \underline{560} \\ 35 \end{array} \quad 80 \times 7$$

---

11 Suppose you want to divide 424 by 8. 42 ÷ 8 is between ○ and _. Complete the long division:

$$\begin{array}{r} \bigcirc\,\_ \\ 8\overline{)4\,2\,4} \\ \underline{\_\,\_\,\_} \\ \_\,\_ \end{array} \quad \bigcirc 0 \times 8$$

---

5 6

$$\begin{array}{r} 53 \\ 8\overline{)424} \\ \underline{400} \\ 24 \end{array} \quad 50 \times 8$$

---

12 Suppose you want to divide 352 by 4. Find the first digit of the answer and complete the long division:

$$\begin{array}{r} \bigcirc\,\_ \\ 4\overline{)3\,5\,2} \\ \underline{\_\,\_\,\_} \\ \_\,\_ \end{array} \quad \bigcirc 0 \times 4$$

$$\begin{array}{r} 88 \\ 4\overline{)352} \\ \underline{320} \\ 32 \end{array} \quad 80 \times 4$$

---

13 Use long division to divide:

a. 228 by 6:

$$\begin{array}{r} \bigcirc\,\_ \\ 6\overline{)2\,2\,8} \\ \underline{\_\,\_\,\_} \\ \_\,\_ \end{array} \quad \bigcirc 0 \times 6$$

b. 532 by 7:

$$\begin{array}{r} \_\,\_ \\ 7\overline{)5\,3\,2} \\ \underline{\_\,\_\,\_} \\ \_\,\_ \end{array}$$

---

a.

$$\begin{array}{r} 38 \\ 6\overline{)228} \\ \underline{180} \\ 48 \end{array} \quad 30 \times 6$$

b.

$$\begin{array}{r} 76 \\ 7\overline{)532} \\ \underline{490} \\ 42 \end{array}$$

---

14 *a.* Use long division to divide 465 by 5:

```
    _ _
5)4 6 5
  _ _ _
    _ _
```

b. Use long division to divide 387 by 9.

---

a.
```
    93
 5)465
   450
    15
```
b.
```
    43
 9)387
   360
    27
```

---

15 *a.* Use long division to divide 222 by 3.

b. Use long division to divide 360 by 8.

---

a.
```
    74
 3)222
   210
    12
```
b.
```
    45
 8)360
   320
    40
```

---

16 So far we have dealt only with cases which are divisible exactly. When there is a remainder, the procedure is just the same, but the division is a bit longer. Suppose we divide 469 by 9. Starting as before, we get

```
    5
 9)469
   450
    19
```

Now 19 is not divisible by 9, but the next lowest number divisible by 9 is 18 and 18 ÷ 9 = _____, so the next digit is clearly _____. We can then obtain the remainder by subtraction.

```
    5○
 9)4 6 9
   4 5 0
     1 9
     _ _   ○ × 9
       _   Remainder
```

So 469 ÷ 9 = ____ remainder ____.

---

2 2

```
     52
 9)469
   450
    19
    18   2 × 9
     1
```

$469 \div 9 = 52$ remainder 1 (also written $52\frac{1}{9}$).

---

17 Complete the following long division:

```
    7○
 7)528
   490
    38
    __   ○ × 7
     _   Remainder
```

$528 \div 7 =$ ____ remainder ____.

```
    75
 7)528
   490
    38
    35   5 × 7
     3
```

75 remainder 3

---

18 Use long division to divide 579 by 6:

```
    □○
 6)579
   ___   □0 × 6
    __
    __   ○ × 6
     _   Remainder
```

$579 \div 6 =$ ____ remainder ____.

```
    96
 6)579
   540   90 × 6
    39
    36   6 × 6
     3
```

96 remainder 3

---

19 The distributive law is being used twice in Frame 18. First 579 ÷ 6 = (_____ + 39) ÷ 6 = _____ + (39 ÷ 6); then 39 ÷ 6 = (__ + 3) ÷ 6 = _____ remainder _____. So 579 ÷ 6 = ____ remainder ____.

---

540 90 36 6 3 96 3

---

20 Use long division to divide:

a. 472 by 7:

```
    □○
7)472
  ___   □0 × 7
   __
   __   ○ × 7
    _   Remainder
```

b. 517 by 8:

```
   __
8)517
  ___
   __
   __
    _
```

---

a.

```
   67
7)472
  420
   52
   49
    3
```

b.

```
   64
8)517
  480
   37
   32
    5
```

---

21 *a.* Use long division to divide 395 by 9:

```
   __
9)395
  ___
   __
   __
    _
```

b. Use long division to divide 424 by 5.

---

a.

```
   43
9)395
  360
   35
   27
    8
```

b.

```
   84
5)424
  400
   24
   20
    4
```

---

22 Sometimes the second digit of the answer is a 0, and the first subtraction gives the remainder.

a.

```
       _ _
  8 ) 4 8 7

     _ _ _
         _   Remainder
```

487 ÷ 8 = _____ remainder _____.

b. Use long division to divide 285 by 7:

285 ÷ 7 = _____ remainder ____.

---

a.

```
      60
  8 ) 487
      480
        7
```

b.

```
      40
  7 ) 285
      280
        5
```

60 remainder 7   40 remainder 5

---

23 If the answer is greater than 100, the procedure is similar but a bit longer. The answer is worked out digit by digit:

```
     ○□△
  3) 5 9 5
     _ _ _    ○00 × 3
     _ _ _
     _ _ _    □0 × 3
       _ _
       _ _    △ × 3
         _
```

595 ÷ 3 = _____ remainder ____.

```
     198
  3) 595
     300   100 × 3
     295
     270    90 × 3
      25
      24     8 × 3
       1
```

198 remainder 1

---

24 *a.*

```
     _ _ _
  4)9 2 7
    _ _ _    ○00 × 4
    _ _ _
    _ _ _    □0 × 4
        _
        _    △ × 4
        _
```

927 ÷ 4 = _____ remainder ____.

b.

```
     _ _ _
  5)7 3 6
    _ _ _
    _ _ _
    _ _ _
      _ _
      _ _
        _
```

736 ÷ 5 = _____ remainder ____.

---

a.

```
     231
  4)927
     800   200 × 4
     127
     120    30 × 4
       7
       4     1 × 4
       3
```

b.

```
     147
  5)736
     500
     236
     200
      36
      35
       1
```

231 remainder 3 147 remainder 1

---

25 *a.*

```
       _ _ _
  9)3 4 5 8
    _ _ _ _    ○00 × 9
    _ _ _ _
      _ _ _    □0 × 9
        _ _
        _ _    △ × 9
          _
```

3,458 ÷ 9 = _____ remainder _____.

b.

```
       _ _ _
  6)4 7 9 3
    _ _ _ _
      _ _ _
      _ _ _
        _ _
        _ _
          _
```

4,793 ÷ 6 = _____ remainder ____.

---

a.

```
     384
 9)3458
   2700   300 × 9
    758
    720    80 × 9
     38
     36     4 × 9
      2
```

384 remainder 2

b.

```
     798
 6)4793
   4200
    593
    540
     53
     48
      5
```

798 remainder 5

---

26 It is easiest to get the right digit in the right place if you remember that long division is a method of successive approximations. Thus in Frame 25*a* when we put up 3 as the first digit, and we put 3 in the (hundreds/tens/units) ______ column, we are saying that 300 is an approximate answer to 3,458 ÷ 9. When we put up 8 as the second digit and put it in the ______ column, we are saying that □0 is an approximate answer to 758 ÷ 9 and hence that 3□0 is an approximate answer to 3,458 ÷ 9. And when we put up 4 as the third digit and put it in the ______ column, we are saying that △ is an approximate answer to 38 ÷ 9 and hence that 38△ is an approximate answer to 3,458 ÷ 9.

---

hundreds tens 80 380 units 4 384

---

27 It is important to remember these approximations when you get a situation like

```
    7
5)3528
  3500   Don't finish this yet.
    28
```

Our first approximation for 3,528 ÷ 5 is __00, and we now need an approximation for 28 ÷ 5. This approximation is (5/50) ______, so we put a 5 in the (units/tens) ______ column of the answer. Complete the long division now:

```
    7
5)3 5 2 8
  3 5 0 0
      2 8
      _ _
        _
```

Hint: You may need a 0 somewhere.

---

700 5 units 705 705 remainder 3

$$\begin{array}{r} 705 \\ 5\overline{)3528} \\ \underline{3500} \\ 28 \\ \underline{25} \\ 3 \end{array}$$

---

28 *a.*

$$\begin{array}{r} \_\,\_\,\_ \\ 6\overline{)4\,2\,3\,8} \\ \underline{\_\,\_\,\_\,\_} \\ \_\,\_ \\ \underline{\_\,\_} \\ \_ \end{array}$$

4,238 ÷ 6 = _____ remainder _____.

b. Use long division to divide 7,259 by 9.

---

a.

$$\begin{array}{r} 706 \\ 6\overline{)4238} \\ \underline{4200} \\ 38 \\ \underline{36} \\ 2 \end{array}$$

b.

$$\begin{array}{r} 806 \\ 9\overline{)7259} \\ \underline{7200} \\ 59 \\ \underline{54} \\ 5 \end{array}$$

706 remainder 2 806 remainder 5

---

29 *a.*

$$\begin{array}{r} \_\,\_\,\_ \\ 4\overline{)3\,1\,9\,1} \\ \underline{\_\,\_\,\_\,\_} \\ \_\,\_\,\_ \\ \underline{\_\,\_\,\_} \\ \_\,\_ \\ \underline{\_\,\_} \\ \_ \end{array}$$

3,191 ÷ 4 = _____ remainder _____.

b. Use long division to divide 6,238 by 7.

```
a.      797          b.      891
     4)3191               7)6238
       2800                 5600
        391                  638
        360                  630
         31                    8
         28                    7
          3                    1

797 remainder 3      891 remainder 1
```

30 *a.* Use long division to divide 5,823 by 8 and check your answer by multiplication.

b. Use long division to divide 3,219 by 4 and check your answer by multiplication.

```
a.      727        CHECK
     8)5823
       5600         727
        223          ×8
        160        5816
         63
         56        5,816 + 7 (remainder) = 5,823
          7

b.      804        CHECK
     4)3219
       3200         804
         19          ×4
         16        3216
          3
                   3,216 + 3 = 3,219
```

31 *a.* Use long division to divide 937 by 5 and check your answer by multiplication.

b. Use long division to divide 5,278 by 9 and check your answer by multiplication.

a.

```
   187
5)937
  500
  437
  400
   37
   35
    2
```

CHECK

```
 187
  ×5
 935
```

$935 + 2 = 937$

b.

```
   586
9)5278
  4500
   778
   720
    58
    54
     4
```

CHECK

```
 586
  ×9
5274
```

$5{,}274 + 4 = 5{,}278$

You have finished the program for Sec. 3-6. Now take the posttest to see how much you have learned.

POSTTEST 3-6

Divide the following numbers:

1. 593 by 8
2. 670 by 6
3. 8,237 by 9
4. 4,183 by 5
5. 8,425 by 7

When you have finished the posttest, correct it yourself using the answers on page 161. Then check all the questions you got wrong on the list below:

1. ______ 3. ______ 5. ______

2. ______ 4. ______

If you compare your performance with that on Pretest 3-6, you will see how much you have improved. If you made more than one mistake on this posttest, review the material and take the pretest again. Don't move on to the next section until you can get all the questions right.

Now proceed to Sec. 3-7.

Section 3-7
Long Division

Do this section if you made a mistake in Questions 8 to 10 of Diagnostic Test 3. Begin by taking the pretest below, which will show you how much of Sec. 3-7 you already know and what you will be able to do by the time you have finished it.

PRETEST 3-7

Divide the following numbers:

1. 3,469 by 67
2. 56,703 by 29
3. 42,163 by 87
4. 62,531 by 532
5. 88,182 by 426

When you have finished the pretest, correct it yourself using the answers on page 161. Then check all the questions you got wrong on the list below:

1. ______ 3. ______ 5. ______

2. ______ 4. ______

If you made any mistakes (even if it was only *one* mistake), read the introduction and go through the program.

If you made no mistakes at all, skip Sec. 3-7, because you already know its material, and go on to take Final Test 3 (page 159).

Introduction

This section shows you how to divide large numbers by other large numbers. It assumes that you can now divide large numbers by one-digit numbers and extends the method of long division to division by two-digit and three-digit numbers.

FRAMES FOR SECTION 3-7

1 We divide large natural numbers by using a method of successive approximations, called long division. In this method we find the answer one digit at a time. Suppose we want to divide 278 by 17. You can see that the answer will be more than 10 ($17 \times 10 =$ ______) and less than 20 ($17 \times 20 =$ ______). So the first digit of the answer must be □, and a very approximate answer will be □0. We therefore write the first digit in the (units/tens/hundreds) ______ column of the answer in the position indicated:

$$\begin{array}{r} \square \\ 17\overline{)278} \end{array}$$

Note: We use a square instead of a dash to show that the same digit is involved each time.

---

170 340 1 10 tens

$$\begin{array}{r} 1 \\ 17\overline{)278} \end{array}$$

---

2 We now wish to get a closer approximation to the answer and to find its second digit. Since the first digit is 1, we can write

$278 \div 17 = 1A$ with remainder R

where A is the second digit. So

$$\begin{aligned} 278 &= [1A \times 17] + R \\ &= [(10 + A) \times 17] + R \\ &= (10 \times 17) + (A \times 17) + R \\ &= \_\_\_\_\_\_ + (A \times 17) + R \end{aligned}$$

We can find $(A \times 17) + R$ by subtraction:

$$\begin{array}{rl} 1 & \\ 17\overline{)2\,7\,8} & \\ \underline{\_\ \_\ \_} & 10 \times 17 \\ \_\ \_\ \_ & (A \times 17) + R \end{array}$$

170

$$\begin{array}{r} 1 \\ 17\overline{)278} \\ \underline{170} \\ 108 \end{array}$$

---

3 Now $108 = (A \times 17) + R$, and since R is the remainder, it must be less than ______. We have to guess possible values for A and try them out

$5 \times 17 =$ ______

$6 \times 17 =$ ______

so $A = \square$. We now write in the second digit of the answer and subtract to find the remainder:

```
      1□
17)278
    170    10 × 17
    108
    ___    □ × 17
      _    Remainder
```

So $278 \div 17 =$ ______ remainder ______.

---

17 85 102 6

```
      16
17)278
    170    10 × 17
    108
    102    6 × 17
      6
```

16 6

---

4 Now let us use the same method for $557 \div 19$. The first digit of the answer must be ○ since $19 \times 20 =$ ______ and $19 \times 30 =$ ______. So we try ○0 as our first approximation and subtract to prepare for the next stage:

```
      ○
19)557
    ___    ○0 × 19
    ___
```

---

2 380 570 20

```
      2
19)557
    380    20 × 19
    177
```

---

5 We now need an approximation for $177 \div 19$ in order to obtain the second digit of the answer. It obviously won't be much under 10 because $19 \times 10 = 190$; but it must not multiply to more than 177 if we are to subtract to get the remainder.

```
     2□
19)557
    380     20 × 19
    ---
    177
    ___     □ × 19
    ---
      _     Remainder
```

So $557 \div 19 =$ _____ remainder _____.

```
     29
19)557
    380   20 × 19
    ---
    177
    171    9 × 19
    ---
      6   Remainder
```

29 6

---

6 So we have $557 \div 19 = 29$ remainder 6. The contribution of the first digit of the answer is $20 \times 19 =$ _____ (the third line of the division, counting the answer as the first line). The contribution of the second digit of the answer is _____ $\times 19 =$ _____ (the fifth line of the division), and the remainder is 6. Adding, we get _____ (third line) + _____ (fifth line) + 6 = _____.

---

380 9 171 380 171 557

---

7 Now let's divide 894 by 23.

$23 \times 30 =$ _____ and $23 \times 40 =$ _____

so the first digit of the answer must be ○. Complete the first half of the division:

```
     ○
23)894
   ___    ○0 × 23
   ---
   ___
```

---

690 920 3

```
     3
23)894
   690   30 × 23
   ---
   204
```

---

8 We now need to find the second digit. Let's try 9: $23 \times 9 =$ ______. This is a close approximation, but we cannot subtract this number from 204, so we have to try 8. Complete the division:

```
      3□
23)894
    690   30 × 23
    204
    ___   □ × 23
     __   Remainder
```

So $894 \div 23 =$ ______ remainder ______.

CHECK

Third row + fifth row + remainder = 690 + ______ + ______ = ______.

207

```
     38
23)894
   690  30 × 23
   204
   184   8 × 23
    20
```

38 20 184 20 894

---

9 Divide 738 by 31. Remember to find one digit of the answer at a time and refer back to earlier frames if you need to.

```
      ○□
31)738
   ___   ○0 × 31

   ___
   ___   □ × 31
    __   Remainder
```

So $738 \div 31 =$ ______ remainder ______.

```
     23
31)738
   620   20 × 31
   118
    93    3 × 31
    25
```

23 25

10 When you remember that the first digit followed by an approximate number of 0s is an approximate answer, you will put it in the right place. Consider 4,329 ÷ 57. 4,000 ÷ 50 = 80, so we try 80 as an approximate answer. 57 × 80 = ______, which is too large, so we have to try 70: 57 × 70 = ______. So the first digit is a __ and should be placed in the (units/tens/hundreds) ______ column of the answer.

4,560   3,990   7   tens

---

11 Complete the division:

```
       ⑦□
57)4 3 2 9
   _ _ _ _    ○0 × 57
     _ _ _
     _ _ _    □ × 57
       _ _    Remainder
```

So 4,329 ÷ 57 = ______ remainder ______.

CHECK

Third row + fifth row + remainder = ______ + ______ + ______ = ______.

```
       75
57)4,329
   3,990   70 × 57
     339
     285   5 × 57
      54
```

75   54   3,990 + 285 + 54 = 4,329

---

12 *a.* Divide 3,184 by 47.

```
     ○□          answer = _____ remainder _____
47)3184
   ----   ○0 × 47
    ---
    ---   □ × 47
     --
```

b. Divide 927 by 43.

```
     --          answer = _____ remainder _____
43)927
   ---
    --
    --
    --
```

---

a.

```
      67
47)3184
   2820   60 × 47
    364
    329    7 × 47
     35
```

answer = 67 remainder 35

b.

```
     21
43)927
   860
    67
    43
    24
```

answer = 21 remainder 24

---

13 *a.* Divide 7,253 by 85.

```
     --          answer = _____ remainder _____
85)7253
   ----
    ---
    ---
     --
```

b. Divide 823 by 37.

---

a.

```
      85
85)7253
   6800
    453
    425
     28
```

answer = 85 remainder 28

b.

```
     22
37)823
   740
    83
    74
     9
```

answer = 22 remainder 9

---

14 Sometimes the second digit of the answer is a 0 and the first subtraction gives the remainder.

a. Divide 2,338 by 29:

```
      _ _
29)2 3 3 8
   _ _ _ _
       _ _
```

answer = ____ remainder ____

b. Divide 4,387 by 62.

---

a.

```
      80
29)2338
   2320
     18
```

answer = 80 remainder 18

b.

```
      70
62)4387
   4340
     47
```

answer = 70 remainder 47

---

15 If the answer is greater than 100, the procedure is similar but a little longer. The answer is worked out digit by digit. With 4,994 ÷ 41, the first approximation is ○00, and the first digit is ○ in the (hundreds/tens/units) _____ column of the answer. Now complete the division:

```
    ○□△
41)4 9 9 4
   _ _ _ _     ○00 × 41
     _ _ _
     _ _ _     □0 × 41
       _ _
       _ _     △ × 41
       _ _
```

answer = _____ remainder _____

CHECK

Third row + fifth row + seventh row + remainder = _____ + _____ + _____ + _____ = _____.

---

100 1 hundreds

```
     121
41)4994
   4100   100 × 41
    894
    820    20 × 41
     74
     41     1 × 41
     33
```

answer = 121 remainder 33

CHECK

4,100 + 820 + 41 + 33 = 4,994

---

16 *a.* Divide 5,329 by 37:

```
      ○□△          answer = ______ remainder ______
 37)5 3 2 9
    _ _ _ _    ○00 × 37
    _ _ _ _
    _ _ _ _    □0 × 37
      _ _ _
      _ _ _     △ × 37
          _
```

b. Divide 7,315 by 22:

```
      _ _ _     answer = ______ remainder ______
 22)7 3 1 5
    _ _ _ _
      _ _ _
      _ _ _
        _ _
        _ _
          _
```

---

a.

```
        144
 37)5329
    3700    100 × 37
    1629
    1480     40 × 37
     149
     148      4 × 37
       1
```

answer = 144 remainder 1

b.

```
        332
 22)7315
    6600
     715
     660
      55
      44
      11
```

answer = 332 remainder 11

---

17 *a.* Divide 28,347 by 76:

```
          _ _ _     answer = ______ remainder ______
 76)2 8 3 4 7
    _ _ _ _ _
      _ _ _ _
      _ _ _ _
        _ _ _
        _ _ _
          _ _
```

b. Divide 7,898 by 56.

---

a.

```
      372
76)28347
   22800
    5547
    5320
     227
     152
      75
```

answer = 372 remainder 75

b.

```
     141
56)7898
   5600
   2298
   2240
     58
     56
      2
```

answer = 141 remainder 2

---

18 If the answer is greater than 1,000 and less than 10,000, the division can take up to four stages.

a. answer = _____ remainder _____

```
      _ _ _ _
34)8 2 3 5 1
   _ _ _ _ _    _000 × 34
   _ _ _ _ _
   _ _ _ _ _    _00 × 34
       _ _ _
       _ _ _    _0 × 34
         _ _
         _ _    _ × 34
           _
```

b. answer = _____

```
   _ _ _ _
4)5 1 5 6
  _ _ _ _
  _ _ _ _
    _ _ _
    _ _ _
    _ _ _
      _ _
```

---

a.

```
     2422
34)82351
   68000
   14351
   13600
     751
     680
      71
      68
       3
```

answer = 2,422 remainder 3

b.

```
   1289
4)5156
  4000
  1156
   800
   356
   320
    36
    36
     0
```

answer = 1,289

---

19 It is easiest to get the right digit in the right place if you remember that long division is a method of successive approximations. Thus, in Frame 18*a* when we put 2 as the first digit and put it in the (thousands/hundreds/units) ______ column, we are saying that 2,000 is an approximate answer to 82,351 ÷ 34. When we put up 4 as the second digit and put it in the ______ column, we are saying that ○00 is an approximate answer to 14,351 ÷ 34 and hence that 2○00 is an approximate answer to 82,351 ÷ 34. When we put up 2 as the third digit and put it in the ______ column, we are saying that __0 is an approximate answer to 751 ÷ 34 and hence that ___0 is an approximate answer to 82,351 ÷ 34. And when we put up 2 as the fourth digit and put it in the ______ column, we are saying that __ is an approximate answer to 71 ÷ 34 and hence that ____ is an approximate answer to 82,531 ÷ 34.

---

thousands hundreds 400 2,400 tens 20 2,420
units 2 2,422

---

20 It is important to remember these approximations when you get a situation like

```
     7
51)36138
   35700   Don't finish this yet.
   -----
     438
```

Our first approximation for 36,138 ÷ 51 is __00, and we now need an approximation for 438 ÷ 51. This approximation is (8/80) ______, so we put an 8 in the (units/tens) ______ column of the answer. Complete the long division now:

```
     7
51)3 6 1 3 8
   - - - - -
   =========
       - - -
       =====
         - -
```

Hint: You may need a 0 somewhere.

---

700 8 units

```
     708
51)36138
   35700
   -----
     438
     408
     ---
      30
```

---

21 *a.*

```
   ____        answer = _____ remainder _____
5)7542
  ____
  ____
  ____
  ____
    __
    __
     _
```

b. Divide 4,258 by 4. Answer = _____ remainder _____.

Hint: If there is a 0 and you miss it, you will run out of digits before the end.

---

a.

```
   1508
5)7542
  5000
  2542
  2500
    42
    40
     2
```

answer = 1,508 remainder 2

b.

```
   1064
4)4258
  4000
   258
   240
    18
    16
     2
```

answer = 1,064 remainder 2

---

22 *a.* Divide 79,346 by 28:

```
     ____      answer = _____ remainder _____
28)79346
   _____
   _____
   _____
     ___
     ___
     ___
      __
      __
```

b. Divide 52,634 by 43. Answer = _____ remainder _____.

```
a.       2833            b.        1224
     28)79346                  43)52634
        56000                     43000
        23346                      9634
        22400                      8600
          946                      1034
          840                       860
          106                       174
           84                       172
           22                         2
```

answer = 2,833 remainder 22 answer = 1,224 remainder 2

---

23 The same principles apply when dividing by three-figure numbers.

```
a.          _ _ _        answer = ____ remainder ____
   259)4 6 5 3 2
       _ _ _ _ _   _00 × 259
       _ _ _ _ _
       _ _ _ _ _    _0 × 259
         _ _ _ _
         _ _ _ _     _ × 259
           _ _ _
```

b. Divide 63,928 by 125. Answer = ____ remainder ____.

```
a.        179            b.         511
     259)46532                125)63928
         25900                    62500
         20632                     1428
         18130                     1250
          2502                      178
          2331                      125
           171                       53
```

answer = 179 remainder 171 answer = 511 remainder 53

---

24 *a.* Divide 52,317 by 226. Answer = ____ remainder ____.

b. Divide 73,614 by 89. Answer = ____ remainder ____.

a.
```
        231
226)52317
    45200
     7117
     6780
      337
      226
      111
```
answer = 231 remainder 111

b.
```
       827
89)73614
   71200
    2414
    1780
     634
     623
      11
```
answer = 827 remainder 11

25 *a.* Divide 67,395 by 327. Answer = _____ remainder _____.

b. Divide 79,678 by 34. Answer = _____ remainder _____.

a.
```
        206
327)67395
    65400
     1995
     1962
       33
```
answer = 206 remainder 33

b.
```
      2343
34)79678
   68000
   11678
   10200
    1478
    1360
     118
     102
      16
```
answer = 2,343 remainder 16

You have finished the program for Sec. 3-7. Now take the posttest to see how much you have learned.

POSTTEST 3-7

Divide the following numbers:

1. 4,178 by 73
2. 62,517 by 37
3. 23,724 by 69
4. 57,615 by 471
5. 97,614 by 319

When you have finished the posttest, correct it yourself using the answers on page 161. Then check all the questions you got wrong on the list below:

1. _____ 3. _____ 5. _____

2. _____ 4. _____

If you compare your performance with that on Pretest 3-7, you will see how much you have improved. If you made more than one mistake on this post-test, review the material and take the pretest again.

Now proceed to Final Test 3.

FINAL TEST 3

1. Add 4,348, 2,861, and 2,895.
2. Add 2,060, 7, 437, and 5,596.
3. Subtract 6,738 from 7,705.
4. Subtract 5,918 from 9,000.
5. Multiply 8,050 by 760.
6. Multiply 9,178 by 349.
7. Multiply 4,082 by 805.
8. Divide 24,073 by 86.
9. Divide 80,714 by 47.
10. Divide 87,327 by 172.

The answers to this test are on page 161. Check the questions you got wrong on the list below and compare your results with those on Diagnostic Test 3 to see how much you have improved.

1. ______	3. ______	5. ______	8. ______
2. ______	4. ______	6. ______	9. ______
		7. ______	10. ______

If you made more than one mistake, review the material and then take the diagnostic test again. Make sure that you can get all the questions right before you move on to the next chapter.

Now proceed to Chap. 4, Negative Numbers.

ANSWERS TO TESTS FOR CHAPTER 3

DIAGNOSTIC TEST 3

1. 10,095
2. 9,740
3. 947
4. 3,068
5. 2,806,000
6. 2,780,159
7. 1,566,672
8. 411 remainder 3
9. 1,903 remainder 21
10. 409 remainder 60

PRETEST 3-1

1. 1,110
2. 3,000
3. 20,744
4. 14,004
5. 3,505

POSTTEST 3-1

1. 1,110
2. 4,000
3. 30,173
4. 11,002
5. 7,602

PRETEST 3-2

1. 70
2. 993
3. 3,507
4. 683
5. 2,347

POSTTEST 3-2

1. 80
2. 998
3. 4,504
4. 495
5. 1,713

PRETEST 3-3

1. 204
2. 6,312
3. 4,690
4. 6,804
5. 1,030

POSTTEST 3-3

1. 301
2. 8,622
3. 3,540
4. 3,008
5. 2,040

PRETEST 3-4

1. 569,687
2. 261,612
3. 2,536,794
4. 813,240
5. 5,082,000

POSTTEST 3-4

1. 155,013
2. 268,956
3. 3,747,966
4. 3,224,830
5. 6,768,000

PRETEST 3-5

1. $8\frac{7}{8}$ or 8 remainder 7
2. $9\frac{3}{6}$ or 9 remainder 3
3. $7\frac{4}{5}$ or 7 remainder 4
4. $8\frac{1}{9}$ or 8 remainder 1
5. $6\frac{2}{7}$ or 6 remainder 2

POSTTEST 3-5

1. $7\frac{5}{8}$ or 7 remainder 5
2. $8\frac{5}{6}$ or 8 remainder 5
3. $9\frac{2}{5}$ or 9 remainder 2
4. $9\frac{7}{9}$ or 9 remainder 7
5. $9\frac{1}{7}$ or 9 remainder 1

PRETEST 3-6

1. $60\frac{7}{8}$ or 60 remainder 7
2. $123\frac{2}{6}$ or 123 remainder 2
3. $680\frac{8}{9}$ or 680 remainder 8
4. $1{,}406\frac{4}{5}$ or 1,406 remainder 4
5. $671\frac{6}{7}$ or 671 remainder 6

POSTTEST 3-6

1. $74\frac{1}{8}$ or 74 remainder 1
2. $111\frac{4}{6}$ or 111 remainder 4
3. $915\frac{2}{9}$ or 915 remainder 2
4. $836\frac{3}{5}$ or 836 remainder 3
5. $1{,}203\frac{4}{7}$ or 1,203 remainder 4

PRETEST 3-7

1. 51 remainder 52
2. 1,955 remainder 8
3. 484 remainder 55
4. 117 remainder 287
5. 207

POSTTEST 3-7

1. 57 remainder 17
2. 1,689 remainder 24
3. 343 remainder 57
4. 122 remainder 153
5. 306

FINAL TEST 3

1. 10,104
2. 8,100
3. 967
4. 3,082
5. 6,118,000
6. 3,203,122
7. 3,286,010
8. 279 remainder 79
9. 1,717 remainder 15
10. 507 remainder 123

Chapter 4
Negative Numbers

This chapter introduces the set of integers (positive, negative, and zero) and defines the four operations of addition, subtraction, multiplication, and division within the set of integers. You learn how to evaluate expressions containing multiple operations with integers.

In Sec. 4-1 negative integers are introduced through the one-to-one correspondence of points on the integer number line with miles on a road, characterized by their distance and direction from a fixed reference point. Addition is defined in terms of trips along the road or along the number line. In Sec. 4-2 a simple situation involving a water tank and a movie helps you understand the operations of multiplying and dividing integers. In Sec. 4-3 the operation "negative of" is defined in terms of the number line, and subtraction is defined as "addition of the negative." This leads in Sec. 4-4 to a definition of multiple subtractions and to the evaluation of expressions containing additions, subtractions, and multiplications of integers.

This chapter requires completion of Chaps. 1 and 2.

DIAGNOSTIC TEST 4

1. $(-53) + 75 =$ ______
2. $0 + (-33) =$ ______
3. $(-16) + (-46) =$ ______
4. $32 + (-14) =$ ______
5. $303 + (-303) =$ ______
6. $(-6) \div (-2) =$ ______
7. $6 \div (-3) =$ ______
8. $0 \div (-3) =$ ______
9. $(-5) \times (-6) =$ ______
10. $(-22) - 37 =$ ______
11. $0 - 43 =$ ______
12. $(-16) - (-34) =$ ______
13. $19 - (-16) =$ ______
14. $23 + (-2) - 6 - (-13) + 37 =$ ______
15. $(-18) - 2 + 15 + (-23) - (-19) =$ ______

16. $-[(-3) - (-10)] =$ ______

17. $-6[(-7) + 9 - (-3) - 16] =$ ______

When you have finished the diagnostic test, correct it yourself using the answers on page 199. Then check all the questions you got wrong on the list below:

1. ______	6. ______	10. ______	14. ______
2. ______	7. ______	11. ______	15. ______
3. ______	8. ______	12. ______	16. ______
4. ______	9. ______	13. ______	17. ______
5. ______			

If you made no mistakes at all, you need not do any work in this chapter, so go straight on to Chap. 5. If you made any mistakes, follow these instructions:

	section	page
If you made any mistakes in Questions 1 to 5, circle Sec. 4-1.	4-1	164
If you made any mistakes in Questions 6 to 9, circle Sec. 4-2.	4-2	175
If you made any mistakes in Questions 10 to 13, circle Sec. 4-3.	4-3	183
If you made any mistakes in Questions 14 to 17, circle Sec. 4-4.	4-4	189

You should now work through the sections whose numbers you have circled, taking them in numerical order. The sections begin on the pages indicated.

Section 4-1
Definition and Addition of Negative Numbers

Do this section if you made a mistake in Questions 1 to 5 of Diagnostic Test 4. Begin by taking the pretest below, which will show you how much of Sec. 4-1 you already know and what you will be able to do by the time you have finished it.

PRETEST 4-1

1. $(-11) + 11 =$ _____
2. $(-30) + 71 =$ _____
3. $(-15) + 0 =$ _____
4. $32 + (-18) =$ _____
5. $(-17) + (-35) =$ _____
6. $0 + (-23) =$ _____
7. $115 + (-115) =$ _____

When you have finished the pretest, correct it yourself using the answers on page 199. Then check all the questions you got wrong on the list below:

1. _____
2. _____
3. _____
4. _____
5. _____
6. _____
7. _____

If you made any mistakes (even if it was only *one* mistake), read the introduction and go through the program.

If you made no mistakes at all, skip Sec. 4-1, because you already know its material, and go on to the next section whose number you circled on page 163. If no other numbers were circled, take Final Test 4 (page 198).

Introduction

Section 4-1 shows how points along a road can be described in terms of their distance from a fixed reference point. You will see that for a complete description of the position of a point some indication of direction is also needed. The directions are labeled *east* and *west* on one map and *positive* and *negative* on another map. The east-west map of mile points is shown to be in one-to-one correspondence with the positive-negative map, which is identified with the integer number line.

Trips along the road correspond with trips along the number line and can be represented by additions. This process is used to define the operation of addition for all integers (including negative integers and 0), and you practice addition of integers in several examples.

This section requires the ability to add and subtract small integers (Chap. 1) and the ability to understand the representation of points along a road as points along a line.

FRAMES FOR SECTION 4-1

1 Suppose you live on a long road which lies in an east-west direction. If you take your home as a reference point, you can represent points along the road by numbers which indicate the distance from your home in miles.

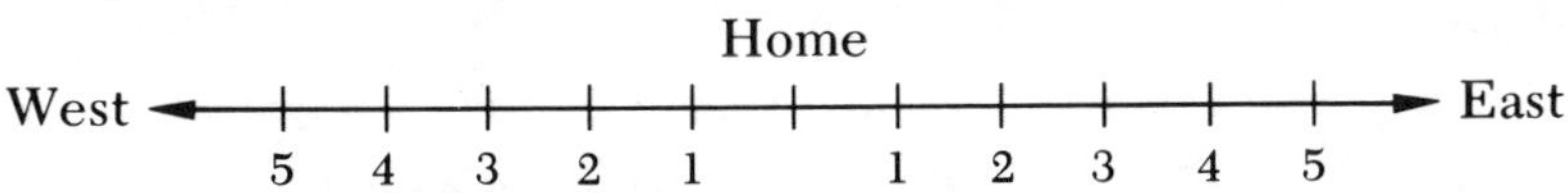

a. How many points are there on the road at a distance of 3 miles from your home?

b. If a friend phoned to say that his car had broken down 3 miles away from your home, would you know where he was?

---

a. 2

b. Not unless you knew which direction he was coming from; he could be at either of the points marked 3.

---

2 You can remove this doubt by labeling points east of your home with an E and points west of your home with a W.

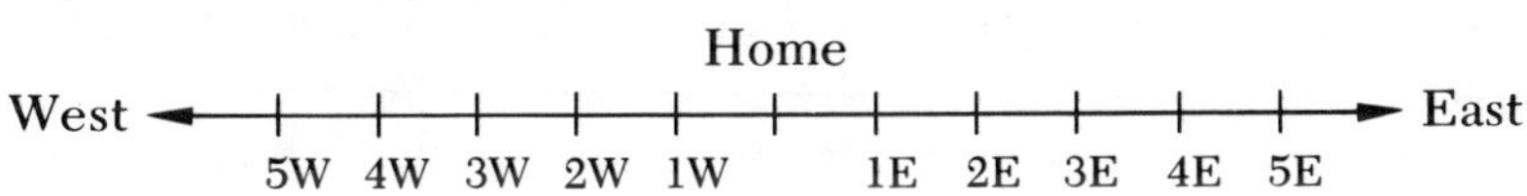

The point labeled 4W is ______ miles (east/west) ______ of your home; and the point labeled 2E is ______ miles ______ of your home.

---

4 west 2 east

---

3 *a.* If you start 2 miles east of home and travel 3 miles east, where do you finish?

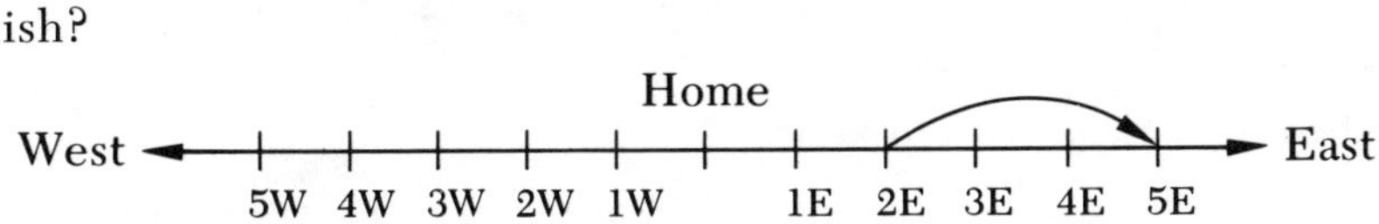

b. If you start 1 mile *west* of home and travel 2 miles *west*, where do you finish?

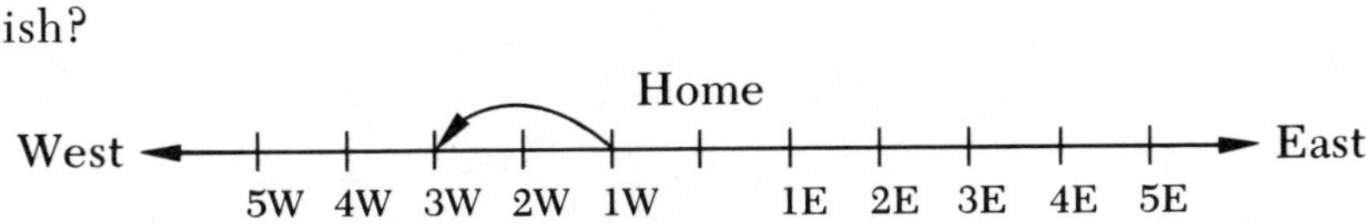

---

a. point 5E, 5 miles east of home *b.* point 3W, 3 miles west of home

---

4 While you travel, you notice that your friend uses a diagram rather like your own, but he marks his points + and − instead of E and W. Your diagram:

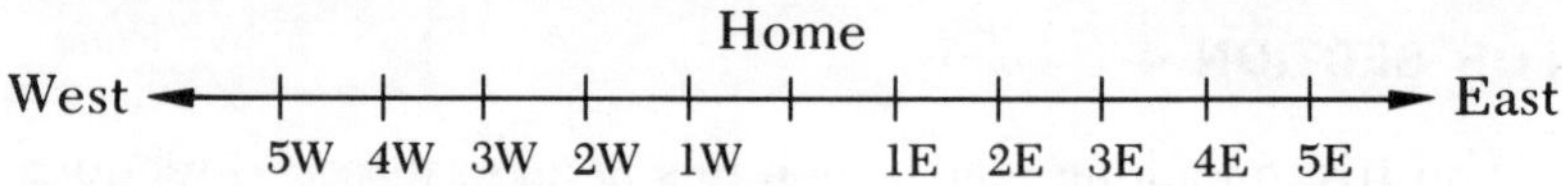

Your friend's diagram:

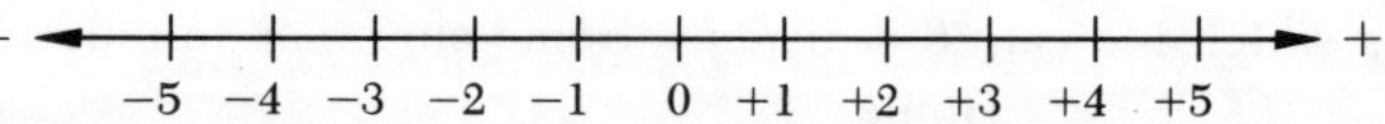

He tells you that he can use *his* diagram for roads that go in any direction and that for this particular road he uses the + side of his diagram for points east of your home and the − side of his diagram for points west of your home. He uses the point 0 to represent your home and also measures his distances in miles. (−3) on his diagram is equivalent to ______ on your diagram, and (+4) on his diagram is equivalent to ______ on your diagram.

3W 4E

---

5 *a.*

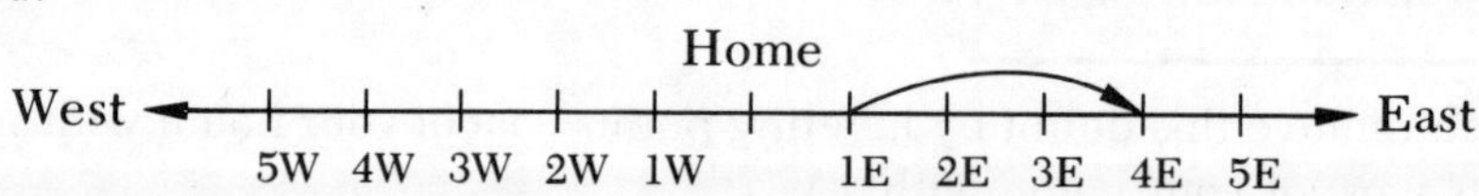

If you start 1 mile east of home and travel 3 miles east, where do you finish on your diagram?

b.

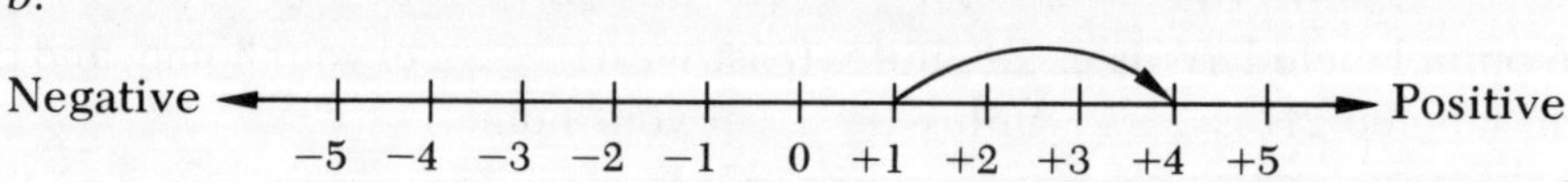

If your friend starts at point (+1) and goes 3 miles in the positive direction, where does he finish on his diagram?

---

a. point 4E, 4 miles east of home *b.* point (+4)

---

6 These trips can be represented as:

a. 1E + 3E = 4E, where +3E means "move ______ miles to the east."

b. (+1) + (+3) = ______, where + (+3) means "move ______ miles in the + direction."

---

a. 3 *b.* (+4); 3

---

7 *a.*

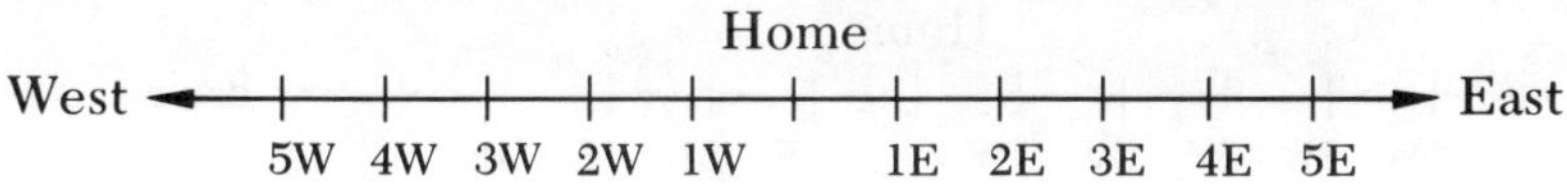

If you start 2 miles east of home and travel 1 mile east, where do you finish on your diagram?

b.

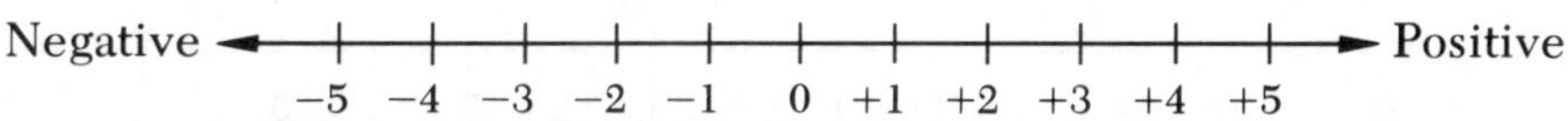

If your friend starts at point (+2) and goes 1 mile in the positive direction, where does he finish on his diagram?

c. These trips can be represented as 2E + 1E = ______, where +1E means "move ______ miles to the ______" and (+2) + (+1) = ______, where + (+1) means "move ______ miles in the ______ direction."

---

a. point 3E, 3 miles east of home *b.* point (+3)
c. 3E; 1, east; (+3); 1, +

---

8 *a.*

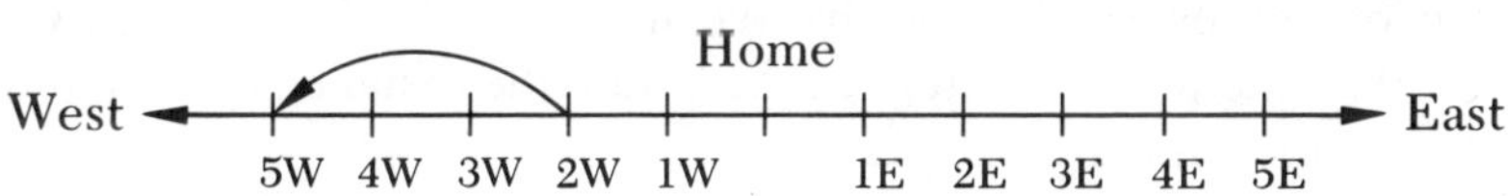

If you start 2 miles *west* of home and travel 3 miles *west,* where do you finish on your diagram?

b.

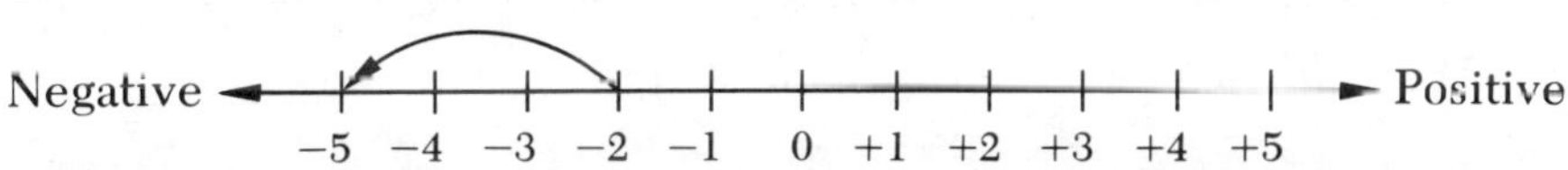

If your friend starts at point (−2) and goes 3 miles in the negative direction, where does he finish on his diagram?

c. These trips can be represented as 2W + 3W = ______, where +3W means "move ______ miles to the ______," and (−2) + (−3) = ______, where + (−3) means "move ______ miles in the ______ direction."

---

a. point 5W, 5 miles west of home *b.* point (−5)
c. 5W; 3, west; (−5); 3, −

---

9 *a.*

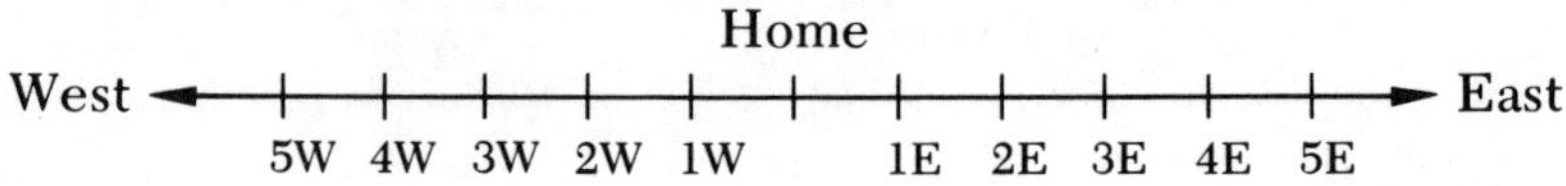

If you start 2 miles *west* of home and travel 2 miles *west,* where do you finish on your diagram?

b.

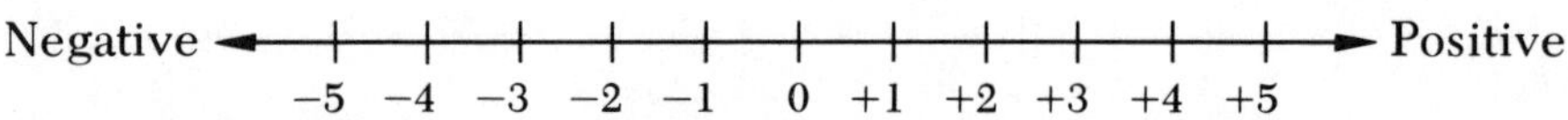

If your friend starts at point (−2) and goes 2 miles in the negative direction, where does he finish on his diagram?

c. These trips can be represented as 2W + 2W = _____ and (−2) + (−2) = _____.

---

a. point 4W, 4 miles *west* of home *b.* point (−4) *c.* 4W; (−4)

---

10 The numbers 1, 2, 3, 4, 5, . . . are called *positive integers,* and the distance of a point from home (on your diagram) or from 0 (on your friend's diagram) can be represented by a positive integer. So the distance corresponding to 3W is the positive integer _____, and the distance corresponding to (−2) is the positive integer _____.

---

3 2

---

11 Give the distances corresponding to the following symbols:

a. 5E *b.* 4W *c.* (+3) *d.* (−1)

---

a. 5 *b.* 4 *c.* 3 *d.* 1

---

12 Our trips in Frames 5 to 9 were represented by the following equations:

1E + 3E = 4E or (+1) + (+3) = (+4)

2E + 1E = 3E or (+2) + (+1) = (+3)

2W + 3W = 5W or (−2) + (−3) = (−5)

2W + 2W = 4W or (−2) + (−2) = (−4)

These suggest that whenever the directions on the left-hand side of the equation are the same, the direction of the right-hand side is (also the same/opposite) ______ and that the distance of the end point from home is obtained by (adding/subtracting) ______ the distances on the left-hand side of the equation. Applying this rule to other trips without looking at the diagram, find the following end points:

7E + 4E = ______ (+7) + (+4) = ______

3W + 4W = ______ (−3) + (−4) = ______

---

also the same; adding 11E (+11) 7W (−7)

---

13 (+13) + (+17) = ______

(−15) + (−14) = ______

(−25) + (−37) = ______

(+30) (−29) (−62)

---

14 *a.*

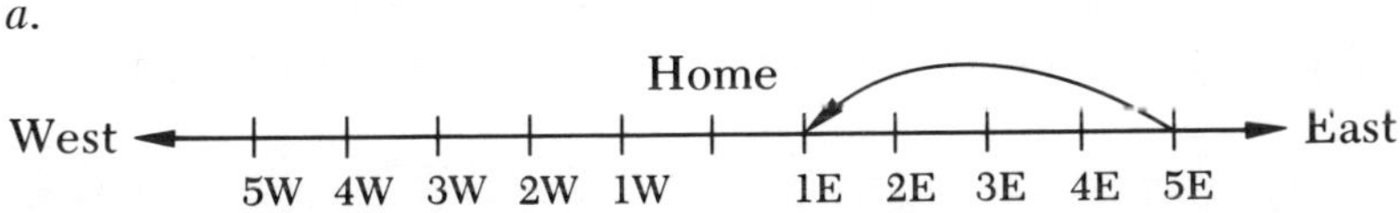

If you start 5 miles *east* of home and travel 4 miles *west*, where do you finish?

b.

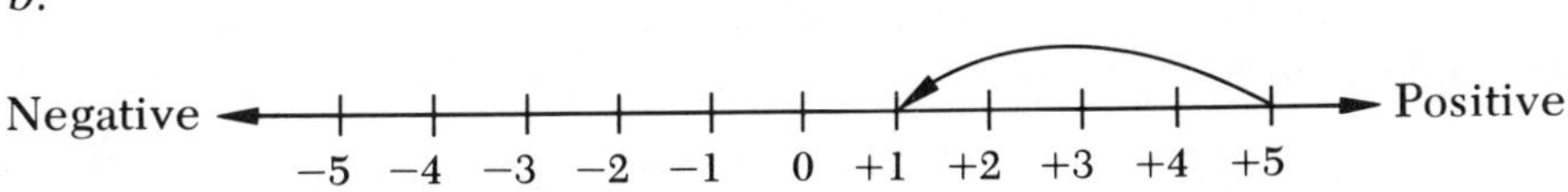

If your friend starts at point (+5) and travels 4 miles in the negative direction, where does he finish?

c. These trips can be represented as 5E + 4W = ______ and (+5) + (−4) = ______.

a. point 1E, 1 mile east of home *b.* point (+1) *c.* 1E; (+1)

15 *a.*

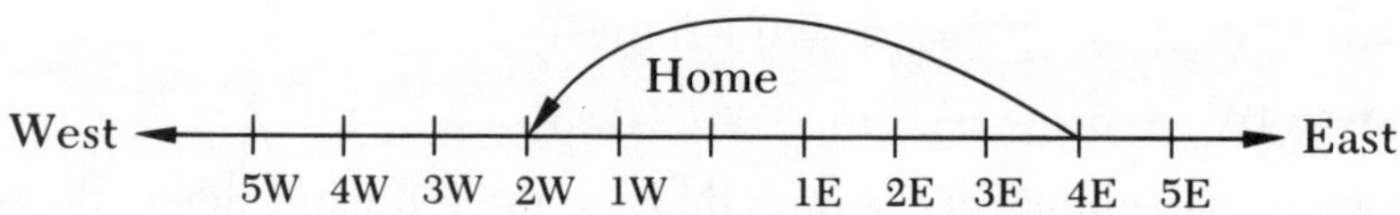

If you start 4 miles *east* of home and travel 6 miles *west*, where do you finish?

b.

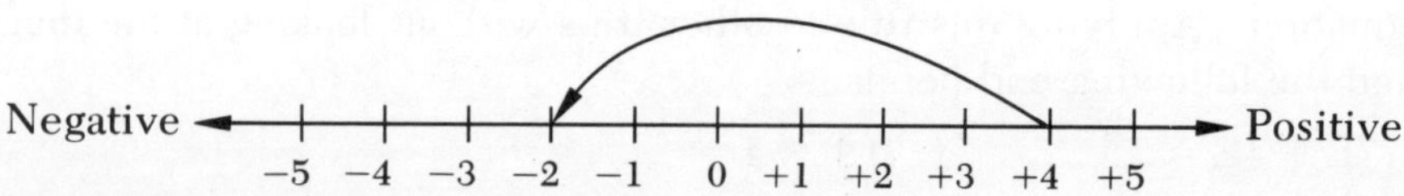

If your friend starts at point (+4) and travels 6 miles in the negative direction, where does he finish?

c. These trips can be represented as 4E + 6W = _____ and (+4) + (−6) = _____.

a. point 2W, 2 miles west of home *b.* point (−2) *c.* 2W; (−2)

16 *a.*

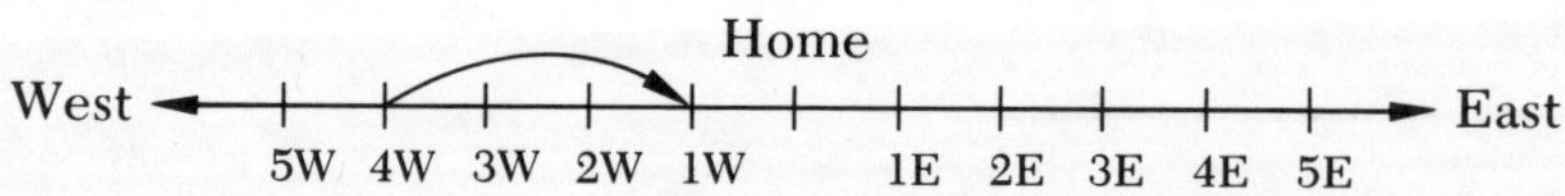

If you start 4 miles *west* of home and travel 3 miles *east*, where do you finish?

b.

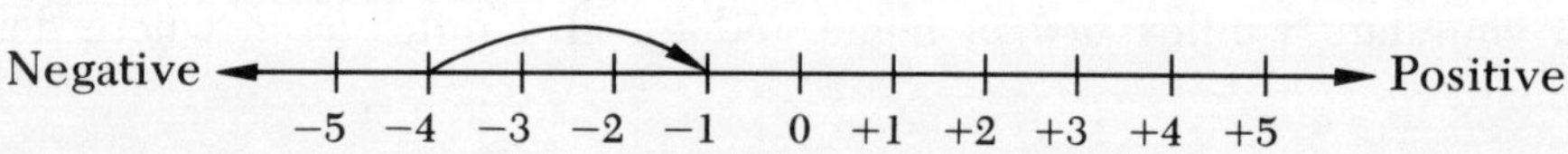

If your friend starts at point (−4) and travels 3 miles in the positive direction, where does he finish?

c. These trips can be represented as 4W + 3E = _____ and (−4) + (+3) = _____.

a. point 1W, 1 mile west of home *b.* point (−1) *c.* 1W; (−1)

17 *a.*

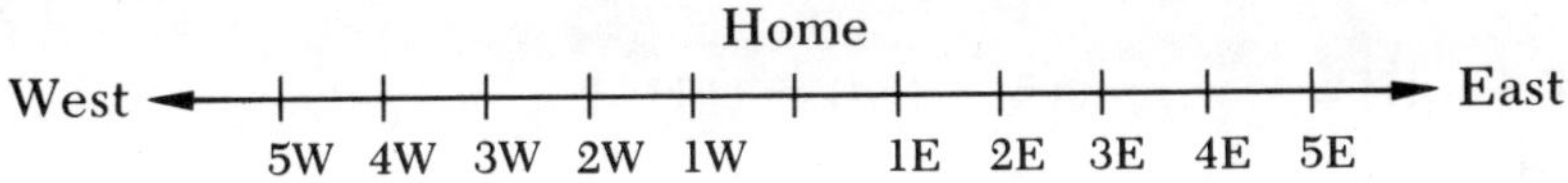

If you start 3 miles west of home and travel 7 miles east, where do you finish?

b.

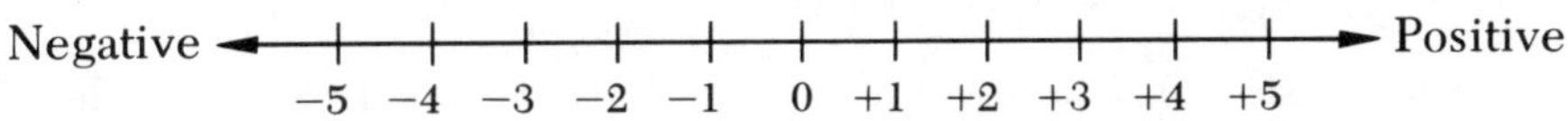

If your friend starts at point (−3) and travels 7 miles in the positive direction, where does he finish?

c. These trips can be represented as 3W + 7E = ______ and (−3) + (+7) = ______.

---

a. point 4E, 4 miles east of home *b.* point (+4) *c.* 4E; (+4)

---

18 The points on your friend's diagram all belong to the set of numbers which we call *integers*. These consist of

1. The positive integers: 1, 2, 3, 4, 5, . . . , also written as (+1), (+2), (+3), (+4), (+5), ...
2. The negative integers: (−1), (−2), (−3), (−4), (−5), ...
3. The number 0.

Which of the following numbers are:

a. Integers *b.* Positive integers *c.* Negative integers

(+5) 0 (−11) 12 (−19)

---

a. all are integers *b.* (+5) and 12 are positive integers
c. (−11) and (−19) are negative integers

---

19 Distances can be represented only by 0 or by positive integers. Give the distances corresponding to 3E, 2W, and (−5).

---

3 2 5

---

20 Our trips in Frames 14 to 17 were represented by the following equations:

5E + 4W = 1E or (+5) + (−4) = (+1)

4E + 6W = 2W or (+4) + (−6) = (−2)

4W + 3E = 1W or (−4) + (+3) = (−1)

3W + 7E = 4E or (−3) + (+7) = (+4)

These suggest that whenever the directions on the left-hand side of the equation are opposite, the distance of the end point is obtained by taking the distances from the left-hand side and (adding/subtracting) _____ the smaller (to/from) _____ the larger and that the direction of the end point will be the same as that corresponding to the (smaller/larger) _____ of the two distances. Applying this rule to other trips without looking at the diagram,

8E + 5W = _____ (+8) + (−5) = _____

3E + 9W = _____ (+3) + (−9) = _____

5W + 7E = _____ (−5) + (+7) = _____

12W + 3E = _____ (−12) + (+3) = _____

---

subtracting; from; larger 3E (+3) 6W (−6) 2E (+2) 9W (−9)

---

21 (−13) + (+11) = _____ (+15) + (−23) = _____

(+18) + (−7) = _____ (−27) + (+16) = _____

(Don't try to do everything in your head if you aren't very good at subtracting!)

(−2) (−8) (+11) (−11)

---

22 *a.* When the signs are the same, distances are (added/subtracted) _____ and the common sign is retained. Thus

b. (+5) + (+17) = _____

c. (−11) + (−13) = _____

d. When the signs are different, distances are (added/subtracted) _____ and the sign of the (larger/smaller) _____ is retained. Thus

e. (+15) + (−26) = _____

f. (−11) + (+27) = _____

---

a. added *b.* (+22) *c.* (−24) *d.* subtracted; larger *e.* (−11) *f.* (+16)

---

23 (+32) + (−17) = ______

(−11) + (−15) = ______

(−23) + (+9) = ______

(Don't try to do too much in your head!)

(+15) (−26) (−14)

---

24 Remembering that (+12) can also be written as 12, we have:

12 + (−40) = ______ 13 + 19 = ______ (−13) + 29 = ______

(−28) 32 16

25

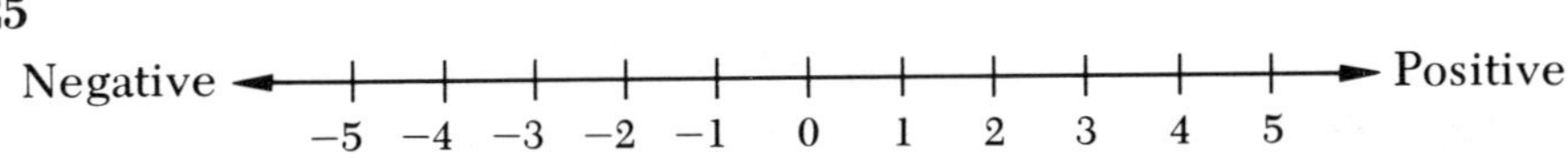

a. If you start at point (−5) and travel 5 spaces in a positive direction, where do you finish? This trip can be represented as (−5) + 5 = ______.

b. If you start at point 4 and travel 4 spaces in a negative direction, where do you finish? This trip can be represented as 4 + (−4) = ______.

---

a. point 0; 0 *b.* point 0; 0

---

26 (−13) + 13 = ______ 8 + (−8) = ______ (−105) + 105 = ______

---

0 0 0

27

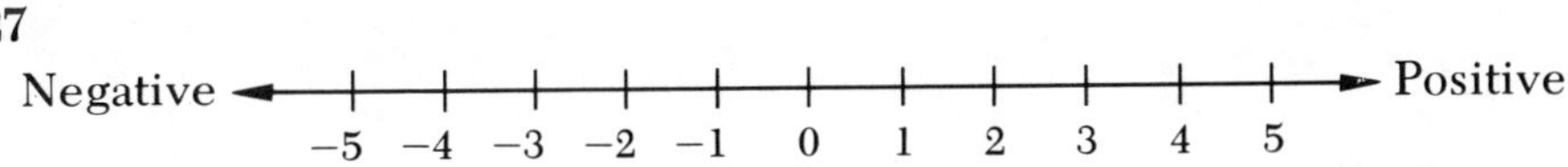

a. If you start at point (−3) and travel 0 spaces, where do you finish? This trip can be represented as (−3) + 0 = ______.

b. If you start at point 0 and travel 3 spaces in a positive direction, where do you finish? This trip can be represented as 0 + 3 = ______.

c. If you start at point 0 and travel 4 spaces in a negative direction, where do you finish? This can be represented as 0 + (−4) = ______.

---

a. point (−3); (−3) *b.* point 3; 3 *c.* point (−4); (−4)

---

28 $(-4) + 0 =$ _____ $\quad 15 + 0 =$ _____

$0 + 17 =$ _____ $\quad 0 + (-23) =$ _____

(−4) 15 17 (−23)

---

29 $(-23) + 23 =$ _____

$0 + (-17) =$ _____

$(-8) + 0 =$ _____

---

0 (−17) (−8)

You have finished the program for Sec. 4-1. Now take the posttest to see how much you have learned.

POSTTEST 4-1

1. $(-17) + 17 =$ _____
2. $(-40) + 61 =$ _____
3. $(-25) + 0 =$ _____
4. $44 + (-16) =$ _____
5. $(-36) + (-19) =$ _____
6. $0 + (-42) =$ _____
7. $134 + (-134) =$ _____

When you have finished the posttest, mark it yourself using the answers on page 199. Then check all the questions you got wrong on the list below:

1. _____ 5. _____

2. _____ 6. _____

3. _____ 7. _____

4. _____

If you compare your performance with that on Pretest 4-1, you will see how much you have improved. If you made more than one mistake on this posttest, review the material and take the pretest again. Don't move on to the next section until you can get all the questions right.

Now proceed to the next section whose number you circled on page 163. If no other sections were circled, take Final Test 4 (page 198).

Section 4-2
Multiplication and Division with Negative Numbers

Do this section if you made a mistake in Questions 6 to 9 of Diagnostic Test 4. Begin by taking the pretest below, which will show you how much of Sec. 4-2 you already know and what you will be able to do by the time you have finished it.

PRETEST 4-2

1. $4 \times (-5) =$ ______
2. $(-10) \div 2 =$ ______
3. $(-9) \div (-3) =$ ______
4. $6 \times 0 =$ ______
5. $(-2) \times (-3) =$ ______
6. $14 \div (-7) =$ ______
7. $0 \div (-3) =$ ______

When you have finished the pretest, correct it yourself using the answers on page 199. Then check all the questions you got wrong on the list below:

1. ______
2. ______
3. ______
4. ______
5. ______
6. ______
7. ______

If you made any mistakes (even if it was only *one* mistake), read the introduction and go through the program.

If you made no mistakes at all, skip Sec. 4-2, because you already know its material, and go on to the next section whose number you circled on page 163. If no other numbers were circled, take Final Test 4 (page 198).

Introduction

Section 4-2 introduces a simple concrete situation. A water tank is filled or emptied at various rates (represented by signed numbers), and a camera takes movies of this process. The developed film can be run forward or backward at various rates (represented by signed numbers). The resultant movie shows a tank filling or emptying at various rates (represented by signed numbers). This example is used to derive definitions of the multiplication of integers (including negative integers and zero). The division of integers is then derived, and you see why it is impossible to arrive at a consistent definition for division by 0.

FRAMES FOR SECTION 4-2

1 Suppose we have a glass tank and a pump which can pump water into or out of the tank. If 4 gallons of water flows through the pipe each minute and the pump is run for 3 minutes, the change in the volume of water in the tank will be _____ gallons.

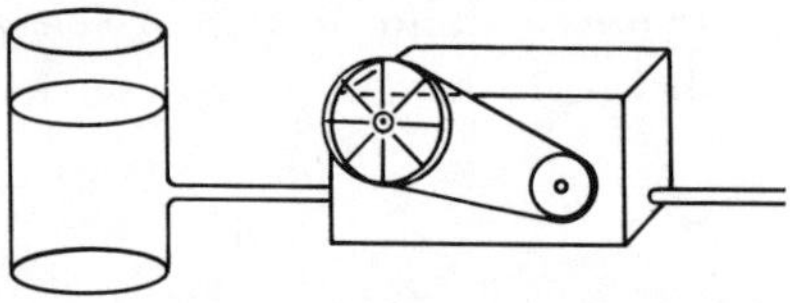

12

2 This change could be in either of two directions. If water is being pumped *into* the tank, the volume will be __creasing. If water is being pumped *out of* the tank, the volume will be __creasing.

increasing decreasing

3 We can distinguish between an increase and a decrease in the volume of water in the tank by using signed numbers. We call an increase of 12 gallons a change of +12 gallons and a decrease of 12 gallons a change of (−12) gallons. If 2 gallons of water is pumped *out* of the tank every minute and the pump is run for 4 minutes, the change in the volume of water in the tank will be _____ gallons. (Give a signed number as your answer.)

(−8)

4 If 5 gallons of water is pumped *into* the tank every minute for 3 minutes, the change in volume of the water in the tank will be _____ gallons. (Give a signed number as your answer.)

+15

5 Suppose we add a movie camera which takes a movie of the tank while the pump is operating.

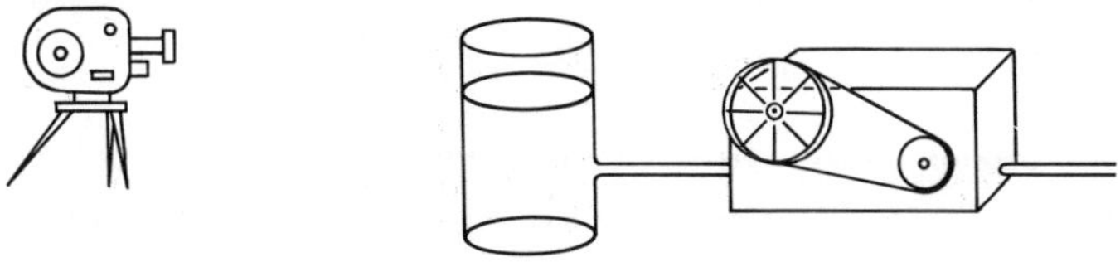

Suppose that the pump and camera are turned on at the same time and that 3 gallons of water is pumped *into* the tank each minute. After 3 minutes the pump and camera are stopped and the film is developed. The film is then projected on the screen. What change in water volume do you observe on the screen? (Give a signed number as your answer.)

+9

---

6 Suppose only 2 minutes of the film is shown, instead of all 3 minutes. What change in water volume do you observe on the screen? (Give a signed number as your answer.)

+6

---

7 If we are not watching all the film, we do not need to know how long the film is in order to predict the change on the screen. We need only know:

a. The change on the screen each minute and

b. The ______ ____ ______ the film is being shown.

---

number of minutes

---

8 If water is pumped out of the tank at 2 gallons a minute and the film is shown for 3 minutes, what change in volume of water is observed on the screen? (Give a signed number as your answer.)

(−6)

---

9 When we describe the change observed on the screen by a positive number, we mean that the water volume appears to __crease. When we describe the change observed on the screen by a negative number, we mean that the water volume appears to __crease.

---

increase; decrease

---

10 If the change in the tank is positive, the change on the screen is ____tive. If the change in the tank is negative, the change on the screen is ____tive.

---

positive; negative

---

11 Suppose that the tank is being filled up and we run the film backward through the projector. On the screen it will look as though the tank is (filling up/emptying) ______.

---

emptying

---

12 When the film is run backward, a positive change in the tank will appear as a ____tive change on the screen and a negative change in the tank will appear as a ____tive change on the screen.

---

negative; positive

---

13 We can also use signed numbers to indicate whether the film is run forward or backward through the projector and for how many minutes the film is run. We represent the number of minutes the film is run forward by a positive number and the number of minutes the film is run backward by a ____tive number.

---

negative

---

14 If the film is run *forward* for 3 minutes, we say that the film has been run for ______ minutes. (Give a signed number for your answer.) If the film is run *backward* for 6 minutes, we say that the film has been run for ______ minutes. (Give a signed number for your answer.)

$+3$ (-6)

---

15 What we see is affected both by what happens to the tank and by what happens to the film. The screen effect is the *product* of the tank effect and the film effect. We can write this as

Tank effect × film effect = screen effect

For example, if water is pumped *out* of the tank at 2 gallons a minute and the film is run *forward* for 2 minutes, the change observed on the screen will be ______ (give a signed number). This can be written as:

$$\underset{\text{tank}}{(-2)} \times \underset{\text{film}}{(+2)} = \underset{\text{screen}}{\underline{\qquad}}$$

Hint: Does the tank appear to be emptying or filling on the screen?

(−4) (−4)

---

16 If water is pumped *into* the tank at 3 gallons a minute and the film is run *forward* for 4 minutes, the change observed on the screen will be ______. (Give a signed number.) **Hint:** Does the tank appear to be emptying or filling? Therefore,

_____ × _____ = ______
tank film screen

(Use signed numbers.)

(+12) (+3) × (+4) = (+12)

---

17 If water is pumped *into* the tank at 4 gallons a minute and the film is run *backward* for 3 minutes, the change observed on the screen will be ______. (Give a signed number.) **Hint:** Does the tank appear to be emptying or filling? Therefore

_____ × _____ = ______
tank film screen

(Use signed numbers.)

−12 (+4) × (−3) = (−12)

---

18 If water is pumped *out* of the tank at 3 gallons a minute and the film is run *backward* for 2 minutes, the change observed on the screen will be ______. (Give a signed number.) **Hint:** Does the tank appear to be emptying or filling? Therefore

_____ × _____ = ______
tank film screen

(Use signed numbers.)

(+6) (−3) × (−2) = +6

---

19 The equations we wrote in Frames 15 to 18 were

$(-2) \times (+2) = (-4)$ Frame 15

$(+3) \times (+4) = +12$ Frame 16

$(+4) \times (-3) = (-12)$ Frame 17

$(-3) \times (-2) = +6$ Frame 18

In general,

a. Multiplying by a positive number (always/never) ______ changes the sign.

b. Multiplying by a negative number (always/never) ______ changes the sign.

c. The product of a negative number and a positive number is a ____tive number.

d. The product of two negative numbers is a ____tive number.

---

a. never *b.* always *c.* negative *d.* positive

---

20 *a.* $(-5) \times 4 =$ ______

b. $3 \times (-6) =$ ______

c. $(-7) \times (-5) =$ ______

Hint: Think in movie and tank terms if you have any difficulty.

---

a. (−20) *b.* (−18) *c.* +35

---

21 If the tank is filling at 2 gallons a minute and the film is run for 0 minutes, the change on the screen will be ______.

$\underset{\text{tank}}{+2} \times \underset{\text{film}}{0} = \underset{\text{screen}}{\underline{\qquad}}$

---

0 0

---

22 If the pump is off and the film is run forward for 6 minutes, the change on the screen will be ______.

$0 \times (+6) = \underset{\text{screen}}{\underline{\qquad}}$

---

0 0

---

23 If a is any number, $a \times 0 =$ ______, and if b is any number, $0 \times b =$ ______.

0 0

---

24 The laws of division can easily be deduced from the laws of multiplication. Since $3 \times (-4) = (-12)$, we can deduce that

$$\frac{(-12)}{3} = (-12) \div 3 = \_\_\_\_\_\_$$

and that

$$\frac{(-12)}{(-4)} = (-12) \div (-4) = \_\_\_\_\_\_$$

Dividing a negative number by a positive number gives a ____tive number. Dividing a negative number by a negative number gives a ____tive number.

(−4) 3 negative; positive

---

25 Since $(-6) \times (-3) = 18$, we can deduce that

$$\frac{18}{(-6)} = 18 \div (-6) = \_\_\_\_\_\_$$

and that

$$\frac{18}{(-3)} = 18 \div (-3) = \_\_\_\_\_\_$$

Dividing a positive number by a negative number gives a ____tive number.

(−3) (−6) negative

---

26 $\frac{18}{(-6)} = \_\_\_\_\_\_$ $\frac{(-15)}{3} = \_\_\_\_\_\_$ $\frac{(-14)}{(-7)} = \_\_\_\_\_\_$

(−3) (−5) 2

---

27 To find out the meaning of $0 \div 2$, let $\frac{0}{2} = y$. Then $2 \times y = 0$, so $y = \_\_\_\_\_\_$.

$\frac{0}{2} = \_\_\_\_\_\_$ and $\frac{0}{3} = \_\_\_\_\_\_$

0 0 0

---

28 $\frac{0}{4} = \_\_\_\_\_\_$ and $\frac{0}{(-4)} = \_\_\_\_\_\_$

---

0 0

---

29 To find the meaning of $2 \div 0$, let $\frac{2}{0} = z$, where z is any number. Then $2 = z \times 0$. But we have already shown that $z \times 0 = 0$; therefore $2 = 0$. This is nonsense, so there (can/cannot) _____ be any number z which equals $\frac{2}{0}$.

---

cannot

---

30 Can there be a number which equals $\frac{4}{0}$? Therefore dividing by 0 (can/cannot) _____ be defined.

---

no; cannot

You have finished the program for Sec. 4-2. Now take the posttest to see how much you have learned.

POSTTEST 4-2

1. $3 \times (-7) =$ _____
2. $(-14) \div 2 =$ _____
3. $(-10) \div (-2) =$ _____
4. $4 \times 0 =$ _____
5. $(-6) \times (-3) =$ _____
6. $35 \div (-5) =$ _____
7. $0 \div (-4) =$ _____

When you have finished the posttest, mark it yourself using the answers on page 199. Then check all the questions you got wrong on the list below:

1. _____
2. _____
3. _____
4. _____
5. _____
6. _____
7. _____

If you compare your performance with that on Pretest 4-2, you will see how much you have improved. If you made more than one mistake on this post-test, review the material and take the pretest again. Don't move on to the next section until you can get all the questions right.

Now proceed to the next section whose number you circled on page 163. If no other numbers were circled, take Final Test 4 (page 198).

Section 4-3
Subtraction of Negative Numbers

Do this section if you made a mistake in Questions 10 to 13 of Diagnostic Test 4. Begin by taking the pretest below, which will show you how much of Sec. 4-3 you already know and what you will be able to do by the time you have finished it.

PRETEST 4-3

1. $-(-3) =$ _____ (what integer?)
2. $0 - 37 =$ _____
3. $(-19) - 62 =$ _____
4. $11 - (-17) =$ _____
5. $23 - 48 =$ _____
6. $0 - (-26) =$ _____
7. $(-12) - (-35) =$ _____

When you have finished the pretest, correct it yourself using the answers on page 199. Then check all the questions you got wrong on the list below:

1. _____
2. _____
3. _____
4. _____
5. _____
6.
7. _____

If you made any mistakes (even if it was only *one* mistake), read the introduction and go through the program.

If you made no mistakes at all, skip Sec. 4-3, because you already know its material, and go on to the next section whose number you circled on page 163. If no other numbers were circled, take Final Test 4 (page 198).

Introduction

Section 4-3 begins by defining the operation "negative of" in terms of position on the number line and shows that it is equivalent to the operations of multiplying by (−1) and dividing by (−1). We then derive a definition of subtraction as "addition of the negative" and show it to be applicable to the whole set of integers.

FRAMES FOR SECTION 4-3

1

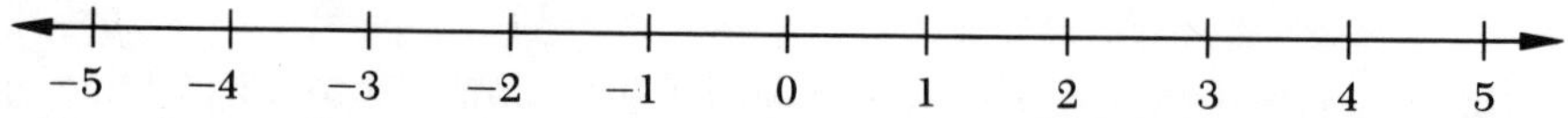

The line shown above is called a number line. All the points we have marked on it represent integers. Positive integers are represented by points on the number line to the (left/right) ______ of 0, and negative integers by points to the (left/right) ______ of 0.

---

right; left

2

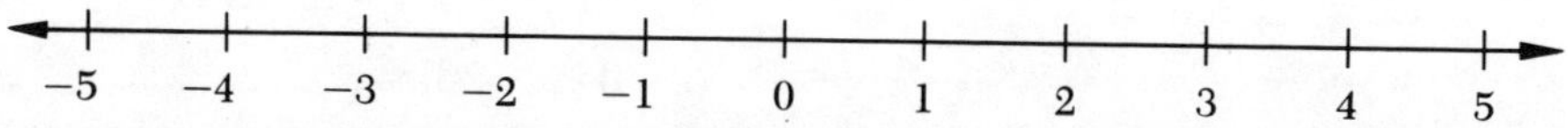

If n is a positive integer, it is located on the (left/right) ______ side of 0 and $(-n)$ is located on the ______ side of 0. The two points representing n and $(-n)$ (are/are not) ______ at the same distance from 0.

---

right; left; are

---

3 We define the negative of a number as the number on the *opposite side* of 0 at the *same distance* from 0. Hence the negative of (-2) [written as $-(-2)$] is ______, the negative of (-5) [written as $-(-5)$] is ______, and $-(-3) =$ ______.

---

2 5 3

---

4 The negative of 3 is ______, the negative of (-4) is ______, and $-(-7) =$ ______.

(-3) 4 7

---

5 Another definition of the operation "negative of" is given by the equations

$$-(n) = (-n) \qquad \text{and} \qquad -(-n) = \_\_\_\_\_\_$$

---

n

---

6 And since $n \times$ ______ $= (-n)$ and $(-n) \times$ ______ $= n$, we can say that the operation "negative of" is equivalent to the operation of "multiplying by ______."

(−1) (−1) (−1)

7 We also know that $n \div$ ______ $= (-n)$ and $(-n) \div$ ______ $= n$, so we can deduce that the operation "negative of" is also equivalent to the operation of "dividing by ______."

(−1) (−1) (−1)

8 There are three different operations which change n to $(-n)$. What are they?

a. ______ *b.* ______ *c.* ______

d. What effect do these operations have on $(-n)$?

a. taking the negative *b.* multiplying by (−1)
c. dividing by (−1) *d.* change $(-n)$ to n

9 *a.* $-(-4) =$ ______ (what integer?)

b. $3 \times (-1) =$ ______ (what integer?)

c. $(-5) \div (-1) =$ ______ (what integer?)

a. 4 *b.* (−3) *c.* 5

10 *a.* $(-5) \times (-1) =$ ______ (what integer?)

b. $-(-13) =$ ______ (what integer?)

c. $18 \div (-1) =$ ______ (what integer?)

a. 5 *b.* 13 *c.* (−18)

11 $(-5) + 8 =$ ______ $12 + (-23) =$ ______ $(-6) + (-18) =$ ______

3 (−11) (−24)

12 $7 + (-2) =$ ______ and $6 + (-3) =$ ______

but we also know that

$7 - 2 =$ ______ and $6 - 3 =$ ______

so adding the negative of a number gives (the same/a different) ______ answer (as/from) ______ subtracting the number.

---

5 3 5 3 the same; as

---

13 We can use this property to define subtraction. If a and b are integers, we can define $a - b$ as $a + (-b)$. So

$8 - 4 = 8 + (-4) =$ _____

$8 - 10 = 8 +$ _____ $=$ _____

$(-3) - 6 = (-3) +$ _____ $=$ _____

---

4 (−10) (−2) (−6) (−9)

---

14 Notice that we are now able to define the subtraction of a larger number from a smaller number, that is, $8 - 10$, whereas this operation was undefined before. Use this new definition of subtraction to replace the following subtractions by additions and then find the answers:

a. $(-7) - 4 = (-7) + (-4) =$ _____ *b.* $4 - 10 =$ __________ $=$ _____

c. $0 - 7 =$ __________ $=$ _____ *d.* $(-8) - 5 =$ __________ $=$ _____

---

a. (−11) *b.* $4 + (-10) = (-6)$ *c.* $0 + (-7) = (-7)$
d. $(-8) + (-5) = (-13)$

---

15 Replace the following subtractions by additions and then calculate the answers:

a. $0 - 32 =$ __________ $=$ _____

b. $11 - 42 =$ __________ $=$ _____

c. $(-17) - 15 =$ __________ $=$ _____

---

a. $0 + (-32) = (-32)$ *b.* $11 + (-42) = (-31)$
c. $(-17) + (-15) = (-32)$

---

16 This new definition of subtraction also applies to the subtraction of negative numbers:

$3 - (-8) = 3 + [-(-8)]$

but

$-(-8) =$ _____

Therefore

$3 - (-8) = 3 +$ _____ $=$ _____

8 8 11

17 $5-(-3)=5+[-(-3)]=5+$ _____ (since $[-(-3)]=$ _____)
$=$ _____

3 3 8

18 *a.* $9-(-5)=9+[-(-5)]=9+$ _____ $=$ _____
b. $3-(-6)=3+$ _____ $=$ _____

a. 5; 14 *b.* 6; 9

19 *a.* $4-(-2)=4+$ _____ $=$ _____
b. $(-3)-(-5)=(-3)+$ _____ $=$ _____
c. $(-8)-(-1)=(-8)+$ _____ $=$ _____
d. $0-(-7)=0+$ _____ $=$ _____

a. 2; 6 *b.* 5; 2 *c.* 1; (−7) *d.* 7; 7

20 Replace the following subtractions by additions and then calculate the answers:

a. $3-(-4)=3+4=$ _____
b. $5-(-2)=$ _____ $=$ _____
c. $(-2)-(-7)=$ _____ $=$ _____
d. $0-(-3)=$ _____ $=$ _____

a. 7 *b.* $5+2=7$ *c.* $(-2)+7=5$ *d.* $0+3=3$

21 Replace the following subtractions by additions and then calculate the answers:

a. $2-(-3)=$ _____ $=$ _____
b. $(-3)-(-4)=$ _____ $=$ _____
c. $(-12)-(-3)=$ _____ $=$ _____
d. $0-(-2)=$ _____ $=$ _____

a. $2+3=5$ *b.* $(-3)+4=1$ *c.* $(-12)+3=(-9)$
d. $0+2=2$

22 Replace the following subtractions by additions and then calculate the answers. (Don't try to do everything in your head if you aren't very good at adding!)

a. $0 - (-58) =$ __________ $=$ ______

b. $23 - (-39) =$ __________ $=$ ______

c. $(-32) - (-47) =$ __________ $=$ ______

d. $(-64) - (-57) =$ __________ $=$ ______

---

a. $0 + 58 = 58$ *b.* $23 + 39 = 62$ *c.* $(-32) + 47 = 15$
d. $(-64) + 57 = (-7)$

---

23 Replace the following subtractions by additions and then calculate the answers. (Notice that you are subtracting positive numbers again.)

a. $15 - 2 =$ __________ $=$ ______

b. $0 - 39 =$ __________ $=$ ______

c. $(-19) - 32 =$ __________ $=$ ______

---

a. $15 + (-2) = 13$ *b.* $0 + (-39) = (-39)$ *c.* $(-19) + (-32) = (-51)$

---

24 Replace the following subtractions by additions and then calculate the answers:

a. $(-15) - 32 =$ __________ $=$ ______

b. $(-31) - (-49) =$ __________ $=$ ______

c. $0 - 52 =$ __________ $=$ ______

d. $0 - (-43) =$ __________ $=$ ______

---

a. $(-15) + (-32) = (-47)$ *b.* $(-31) + 49 = 18$
c. $0 + (-52) = (-52)$ *d.* $0 + 43 = 43$

You have finished the program for Sec. 4-3. Now take the posttest to see how much you have learned.

POSTTEST 4-3

1. $-(-7) =$ ______ (what integer?)
2. $0 - 26 =$ ______
3. $(-43) - 59 =$ ______
4. $25 - 47 =$ ______
5. $21 - (-18) =$ ______
6. $0 - (-33) =$ ______
7. $(-22) - (-46) =$ ______

When you have finished the posttest, mark it yourself using the answers on page 199. Then check all the questions you got wrong on the list below:

1. ______ 5. ______
2. ______ 6. ______
3. ______ 7. ______
4. ______

If you compare your performance with that on Pretest 4-3, you will see how much you have improved. If you made more than one mistake on this posttest, review the material and take the pretest again. Don't move on to the next section until you can get all the questions right.

Now proceed to the next section whose number you circled on page 163. If no other sections were circled, take Final Test 4 (page 198).

Section 4-4 Multiple Operations with Negative Numbers

Do this section if you made a mistake in Questions 14 to 17 of Diagnostic Test 4. Begin by taking the pretest below, which will show you how much of Sec. 4-4 you already know and what you will be able to do by the time you have finished it.

PRETEST 4-4

1. $(-11) - 5 - (-10) + 18 =$ ______
2. $14 + (-13) - 25 - (-2) + 17 =$ ______
3. $(-16) - 1 + 24 + (-2) - (-27) =$ ______
4. $-7[(-6) - (-9)] =$ ______
5. $-[3 + (-7)] =$ ______
6. $-3[(-4) + 5 - (-11) - 8] =$ ______

When you have finished the pretest, correct it yourself using the answers on page 200. Then check all the questions you got wrong on the list below:

1. ______ 4. ______
2. ______ 5. ______
3. ______ 6. ______

If you made any mistakes (even if it was only *one* mistake), read the introduction and go through the program.

If you made no mistakes at all, skip Sec. 4-4, because you already know its material, and go on to take Final Test 4 (page 198).

Introduction

The definition of subtraction as "addition of the negative" enables us to convert subtractions (which are not associative) to additions (which are associative). Thus multiple subtractions can be defined without using parentheses. We then have a number of examples of multiple operations involving addition and subtraction.

Combined multiplications and additions are introduced for the set of integers, leading to combined multiplications and subtractions. The common practice of omitting parentheses from negative numbers in certain types of expressions is adopted.

FRAMES FOR SECTION 4-4

1 In an earlier chapter we showed that the addition of *positive* integers was commutative, i.e., that

$a + b = b + a$

Addition of negative integers can also be shown to be commutative. For example,

$(-7) + (-8) =$ _____ and $(-8) + (-7) =$ _____

(−15) (−15)

---

2 $7 - 9 =$ _____ and $9 - 7 =$ _____

So subtraction of positive integers (is/is not) _____ commutative.

(−2) 2 is not

---

3 We also showed earlier that addition of positive integers was associative, i.e., that

$$(a + b) + c = a + (b + c)$$

Addition of negative integers can also be shown to be associative. For example,

$[(-3) + 4] + (-5) =$ ______ $+ (-5) =$ ______

and

$(-3) + [4 + (-5)] = (-3) +$ ______ $=$ ______

---

1 (−4) (−1) (−4)

---

4 Now if we consider an example with multiple subtractions,

$(14 - 4) - 3 =$ ______ $- 3 =$ ______

and

$14 - (4 - 3) = 14 -$ ______ $=$ ______

So subtraction of positive integers (is/is not) ______ always associative.

---

10 7 1 13 is not

---

5 This means that whereas $14 + 4 + 3$ will always equal ______, $14 - 4 - 3$ could equal ______ or ______, depending upon which subtraction we decided to carry out first.

---

21 7 or 13

---

6 If, however, we stick to the definition of subtraction developed in Sec. 4-3, i.e., that $a - b = a + (-b)$, we find that $14 - 4 - 3$ is defined as

$14 +$ ______ $+$ ______

(−4) (−3)

---

7 Now $14 + (-4) + (-3)$ only has one value because addition is associative. Thus

$[14 + (-4)] + (-3) =$ ______ $+ (-3) =$ ______

and

$14 + [(-4) + (-3)] = 14 +$ ______ $=$ ______

---

10 7 (−7) 7

---

8 For this reason the definition of subtraction as "the addition of the negative" is commonly adopted, and multiple subtractions are written without parentheses. Thus $12 - 7 - 8$ is defined as $12 + \_\_\_\_ + \_\_\_\_ = \_\_\_\_$, and $13 - 14 - 15 = \_\_\_\_ + \_\_\_\_ + \_\_\_\_ = \_\_\_\_$.

(-7) (-8) (-3) $13 + (-14) + (-15) = (-16)$

---

9 Similarly $(-5) - 6 - (-7) = \_\_\_\_ + \_\_\_\_ + \_\_\_\_ = \_\_\_\_$, and $7 - (-6) - (-3) = \_\_\_\_ + \_\_\_\_ + \_\_\_\_ = \_\_\_\_$.

$(-5) + (-6) + 7 = (-4)$ $7 + 6 + 3 = 16$

---

10 *a.* $5 - (-7) - 8 = \_\_\_\_ + \_\_\_\_ + \_\_\_\_ = \_\_\_\_$

b. $(-6) - (-5) - (-7) = \_\_\_\_$

---

a. $5 + 7 + (-8) = 4$ *b.* $(-6) + 5 + 7 = 6$

---

11 Any expression containing subtractions can be replaced in this way by an equivalent expression which contains only additions.

a. $3 + (-1) - 3 + 2 - (-6) - 7 = \_\_\_\_ + \_\_\_\_ + \_\_\_\_ + \_\_\_\_ + \_\_\_\_$

b. $(-2) + 9 + (-6) - 6 + (-2) - (-3)$

$$= \_\_\_\_ + \_\_\_\_ + \_\_\_\_ + \_\_\_\_ + \_\_\_\_ + \_\_\_\_$$

c. $7 - 2 - 3 - (-5) - 6 + 1 = \_\_\_\_ + \_\_\_\_ + \_\_\_\_ + \_\_\_\_ + \_\_\_\_ + \_\_\_\_$

---

a. $3 + (-1) + (-3) + 2 + 6 + (-7)$ *b.* $(-2) + 9 + (-6) + (-6) + (-2) + 3$
c. $7 + (-2) + (-3) + 5 + (-6) + 1$

---

12 Since addition is commutative, these expressions can now be regrouped to bring all the positive integers together and all the negative integers together. Thus

$$3 + (-1) + (-3) + 2 + 6 + (-7) = (3 + 2 + 6) + [(-1) + (-3) + (-7)]$$
$$= \_\_\_\_ + (\_\_\_\_) = \_\_\_\_$$

and

$$(-2) + 9 + (-6) + (-6) + (-2) + 3$$
$$= (\_\_\_\_ + \_\_\_\_) + [(\_\_\_\_) + (\_\_\_\_) + (\_\_\_\_) + (\_\_\_\_)]$$
$$= \_\_\_\_ + (\_\_\_\_) = \_\_\_\_$$

$11 + (-11) = 0$ $(9 + 3) + [(-2) + (-6) + (-6) + (-2)] = 12 + (-16) = (-4)$

---

13 Similarly

$$7 + (-2) + (-3) + 5 + (-6) + 1 = \underset{\text{sum of all positives}}{\underline{\qquad\qquad}} + (\underset{\text{sum of all negatives}}{\underline{\qquad\qquad}}) = \underline{\qquad\qquad}$$

and

$$6 + (-5) + (-4) + (-3) + 2 + (-9) + 7 = \underline{\qquad\qquad} + (\underline{\qquad\qquad}) = \underline{\qquad\qquad}$$

$13 + (-11) = 2 \qquad 15 + (-21) = (-6)$

---

14 This suggests a simple method for simplifying multiple additions and subtractions:

$$1 - 5 - (-7) + (-6) - (-8) + 3 - 11 - (-7)$$

↕ Change all subtractions to additions

$$1 + (-5) + 7 + (-6) + 8 + 3 + (-11) + 7$$

↕ Add all positive numbers and all negative numbers

$$\underline{\qquad\qquad} + (\underline{\qquad\qquad})$$

↕ Complete the problem

$$\underline{\qquad\qquad}$$

---

26 (−22) 4

---

15 How many stages were there in the method used in Frame 14? Simplify the same expression by adding or subtracting one number at a time, that is,

$$1 - 5 = (-4)$$

$$(-4) - (-7) = (-4) + 7 = 3$$

$$3 + (-6) = (-3)$$

and so forth. How many stages did this method take? Which method do you think is the quickest?

---

3 stages; 7 stages; we found the method used in Frame 14 quicker and hope you did too. We shall use this method from now on.

---

16 Use the method of Frame 14 to simplify the expression

$$3 + (-9) - 6 - (-5) + 6 + (-3) - 2 - (-10)$$

↕ Change all subtractions to additions

_____ + _____ + _____ + _____ + _____ + _____ + _____ + _____

↕ Add all positive numbers and all negative numbers

_____ + (_____)

↕ Complete the problem

_____

$3 + (-9) + (-6) + 5 + 6 + (-3) + (-2) + 10$ $\quad 24 + (-20)$ $\quad 4$

---

17 Simplify the following expression by the same method:

$(-7) + 3 - (-10) - 7 + (-8) - 4$

↕ Change all subtractions to additions

_____

↕ Add all positive numbers and all negative numbers

_____

↕ Complete the problem

_____

$(-7) + 3 + 10 + (-7) + (-8) + (-4)$ $\quad 13 + (-26)$ $\quad (-13)$

---

18 Simplify the following expressions:

a. $(-5) - (-6) - 6 + 2$

↕ _____ ↕ _____ ↕ _____

b. $17 - 28 + (-3) - (-16)$

↕ _____ ↕ _____ ↕ _____

---

a. $(-5) + 6 + (-6) + 2$; $8 + (-11)$; (-3)
b. $17 + (-28) + (-3) + 16$; $33 + (-31)$; 2

---

19 Simplify the following expressions:

a. $(-6) + 3 - (-7) + (-8) - (-5) + 2$

b. $(-12) + (-3) - (-27) - 36 + 8 - 12$

c. $1 - (-6) + (-25) - 14 + 3 - (-12) + (-31)$

---

a. $(-6) + 3 + 7 + (-8) + 5 + 2 = (-14) + 17 = 3$
b. $(-12) + (-3) + 27 + (-36) + 8 + (-12) = (-63) + 35 = (-28)$
c. $1 + 6 + (-25) + (-14) + 3 + 12 + (-31) = (-70) + 22 = (-48)$

---

20 Combinations of multiplication and addition are written using parentheses and brackets to indicate which operation should be performed first. Thus

$$3(5 + 7) = 3 \times \_\_\_\_ = \_\_\_\_$$

$$3[5 + (-7)] = 3 \times \_\_\_\_ = \_\_\_\_$$

$$3[(-5) + (-7)] = 3 \times \_\_\_\_ = \_\_\_\_$$

---

12 36 (−2) (−6) (−12) (−36)

---

21 Similarly

$$5(2 - 7) = 5(2 + \_\_\_\_) = 5 \times \_\_\_\_ = \_\_\_\_$$

$$5[2 - (-7)] = 5(2 + \_\_\_\_) = \_\_\_\_$$

$$5[(-2) - (-7)] = 5[(-2) + \_\_\_\_] = \_\_\_\_$$

(−7) (−5) (−25) 7 $5 \times 9 = 45$ 7 $5 \times 5 = 25$

---

22 *a.* $(-4)(3 - 5) = (-4)(3 + \_\_\_\_) = (-4) \times \_\_\_\_ = \_\_\_\_$

b. $(-2)[2 - (-6)] = (-2)(2 + \_\_\_\_) = (-2) \times \_\_\_\_ = \_\_\_\_$

c. $(-5)[(-3) - (-9)] = (-5)[(-3) + \_\_\_\_] = \_\_\_\_$

---

a. (−5); (−2); 8 *b.* 6; 8; (−16) *c.* 9; $(-5) \times 6 = (-30)$

---

23 *a.* $(-6)[(-5) - 3] = (-6)[(-5) + \_\_\_\_] = \_\_\_\_$

b. $(-7)[2 - (-3)] = \_\_\_\_$

c. $(-4)[(-5) + (-6)] = \_\_\_\_$

---

a. (−3); $(-6) \times (-8) = 48$ *b.* $(-7)(2 + 3) = (-7) \times 5 = (-35)$
c. $(-4) \times (-11) = 44$

---

24 The brackets around negative numbers are often omitted in this context, so that:

a. $-5[(-9) + 3] = (-5) \times \_\_\_\_ = \_\_\_\_$

b. $-3[(-7) - 6] = (-3) \times [(-7) + \_\_\_\_] = \_\_\_\_$

c. $-6[(-2) - (-5)] = \_\_\_\_$

---

a. (−6); 30 *b.* (−6); (−3) × (−13) = 39 *c.* (−6) × 3 = (−18)

---

25 *a.* $-4[5+(-6)] =$ ________

b. $-2[4-(-7)] =$ ________

c. $-1[(-3)-8] =$ ________

---

a. (−4) × (−1) = 4 *b.* (−2) × 11 = (−22) *c.* (−1) × (−11) = 11

---

26 Since the operation of "multiplying by (−1)" is equivalent to the operation "negative of," we can write

$-1[(-3)-8]$ or $(-1)\times[(-3)-8]$

as

$-[(-3)-8] = -($____$) =$ ____

Instead of writing $-1[15+(-7)]$ we write $-[15+(-7)]$ and

$-[15+(-7)] = -($____$) =$ ____

−(−11) 11 −(8) (−8)

---

27 *a.* $-[(-3)-(-4)] = -[(-3)+$ ____ $] =$ ____

b. $-[5-(-6)] =$ ________

c. $-[3+(-7)] =$ ________

---

a. 4; −(1) = (−1) *b.* −(5 + 6) = −(11) = (−11) *c.* −(−4) = 4

---

28 *a.* $-6[2+(-7)] =$ ________

b. $-[(-5)+4] =$ ________

c. $-[5-(-3)] =$ ________

---

a. (−6) × (−5) = 30 *b.* −(−1) = 1 *c.* −(8) = (−8)

---

29 *a.* $-4[5+(-7)-3-(-2)] = (-4)\times[5+(-7)+$ ____ $+$ ____ $]$

$= (-4)\times$ ____ $=$ ____

b. $-2[7-8-(-6)+(-3)] = (-2)\times$ ____ $=$ ____

c. $-5[(-3)-7+2-(-6)] =$ ____ $\times$ ____ $=$ ____

---

a. $(-4) \times [5 + (-7) + (-3) + 2] = (-4) \times (-3) = 12$
b. $(-2) \times [7 + (-8) + 6 + (-3)] = (-2) \times 2 = (-4)$
c. $(-5) \times [(-3) + (-7) + 2 + 6] = (-5) \times (-2) = 10$

---

30 *a.* $-3[8 - 7 + (-2)] =$ ______ $\times$ ______ $=$ ______

b. $4[(-8) + 6 - (-7)] =$ ______

c. $-[7 - (-3) + (-5) - 6] =$ ______

---

a. $(-3) \times [8 + (-7) + (-2)] = (-3) \times (-1) = 3$
b. $4 \times [(-8) + 6 + 7] = 4 \times 5 = 20$
c. $-[7 + 3 + (-5) + (-6)] = (-1) \times (-1) = 1$

You have finished the program for Sec. 4-4. Now take the posttest to see how much you have learned.

POSTTEST 4-4

1. $(-9) - 3 - (-4) + 15 =$ ______
2. $12 + (-36) - 8 - (-11) + 11 =$ ______
3. $(-18) - 2 + 27 + (-8) - (-28) =$ ______
4. $-5[(-3) - (-8)] =$ ______
5. $-[6 + (-10)] =$ ______
6. $-4[(-2) + 7 - (-10) - 5] =$ ______

When you have finished the posttest, correct it yourself using the answers on page 200. Then check all the questions you got wrong on the list below:

1.	______	4.	______
2.	______	5.	______
3.	______	6.	______

If you compare your performance with that on Pretest 4-4, you will see how much you have improved. If you made more than one mistake on this posttest, review the material and take the pretest again. Don't move on until you can get all the questions right.

Now take Final Test 4.

FINAL TEST 4

1. $(-62) + 83 =$ _____
2. $0 + (-18) =$ _____
3. $(-32) + (-59) =$ _____
4. $43 + (-16) =$ _____
5. $218 + (-218) =$ _____
6. $(-3) \times (-4) =$ _____
7. $(-9) \div (-3) =$ _____
8. $16 \div (-4) =$ _____
9. $0 \div (-8) =$ _____
10. $(-31) - 45 =$ _____
11. $0 - 32 =$ _____
12. $(-15) - (-43) =$ _____
13. $18 - (-17) =$ _____
14. $24 + (-1) - 5 - (-14) + 33 =$ _____
15. $(-17) - 3 + 16 + (-22) - (-18) =$ _____
16. $-[(-5) - (-13)] =$ _____
17. $-8[(-9) + 7 - (-11) - 14] =$ _____

The answers to this test are on page 200. Check the questions you got wrong on the list below and compare your results with those on Diagnostic Test 4 to see how much you have improved.

1.	_____	6.	_____	10.	_____	14.	_____
2.	_____	7.	_____	11.	_____	15.	_____
3.	_____	8.	_____	12.	_____	16.	_____
4.	_____	9.	_____	13.	_____	17.	_____
5.	_____						

If you made more than one mistake, review the material and then take the diagnostic test again. Make sure that you can get all the questions right before you finish.

Now proceed to Chap. 5.

ANSWERS TO TESTS FOR CHAPTER 4

DIAGNOSTIC TEST 4

1. 22	2. (−33)	3. (−62)	4. 18
5. 0	6. 3	7. (−2)	8. 0
9. 30	10. (−59)	11. (−43)	12. 18
13. 35	14. 65	15. (−9)	16. (−7)
17. 66			

PRETEST 4-1

1. 0
2. 41
3. (−15)
4. 14
5. (−52)
6. (−23)
7. 0

POSTTEST 4-1

1. 0
2. 21
3. (−25)
4. 28
5. (−55)
6. (−42)
7. 0

PRETEST 4-2

1. (−20)
2. (−5)
3. 3
4. 0
5. 6
6. (−2)
7. 0

POSTTEST 4-2

1. (−21)
2. (−7)
3. 5
4. 0
5. 18
6. (−7)
7. 0

PRETEST 4-3

1. 3
2. (−37)
3. (−81)
4. 28
5. (−25)
6. 26
7. 23

POSTTEST 4-3

1. 7
2. (−26)
3. (−102)
4. (−22)
5. 39
6. 33
7. 24

PRETEST 4-4

1. 12
2. (–5)
3. 32
4. (–21)
5. 4
6. (–12)

POSTTEST 4-4

1. 7
2. (–10)
3. 27
4. (–25)
5. 4
6. (–40)

FINAL TEST 4

1. 21	2. (–18)	3. (–91)	4. 27
5. 0	6. 12	7. 3	8. (–4)
9. 0	10. (–76)	11. (–32)	12. 28
13. 35	14. 65	15. (–8)	16. (–8)
17. 40			

Chapter 5 Introduction to Rational Numbers

This chapter introduces rational numbers as numbers which can be represented by fractions of the form a/b with a and b integers and b not equal to 0. You will learn how these numbers can be graphed and written in both fractional and integral or mixed-numeral forms. After seeing how a rational number can be represented by an infinite set of equivalent fractions, you learn the definition of the simplest form of a rational number. Then you practice simplifying fractions.

Section 5-1 introduces rational numbers as satisfying the need for a definition of division for all integers. It shows how integers and fractions with positive denominators can be graphed and defines $-a/b$, the negative of a rational number. Section 5-2 shows how fractions with negative denominators can be graphed and defines equivalent fractions as those which have the same graph. Several examples then lead to the inference that a/b is equivalent to $(a \times n)/(b \times n)$ if b and n are not equal to 0 and a, b, and n are integers. Section 5-3 introduces factoring and simplifying fractions by eliminating factors common to both numerator and denominator. The simplest form of a nonintegral rational number is defined as a/b or $-a/b$ with a and b the smallest possible positive integer. Section 5-4 uses the distributive law to derive a definition of the addition of fractions of common denominator. After you are introduced to the mixed-numeral form of a nonintegral rational number, you practice the interconversion of fractional and mixed-numeral forms.

This chapter requires completion of Chaps. 2 and 4.

DIAGNOSTIC TEST 5

Draw sections of the number line and graph the following fractions:

1. $\dfrac{-11}{3}$

2. $-\dfrac{13}{4}$

Select values for m and n which make the second fraction in each pair equivalent to the first. Fractions which represent the same rational number are equivalent.

3. $-\dfrac{4}{5}$ and $\dfrac{m}{-40}$

4. $\dfrac{8}{7}$ and $\dfrac{-56}{n}$

5. Completely factor the integer −120.

Express the following rational numbers in their simplest forms:

6. $\dfrac{108}{-81}$

7. $\dfrac{-112}{-96}$

8. $\dfrac{-7}{9} + \dfrac{2}{9}$

9. Express the mixed numeral $-7\frac{4}{7}$ in fractional form.

10. Express the fraction $\dfrac{-52}{-11}$ in mixed-numeral form.

When you have finished the diagnostic test, correct it yourself using the answers on page 239. Then check all the questions you got wrong on the list below:

1. ____	3. ____	5. ____	8. ____
2. ____	4. ____	6. ____	9. ____
		7. ____	10. ____

If you made no mistakes at all, you need not do any work in this topic and should go straight on to Chap. 6, Operations with Rational Numbers, which begins on page 243. But if you made any mistakes (even if it was only *one* mistake), follow these instructions:

	section	page
If you made any mistakes in Question 1 or 2, circle Sec. 5-1.	5-1	203
If you made any mistakes in Question 3 or 4, circle Sec. 5-2.	5-2	212
If you made any mistakes in Questions 5 to 7, circle Sec. 5-3.	5-3	223
If you made any mistakes in Questions 8 to 10, circle Sec. 5-4.	5-4	232

You should now work through the sections whose numbers you have circled, taking them in numerical order. The sections begin on the pages indicated.

Section 5-1
Definition and Graphing of Rational Numbers

Do this section if you made a mistake in Question 1 or 2 of Diagnostic Test 5. Begin by taking the pretest below, which will show you how much of Sec. 5-1 you already know and what you will be able to do by the time you have finished it.

PRETEST 5-1

1. Which of the following are rational numbers?

$\frac{3}{5}$ $\quad 7 \quad$ $\frac{-6}{0}$ $\quad$ $\frac{-7}{-9}$ $\quad -5 \quad$ $\frac{0}{8}$ $\quad$ $\frac{4}{-6}$

2. Which of the following are integers?

$\frac{4}{-7}$ $\quad$ $\frac{10}{-5}$ $\quad$ $\frac{-5}{-6}$ $\quad$ $\frac{-8}{2}$ $\quad$ $\frac{-12}{-3}$ $\quad$ $\frac{3}{9}$

Draw sections of the number line and graph the following fractions:

3. $\frac{8}{3}$ $\quad$ 4. $\frac{-7}{4}$ $\quad$ 5. $-\frac{7}{5}$

Give the negatives of the following fractions:

6. $\frac{6}{7}$ $\quad$ 7. $\frac{-9}{8}$ $\quad$ 8. $-\frac{4}{11}$

When you finish the pretest, correct it yourself using the answers on page 240. Then check all the questions you got wrong on the list below:

1. ______ 5. ______
2. ______ 6. ______
3. ______ 7. ______
4. ______ 8. ______

If you made any mistakes (even if it was only *one* mistake), read the introduction and go through the program.

If you made no mistakes at all, skip Sec. 5-1, because you already know its material, and go on to the next section whose number you circled on page 202. If no further numbers were circled, take Final Test 5 (page 238).

Introduction

The set of integers is closed under addition, subtraction, and multiplication but not under division. Divisions such as $5 \div 2$ and $7 \div 3$ can be defined only if we introduce a new kind of number. This section shows how rational numbers, defined as numbers which can be written in the form a/b with a and b integers and b not equal to 0, fill this gap. The graphing of integers and of fractions with positive denominators leads to a definition of the negative of a rational number and to the form $-a/b$.

FRAMES FOR SECTION 5-1

1 Integers can be *graphed* on a number line, as shown below for the integer -3:

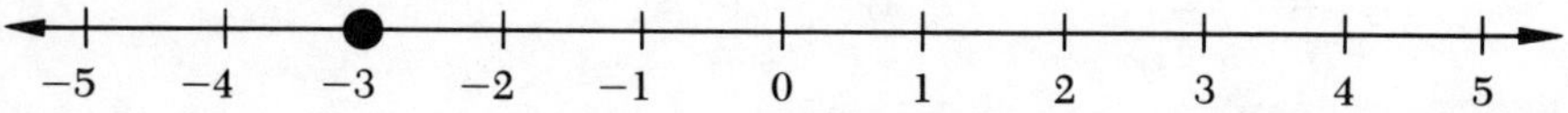

Draw a section of the number line and graph the point representing 2.

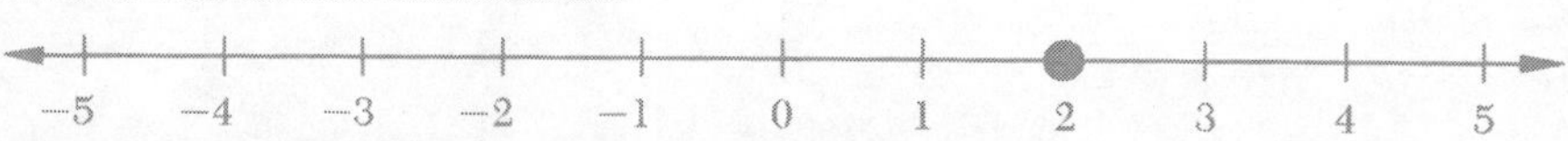

---

2 Draw a section of the number line and graph the points representing -1 and 4.

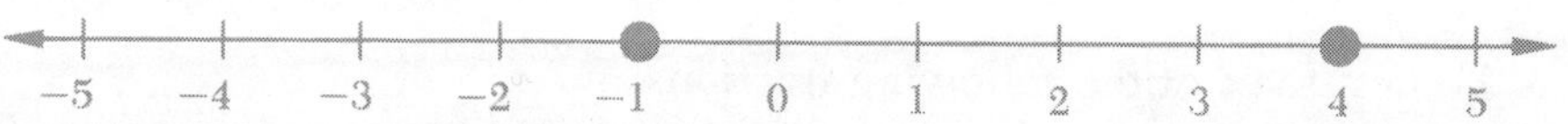

---

3 *a.* Draw a section of the number line (from 7 to 11) and graph the integer 9.

b. Draw another section (from -9 to -3) and graph the integer -7.

---

a.

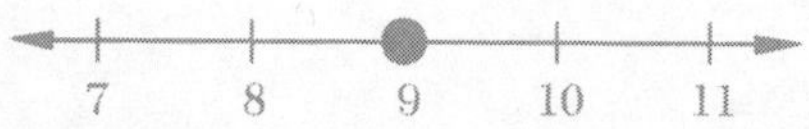

b.

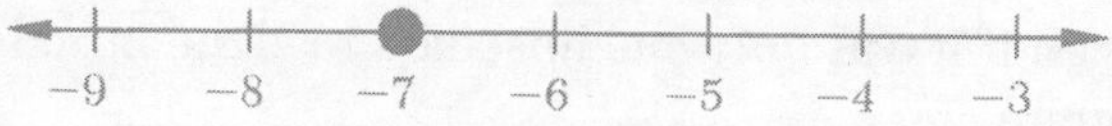

---

4 Draw sections of the number line and graph the integers

a. -10 *b.* 11 *c.* 5

---

a.

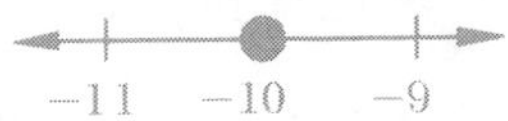

b.

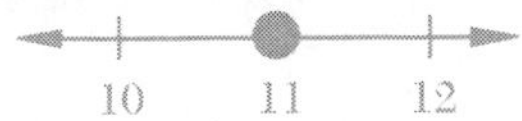

c.

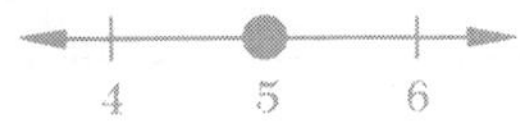

Your sections should each contain at least three integer points.

---

5 If you add any two integers, do you always get another integer?

---

yes

---

6 For this reason the set of integers is said to be *closed under addition.* In the chapter on negative numbers, we defined subtraction as addition of the negative; i.e., we defined $a - b$ as $a + (-b)$.

a. Does this mean that if you subtract any two integers, you always get another integer?

b. Is the set of integers *closed under subtraction?*

---

a. yes, because all subtractions can be converted to additions. *b.* yes

---

7 Is the set of integers *closed under multiplication?*

---

yes

---

8 $6 \div 1 =$ ______ $6 \div 2 =$ ______ $6 \div 3 =$ ______

but $6 \div 4$ (does/does not) ______ give an integer, and $6 \div 5$ (does/does not) ______ give an integer. Is the set of integers *closed under division?*

---

6 3 2 does not; does not; no

---

9 We can introduce a new kind of number to define operations like $6 \div 4$ and $6 \div 5$. The numbers we use for this purpose are called *rational numbers* and are commonly written as fractions, that is, $\frac{3}{8}, \frac{5}{7}$, etc. In general, the fraction $\frac{p}{q}$ is defined by the equation

$$\frac{p}{q} \times q = p \qquad \text{or} \qquad \frac{p}{q} = p \div q \qquad \text{where } q \text{ is not equal to } 0$$

Thus $6 \div 4$ is written as the fraction $\frac{6}{4}$, and $6 \div 5$ is written as the fraction ______, but $6 \div 0$ (is/is not) ______ defined.

$\frac{6}{5}$ is not

---

10 You may remember that the top of a fraction is called its *numerator* and the bottom is called its *denominator.* So the numerator of $\frac{6}{5}$ is _____ and the denominator of $\frac{7}{8}$ is _____.

---

6 8

---

11 A fraction can never have a (numerator/denominator) _____ of 0 because (multiplication/division) _____ by 0 is not defined.

---

denominator; division

---

12 A *rational number* is defined as any number which can be written as a fraction of the form $\frac{p}{q}$ with p and q integers and q not equal to 0. Which of the following are rational numbers?

$\frac{2}{3}$ $\frac{-6}{7}$ $\frac{4}{0}$ $\frac{3}{-5}$ $\frac{-2}{-9}$ $\frac{7}{1}$ $\frac{0}{7}$

$\frac{2}{3}$ $\frac{-6}{7}$ $\frac{3}{-5}$ $\frac{-2}{-9}$ $\frac{7}{1}$ $\frac{0}{7}$ Only $\frac{4}{0}$ is not.

---

13 The set of integers can be written as a set of fractions with denominator 1.

$\ldots$ $\frac{-3}{1}$ $\frac{-2}{1}$ $\frac{-1}{1}$ $\frac{0}{1}$ $\frac{1}{1}$ $\frac{2}{1}$ $\frac{3}{1}$ $\ldots$

So clearly (all/some/no) _____ integers are also rational numbers.

---

all

---

14 Which of the following are rational numbers?

$\frac{5}{6}$ -4 $\frac{-3}{0}$ $\frac{-6}{-7}$ 8 $\frac{0}{9}$ $\frac{6}{-11}$

$\frac{5}{6}$ -4 $\frac{-6}{-7}$ 8 $\frac{0}{9}$ $\frac{6}{-11}$ Only $\frac{-3}{0}$ is not.

---

15 Draw a section of the number line and graph $\frac{-2}{1}$ and $\frac{1}{1}$.

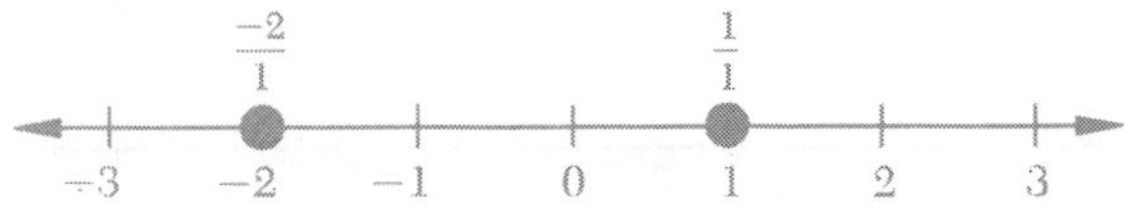

---

16 Fractions with denominator 2 can be graphed on a similar but different number line, as shown for $\frac{5}{2}$ below:

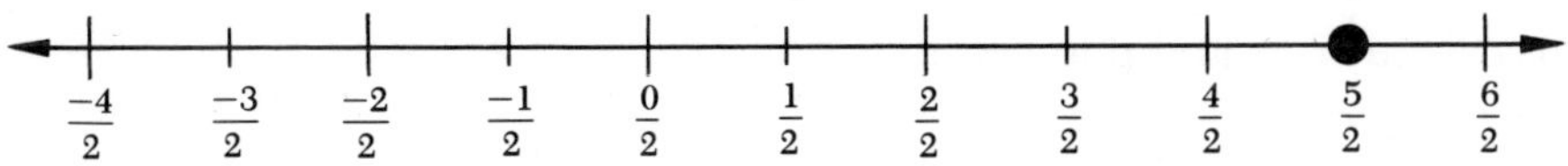

Draw a similar number line and graph $\frac{4}{2}$, $\frac{-1}{2}$, and $\frac{-3}{2}$.

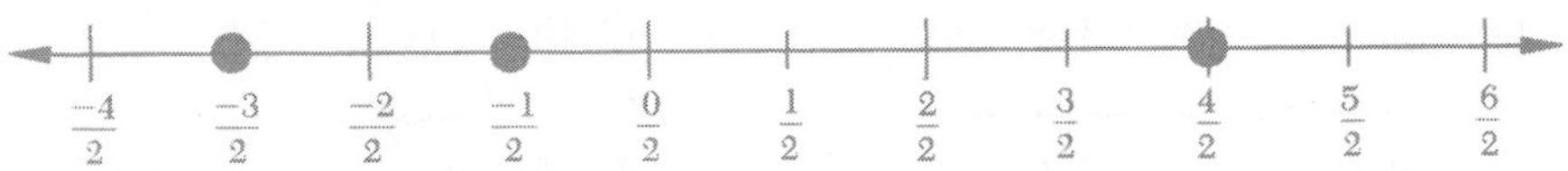

---

17 This number line has a number of points in common with the integer number line because

$\frac{-4}{2} = -4 \div 2 =$ ______

$\frac{-2}{2} =$ ______

$\frac{0}{2} =$ ______

$\frac{2}{2} =$ ______

$\frac{4}{2} =$ ______

−2 −1 0 1 2

---

18 So the number lines can be combined:

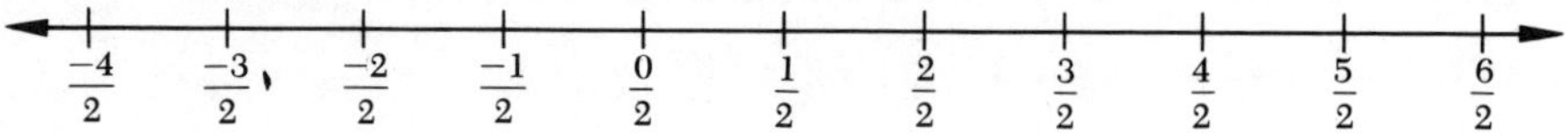

and

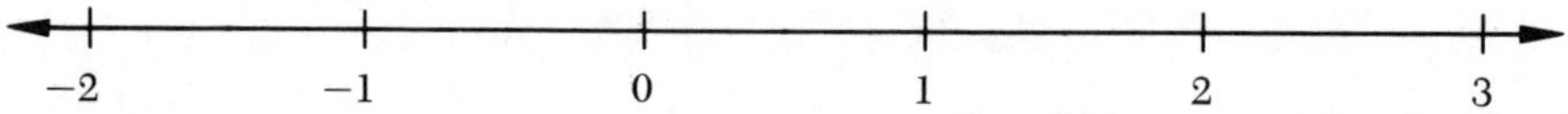

give

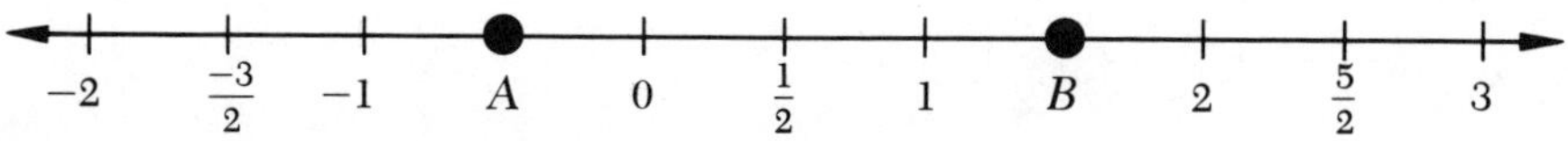

Point A is _____, and point B is _____.

$\frac{-1}{2}$ $\frac{3}{2}$

19 The number line of fractions with denominator 3 is

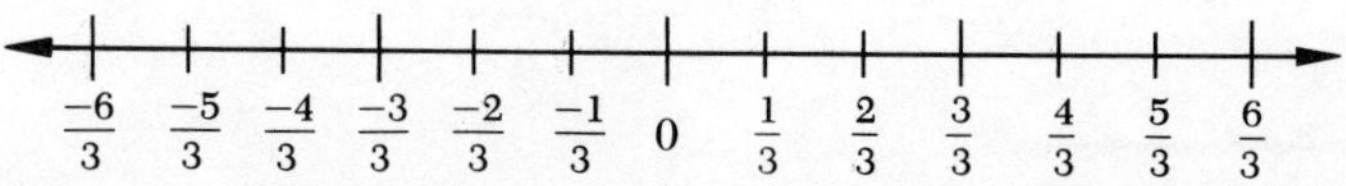

This line can also be combined with the integer number line. Draw the section of this combined number line from $\frac{-7}{3}$ to $\frac{7}{3}$ and graph $\frac{-2}{3}$, $\frac{5}{3}$, and $\frac{6}{3}$.

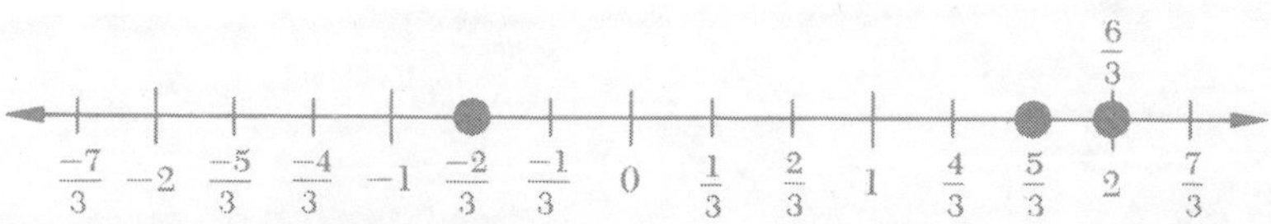

20 Draw the section from -2 to 2 of the combined number line for integers and fractions with denominator 4 and graph $\frac{-7}{4}$, $\frac{-4}{4}$, and $\frac{3}{4}$.

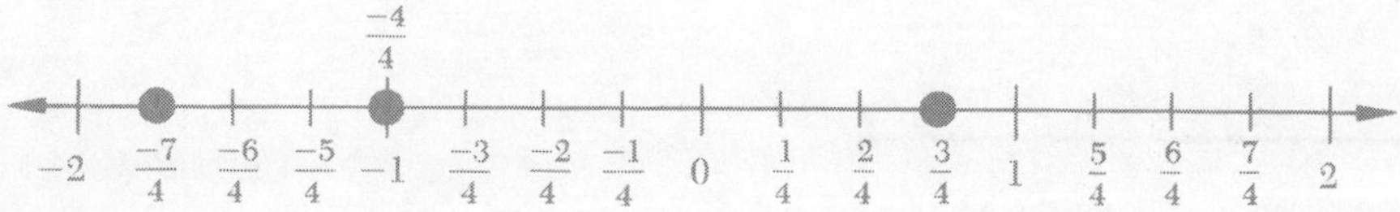

21 All fractions with positive denominators can be graphed in this way, but normally only integer points are marked. Thus $\frac{-7}{4}$, $\frac{0}{4}$, and $\frac{3}{4}$ are graphed as

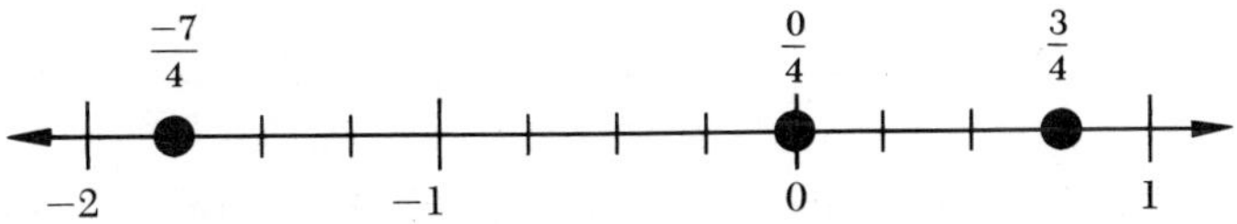

and $\frac{-8}{5}$, $\frac{2}{5}$, and $\frac{5}{5}$ are graphed as

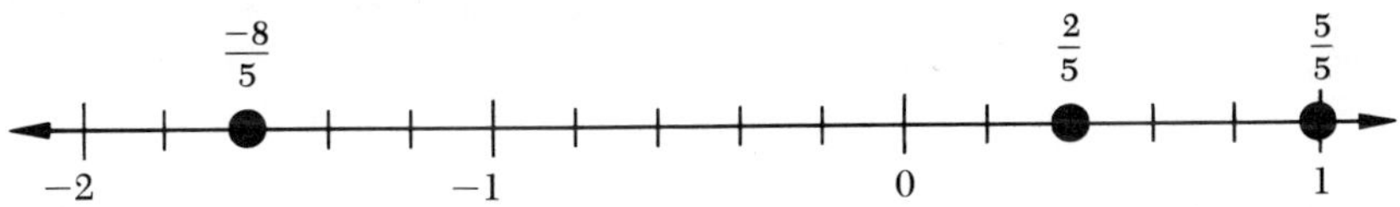

Draw a section of the number line and graph $\frac{-6}{3}$, $\frac{-1}{3}$, and $\frac{5}{3}$.

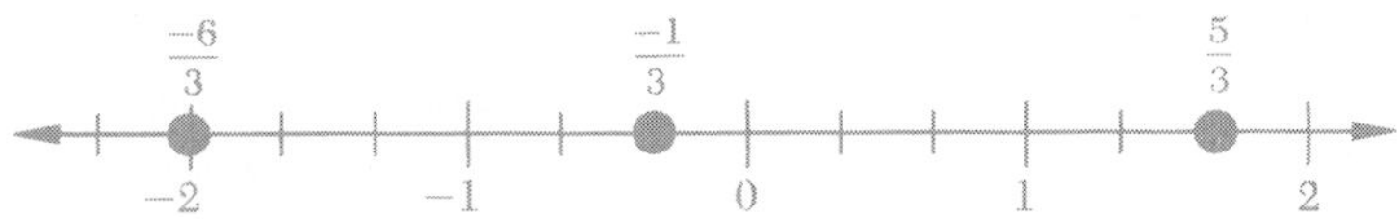

---

22 *a.* Draw a section of the number line and graph $\frac{-6}{5}$ and $\frac{6}{5}$.

b. Draw a section of the number line and graph $\frac{-2}{7}$ and $\frac{2}{7}$.

---

a.

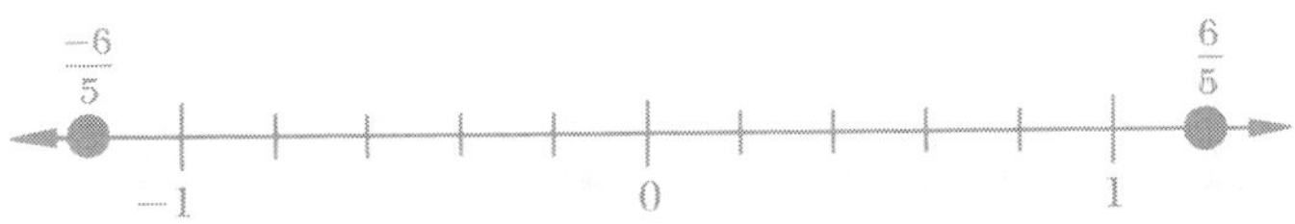

b.

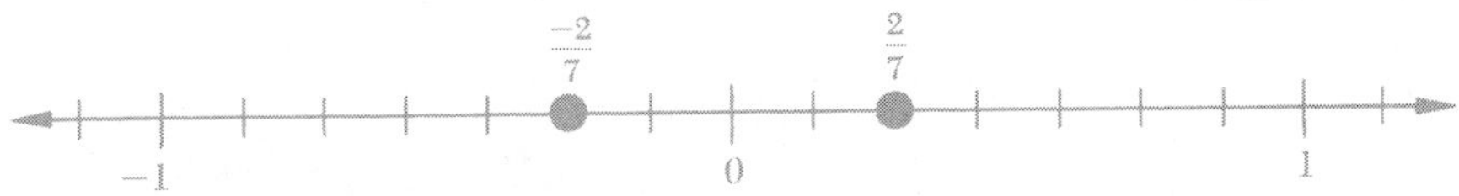

---

23 In the previous chapter we defined the *negative* of an integer as the number which was graphed on the opposite side of 0 and at the same distance from 0. If we define the negative of a rational number in the same way, it follows that the negative of $\frac{-6}{5}$ is _____ and that the negative of $\frac{2}{7}$ is _____.

$\frac{6}{5}$ $\frac{-2}{7}$

24 Just as the negative of 4 is written as −4, the negative of $\frac{2}{7}$ can be written as $-\frac{2}{7}$. This gives us two ways of writing the same number, $\frac{-2}{7}$ and $-\frac{2}{7}$. Give two ways of writing the negative of $\frac{6}{5}$.

$\frac{-6}{5}$ $-\frac{6}{5}$ (either order)

25 Draw sections of the number line and graph $-\frac{2}{3}$ and $-\frac{5}{4}$.

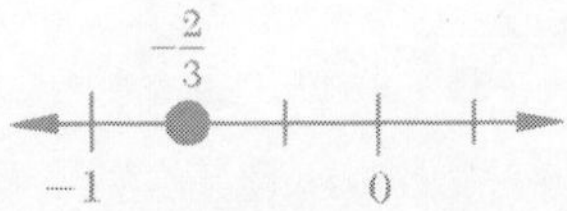

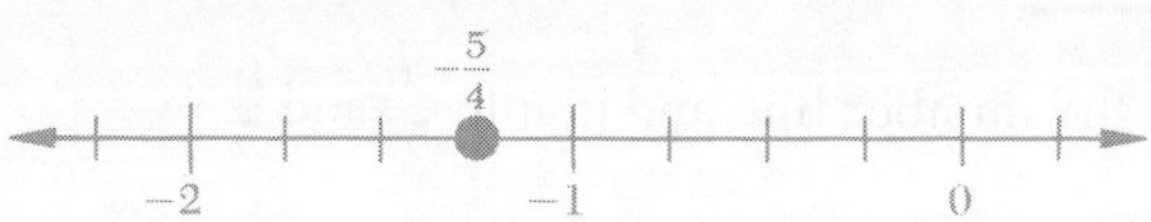

26 Give the negatives of the following numbers:

a. $-\frac{8}{5}$ *b.* $\frac{3}{4}$ *c.* $\frac{-7}{6}$

a. $\frac{8}{5}$ *b.* $-\frac{3}{4}$ or $\frac{-3}{4}$ *c.* $\frac{7}{6}$

27 Draw sections of the number line and graph $-\frac{4}{5}$ and $\frac{-7}{2}$.

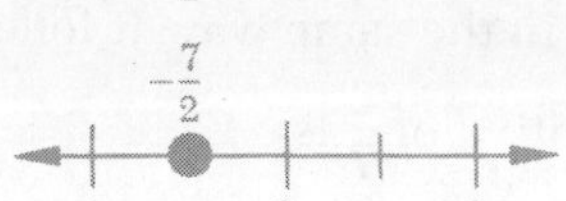

You have finished the program for Sec. 5-1. Now take the posttest to see how much you have learned.

POSTTEST 5-1

1. Which of the following are rational numbers?

$$\frac{8}{9} \quad 5 \quad \frac{-7}{0} \quad \frac{-3}{-8} \quad -6 \quad \frac{0}{5} \quad \frac{3}{-9}$$

2. Which of the following are integers?

$$\frac{5}{-6} \quad \frac{16}{-4} \quad \frac{-3}{-7} \quad \frac{-12}{2} \quad \frac{-15}{-5} \quad \frac{2}{10}$$

Draw sections of the number line and graph the following fractions:

3. $\frac{9}{4}$ 4. $\frac{-5}{3}$ 5. $-\frac{12}{5}$

Give the negatives of the following fractions:

6. $\frac{8}{9}$ 7. $\frac{-7}{11}$ 8. $-\frac{6}{13}$

When you have finished the posttest, correct it yourself using the answers on page 240. Then check all the questions you got wrong on the list below:

1. ______ 5. ______
2. ______ 6. ______
3. ______ 7. ______
4. ______ 8. ______

If you compare your performance with that on Pretest 5-1, you will see how much you have improved. If you made more than one mistake on this posttest, review the material and take the pretest again. Don't move on to the next section until you get all the questions right.

Now proceed to the next section whose number you circled on page 202. If no other sections were circled, take Final Test 5 (page 238).

Section 5-2
Equivalent Rational Numbers

Do this section if you made a mistake in Question 3 or 4 of Diagnostic Test 5. Begin by taking the pretest below, which will show you how much of Sec. 5-2 you already know and what you will be able to do by the time you have finished it.

PRETEST 5-2

Are the following pairs of fractions equivalent, i.e., do they represent the same rational number? Answer yes or no.

1. $\dfrac{-6}{7}$ and $\dfrac{6}{-7}$
2. $\dfrac{-3}{-4}$ and $-\dfrac{3}{4}$
3. $\dfrac{9}{-10}$ and $-\dfrac{9}{10}$
4. $\dfrac{7}{-8}$ and $\dfrac{-7}{-8}$

Select values for a, b, c, d, e, and f which make the second fraction in each pair equivalent to the first.

5. $\dfrac{3}{5}$ and $\dfrac{a}{15}$
6. $\dfrac{-7}{12}$ and $\dfrac{b}{36}$
7. $-\dfrac{7}{8}$ and $\dfrac{35}{c}$
8. $\dfrac{9}{-27}$ and $\dfrac{-3}{d}$
9. $\dfrac{-15}{-3}$ and $\dfrac{-10}{e}$
10. $\dfrac{3}{8}$ and $\dfrac{f}{-32}$

When you have finished the pretest, correct it yourself using the answers on page 241. Then check all the questions you got wrong on the list below:

1. ______	5. ______	8. ______
2. ______	6. ______	9. ______
3. ______	7. ______	10. ______
4. ______		

If you made any mistakes (even if it was only *one* mistake), read the introduction and go through the program.

If you made no mistakes at all, skip Sec. 5-2, because you already know its material, and go on to the next section whose number you circled on page 202. If no further numbers were circled, take Final Test 5 (page 238).

Introduction

Fractions like $\frac{-6}{-3}$ and $\frac{8}{-2}$ can be graphed because they represent integers. This makes it possible for nonintegral fractions with negative denominators to be graphed by interpolation. The term *equivalent* is introduced to describe the relationship between fractions which are graphed onto the same point on the number line. You are given examples of fractions which are equivalent to integers, and led to infer the general rule that k is equivalent to $(k \times n)/n$. Examples of the graphing of nonintegral fractions then lead to the further inference that a/b is equivalent to $(a \times n)/(b \times n)$ for all integral values of a, b, and n provided that neither b nor n is equal to zero.

FRAMES FOR SECTION 5-2

1 In the topic of negative numbers we found that

$4 \div (-1) =$ _____ and $-4 \div (-1) =$ _____

and that

$18 \div (-3) =$ _____ and $-18 \div (-3) =$ _____

-4 4 -6 6

---

2 Draw a number line and graph the following fractions:

a. $\frac{-6}{2}$ *b.* $\frac{-12}{-3}$ *c.* $\frac{14}{-7}$

Hint: First find integers represented by the fractions.

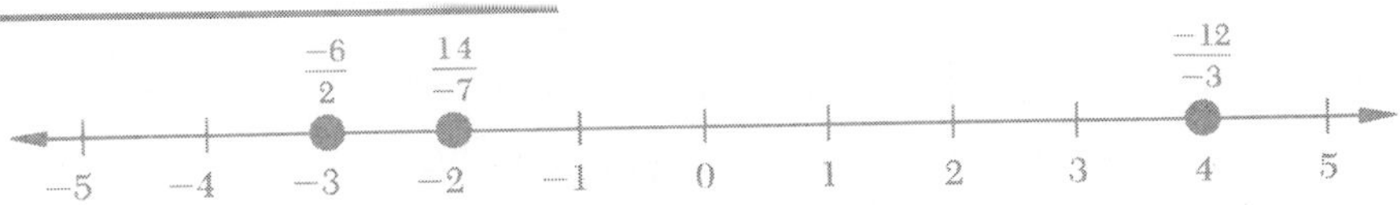

---

3 Draw a section of the number line and graph

$\frac{-6}{-2}$ $\frac{-4}{-2}$ $\frac{-2}{-2}$ $\frac{0}{-2}$ $\frac{2}{-2}$ $\frac{4}{-2}$ $\frac{6}{-2}$

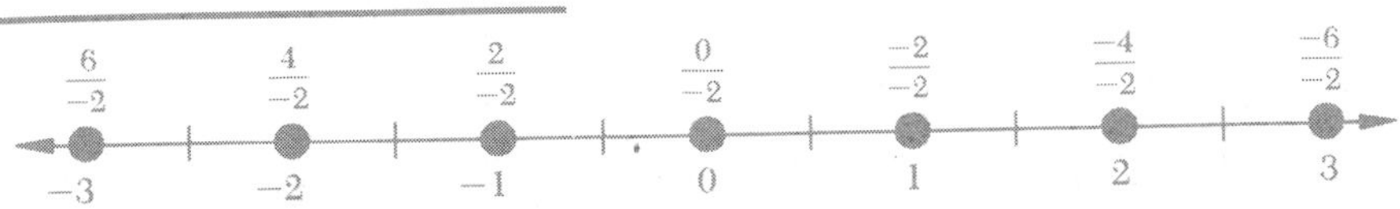

---

4 The graphing of nonintegral rational numbers can be assumed to follow the same pattern, with $\frac{-5}{-2}$ located between $\frac{-6}{-2}$ and $\frac{-4}{-2}$, etc. Use this assumption to draw a section of the number line and graph

$\frac{-5}{-2}$ $\frac{-3}{-2}$ $\frac{-1}{-2}$ $\frac{1}{-2}$ $\frac{3}{-2}$ $\frac{5}{-2}$

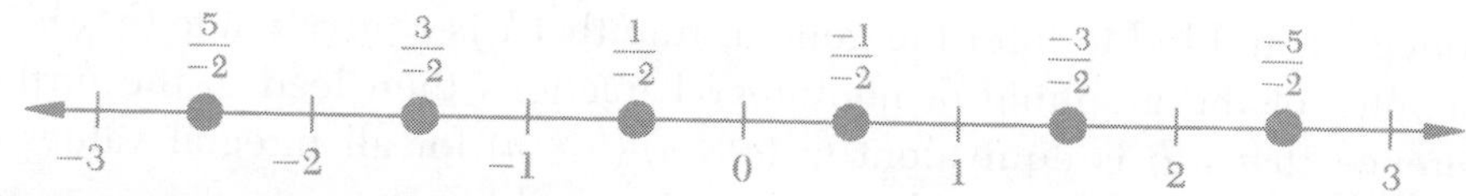

---

5 Now draw a section of the number line and graph $\frac{-5}{2}$ and $\frac{3}{2}$.

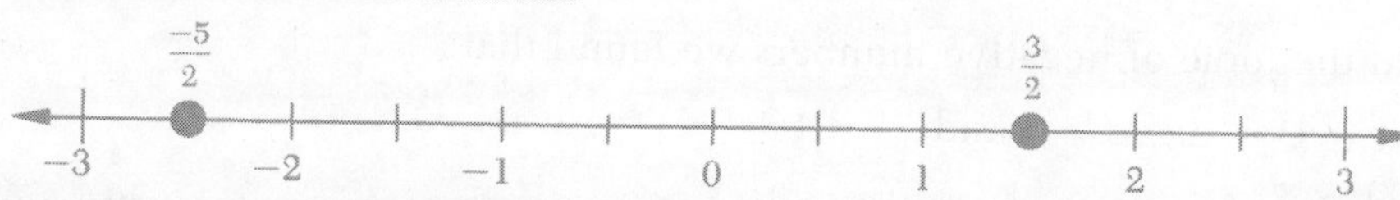

---

6 Compare the graphs in Frames 4 and 5. Notice that the same point on the number line represents $\frac{-5}{2}$ and $\frac{?}{-2}$; and that the same point represents $\frac{3}{2}$ and $\frac{?}{-2}$.

---

5 −3

---

7 Draw a section of the number line and graph

$\frac{-9}{-3}$ $\frac{-6}{-3}$ $\frac{-3}{-3}$ $\frac{0}{-3}$ $\frac{3}{-3}$ $\frac{6}{-3}$ $\frac{9}{-3}$

Then assuming that the graphing of nonintegral rational numbers follows the same pattern, graph $\frac{-7}{-3}$ and $\frac{5}{-3}$.

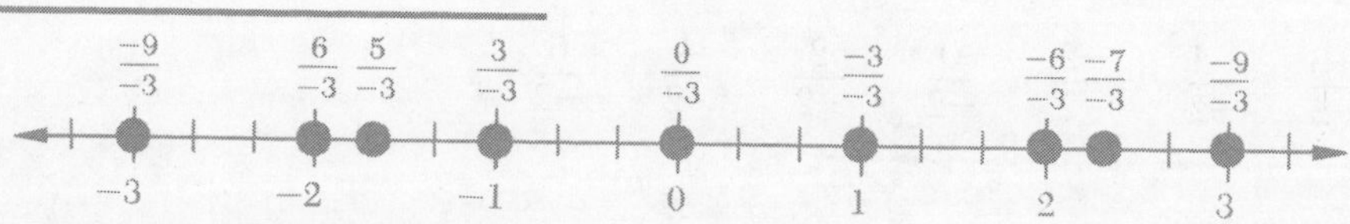

---

8 Notice that $\frac{-7}{-3}$ is represented by the same point as $\frac{?}{3}$ and $\frac{5}{-3}$ is represented by the same point as $\frac{?}{3}$.

---

7 −5

---

9 In general it can be shown that $\frac{-a}{-b}$ is represented by the same point as $\frac{?}{b}$ and $\frac{a}{-b}$ is represented by the same point as $\frac{?}{b}$.

---

a −*a*

---

10 Fractions represented by the same point on the number line are called *equivalent*. Which of the following pairs of fractions are equivalent?

a. $\frac{3}{4}$ and $\frac{-3}{-4}$ *b.* $\frac{-2}{5}$ and $\frac{-2}{-5}$

c. $\frac{7}{-6}$ and $\frac{7}{6}$ *d.* $\frac{5}{-8}$ and $\frac{-5}{8}$

---

a and *d*

---

11 Which of the following pairs of fractions are equivalent?

a. $\frac{3}{-5}$ and $-\frac{3}{5}$ *b.* $\frac{-7}{-8}$ and $-\frac{7}{8}$

c. $\frac{-5}{-6}$ and $\frac{5}{6}$ *d.* $\frac{2}{3}$ and $\frac{2}{3}$

---

a and *c*

---

12 The same definition applies to integers. A fraction and an integer are equivalent if they are both represented by the same point on the number line. Give the integers which are equivalent to the following fractions:

a. $\frac{-10}{5}$ *b.* $\frac{5}{5}$ *c.* $\frac{0}{4}$

d. $\frac{9}{-3}$ *e.* $\frac{-12}{-6}$ *f.* $-\frac{16}{4}$

---

a. −2 *b.* 1 *c.* 0 *d.* −3 *e.* 2 *f.* −4

---

13 By finding equivalent integers, determine which of the following pairs of fractions are equivalent:

a. $\frac{-5}{-5}$ and $-\frac{4}{4}$ *b.* $\frac{-3}{1}$ and $\frac{9}{3}$

c. $\frac{6}{-3}$ and $\frac{-4}{2}$ *d.* $\frac{0}{5}$ and $\frac{0}{-7}$

---

c and *d*

---

14 Give fractions with denominator 2 which are equivalent to the following integers:

a. −5 *b.* 3 *c.* 6

---

a. $\frac{-10}{2}$ *b.* $\frac{6}{2}$ *c.* $\frac{12}{2}$

---

15 Give fractions with denominator 3 which are equivalent to the following integers:

a. −5 *b.* 3 *c.* 6

---

a. $\frac{-15}{3}$ *b.* $\frac{9}{3}$ *c.* $\frac{18}{3}$

---

16 Give fractions with denominator 4 which are equivalent to the following integers:

---

a. $\frac{-20}{4}$ *b.* $\frac{12}{4}$ *c.* $\frac{24}{4}$

---

17 We can use this approach to construct a set of fractions equivalent to any integer. The set of fractions equivalent to −5 includes:

$$\frac{-5}{1} \quad \frac{-10}{2} \quad \frac{-15}{3} \quad \frac{-20}{4} \quad \frac{?}{5} \quad \frac{?}{6} \quad \cdots \quad \frac{?}{-1} \quad \frac{?}{-2} \quad \frac{?}{-3} \quad \cdots$$

−25 −30 5 10 15

---

18 Similarly the set of fractions equivalent to 3 includes

$$\frac{3}{1} \quad \frac{6}{2} \quad \frac{9}{3} \quad \frac{12}{4} \quad \frac{?}{5} \quad \frac{?}{6} \quad \cdots \quad \frac{?}{-1} \quad \frac{?}{-2} \quad \frac{?}{-3} \quad \cdots$$

15 18 −3 −6 −9

19 Notice that

$$\frac{3 \times 1}{1} = \frac{?}{1} \qquad \frac{3 \times (-1)}{-1} = \frac{?}{-1}$$

$$\frac{3 \times 2}{2} = \frac{?}{2} \qquad \frac{3 \times (-2)}{-2} = \frac{?}{-2}$$

$$\frac{3 \times 3}{3} = \frac{?}{3} \qquad \frac{3 \times (-3)}{-3} = \frac{?}{-3}$$

What can you say about these fractions?

3 −3 6 −6 9 −9 They are all equivalent to the integer 3 and belong to the set of fractions described in Frame 18.

20 In general it can be shown that if n is an integer other than 0, $\frac{3 \times n}{n}$ is equivalent to 3. Other fractions equivalent to 3 therefore include

$$\frac{3 \times ?}{7} \quad \text{or} \quad \frac{?}{7} \quad \text{and} \quad \frac{3 \times (-6)}{?} \quad \text{or} \quad \frac{?}{?}$$

7 21 −6 $\frac{-18}{-6}$

21 Complete the following set of fractions so as to make them all equivalent to 3:

a. $\frac{?}{9}$ *b.* $\frac{36}{?}$ *c.* $\frac{-21}{?}$ *d.* $\frac{?}{-8}$

a. 27 *b.* 12 *c.* −7 *d.* −24

22 In a similar manner it can be shown that if n is an integer other than 0, $\frac{-5 \times n}{n}$ is equivalent to −5. Fractions equivalent to −5 therefore include all those in Frame 17 as well as $\frac{-5 \times ?}{8}$ or $\frac{?}{8}$ and $\frac{-5 \times (-5)}{?}$ or $\frac{?}{?}$.

8 −40 −5 $\frac{25}{-5}$

23 Complete the following set of fractions so as to make them all equivalent to –5:

a. $\frac{?}{7}$ *b.* $\frac{55}{?}$ *c.* $\frac{-30}{?}$ *d.* $\frac{?}{-9}$

---

a. –35 *b.* –11 *c.* 6 *d.* 45

---

24 Draw sections of the number line and graph $\frac{1}{2}$ and $\frac{2}{4}$. Are $\frac{1}{2}$ and $\frac{2}{4}$ equivalent?

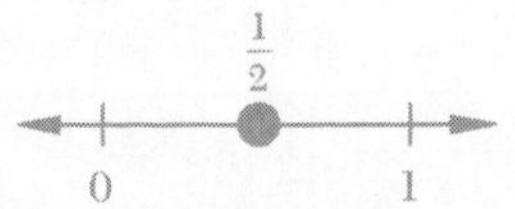

yes

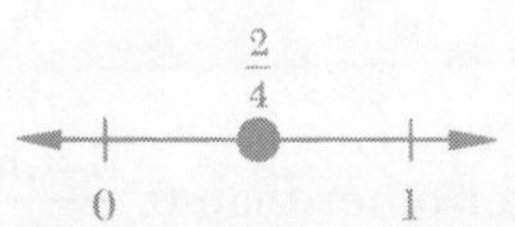

---

25 Draw a section of the number line and graph $-\frac{4}{5}$ and $-\frac{8}{10}$. Are $-\frac{4}{5}$ and $-\frac{8}{10}$ equivalent?

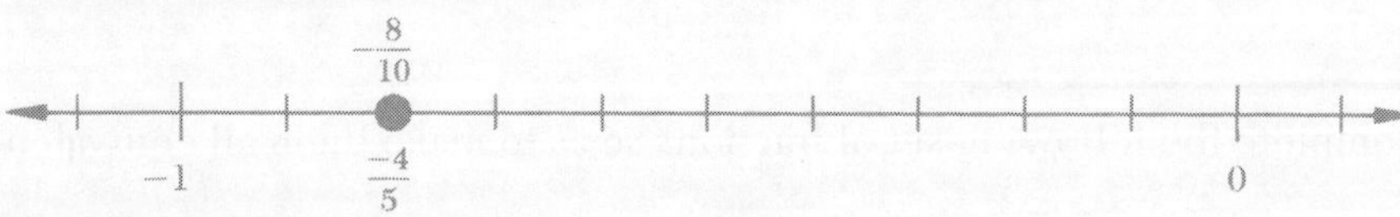

yes

---

26 *a.* Graph $\frac{2}{3}$ and $\frac{4}{6}$ on the same number line (try the technique used just above).

b. Graph $\frac{2}{3}$ and $\frac{6}{9}$ on the same number line.

c. Graph $\frac{2}{3}$ and $\frac{8}{12}$ on the same number line.

d. Graph $\frac{2}{3}$ and $\frac{10}{15}$ on the same number line.

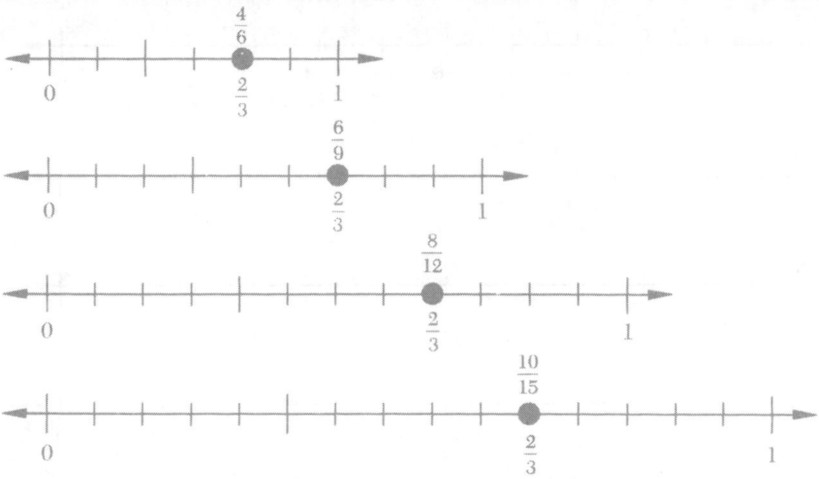

27 Which of the following fractions are equivalent to $\frac{2}{3}$?

a. $\frac{2 \times 2}{3 \times 2}$ *b.* $\frac{2 \times 3}{3 \times 3}$ *c.* $\frac{2 \times 4}{3 \times 4}$ *d.* $\frac{2 \times 5}{3 \times 5}$

They are all equivalent: *a.* $\frac{4}{6}$ *b.* $\frac{6}{9}$ *c.* $\frac{8}{12}$ *d.* $\frac{10}{15}$

28 We therefore know that $\frac{2 \times n}{3 \times n}$ is equivalent to $\frac{2}{3}$ for $n = 1$, ______, ______, ______, and ______, and it can be shown to be true for all integral values of n other than 0. Other fractions equivalent to $\frac{2}{3}$ therefore include $\frac{16}{?}$ and $\frac{?}{-21}$.

2 3 4 5 $\frac{16}{24}$ $\frac{-14}{-21}$

29 Similarly it can be seen from the number lines below

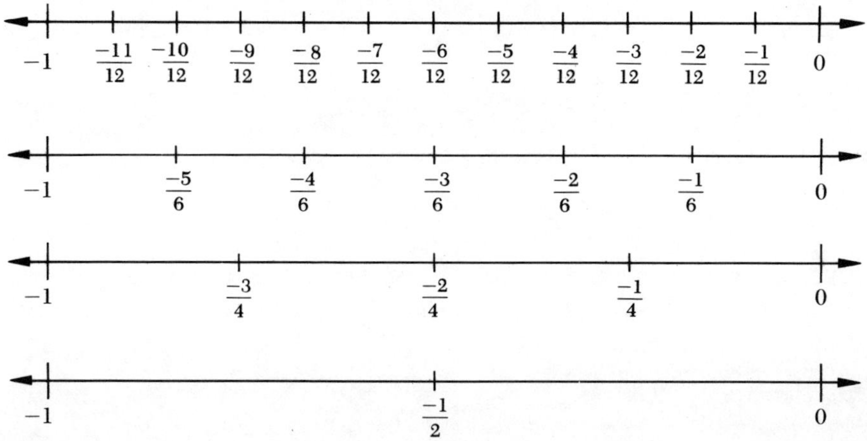

that $\frac{-1}{2}$ is equivalent to $\frac{-1 \times n}{2 \times n}$ for $n = 1, 2, 3$, and 6, that is, that $\frac{-1}{2}$ is equivalent to _____, _____, and _____; and it can be shown that this is true for all positive integral values of n. Other fractions equivalent to $\frac{-1}{2}$ therefore include $\frac{-9}{?}$ and $\frac{?}{-22}$.

$\frac{-2}{4}$ $\frac{-3}{6}$ $\frac{-6}{12}$ $\frac{-9}{18}$ $\frac{11}{-22}$

30 The number lines in Frame 29 also showed that $\frac{-3}{4}$ is equivalent to $\frac{-3 \times n}{4 \times n}$ for $n = 1$ and $n =$ _____, that is, that $\frac{-3}{4}$ is equivalent to _____; and it can be shown to be true for all positive integral values of n. Other fractions equivalent to $\frac{-3}{4}$ therefore include $\frac{?}{20}$ and $\frac{27}{?}$.

3 $\frac{-9}{12}$ $\frac{-15}{20}$ $\frac{27}{-36}$

31 In general it can be shown that $\frac{a \times n}{b \times n}$ is equivalent to $\frac{a}{b}$ for all integral values of a and all integral values of b and n other than 0. Thus $\frac{12}{15}$ is equivalent to $\frac{?}{5}$ and $\frac{7}{42}$ is equivalent to $\frac{-1}{?}$.

$\frac{4}{5}$ $\frac{-1}{-6}$

---

32 Thus fractions equivalent to $\frac{-3}{8}$ include $\frac{?}{56}$ and $\frac{-24}{?}$.

$\frac{-21}{56}$ $\frac{-24}{64}$

---

33 Which of the following pairs of fractions are equivalent?

a. $\frac{4}{5}$ and $\frac{5}{6}$

b. $\frac{5}{-8}$ and $-\frac{25}{40}$

c. $\frac{-7}{9}$ and $\frac{-28}{36}$

d. $\frac{-2}{3}$ and $\frac{-4}{-6}$

---

b and *c*

---

34 Which of the following pairs of fractions are equivalent?

a. $\frac{-3}{7}$ and $\frac{-9}{21}$

b. $-\frac{2}{9}$ and $\frac{-6}{-27}$

c. $\frac{3}{-10}$ and $\frac{-18}{60}$

d. $\frac{-6}{11}$ and $\frac{-7}{12}$

---

a and *c*

---

35 Which of the following pairs of fractions are equivalent?

a. $\frac{-4}{7}$ and $\frac{8}{-21}$

b. $\frac{-8}{-10}$ and $-\frac{40}{50}$

c. $\frac{5}{-20}$ and $-\frac{1}{4}$

d. $\frac{-8}{-5}$ and $\frac{-32}{-20}$

---

c and *d*

You have finished the program for Sec. 5-2. Now take the posttest to see how much you have learned.

POSTTEST 5-2

Are the following pairs of fractions equivalent, i.e., do they represent the same rational number? Answer yes or no.

1. $\frac{-8}{9}$ and $\frac{8}{-9}$
2. $\frac{-2}{-3}$ and $-\frac{2}{3}$
3. $\frac{7}{-12}$ and $-\frac{7}{12}$
4. $\frac{11}{-4}$ and $\frac{-11}{-4}$

Select values for a, b, c, d, e, and f which make the second fraction in each pair equivalent to the first:

5. $\frac{4}{5}$ and $\frac{a}{35}$
6. $\frac{-9}{11}$ and $\frac{b}{44}$
7. $-\frac{7}{9}$ and $\frac{63}{c}$
8. $\frac{12}{-48}$ and $\frac{-3}{d}$
9. $\frac{-18}{-3}$ and $\frac{-12}{e}$
10. $\frac{4}{7}$ and $\frac{f}{-28}$

When you have finished the posttest, correct it yourself using the answers on page 241. Then check all the questions you got wrong on the list below:

1. _____ 5. _____ 8. _____

2. _____ 6. _____ 9. _____

3. _____ 7. _____ 10. _____

4. _____

If you compare your performance with that on Pretest 5-2, you will see how much you have improved. If you made more than one mistake on this posttest, review the material and take the pretest again. Don't move on to the next section until you can get all the questions right.

Now proceed to the next section whose number you circled on page 202. If no further numbers were circled, take Final Test 5 (page 238).

Section 5-3
Simplification of Rational Numbers

Do this section if you made a mistake in Questions 5 to 7 of Diagnostic Test 5. Begin by taking the pretest below, which will show you how much of Sec. 5-3 you already know and what you will be able to do by the time you have finished it.

PRETEST 5-3

Completely factor the following numbers:

1. 96
2. −70

Express the following rational numbers in their simplest forms:

3. $\frac{81}{54}$
4. $\frac{-144}{84}$
5. $\frac{56}{-20}$
6. $\frac{-105}{-33}$

When you have finished the pretest, correct it yourself using the answers on page 241. Then check all the questions you got wrong on the list below.

1. ______	4. ______
2. ______	5. ______
3. ______	6. ______

If you made any mistakes (even if it was only *one* mistake), read the introduction and go through the program.

If you made no mistakes at all, skip Sec. 5-3, because you already know its material, and go on to the next section whose number you circled on page 202. If no further numbers were circled, take Final Test 5 (page 238).

Introduction

After introducing the concept of factoring, this section emphasizes that any rational number can be represented by an infinite number of different but equivalent fractions. You will learn the convenience of using one particular form of a rational number, called the "simplest form." We define it as the integer form, and for nonintegral rational numbers as the form a/b or $-a/b$, with a and b the smallest possible positive integers. You then practice converting rational numbers to their simplest forms by dividing numerator and denominator by a common factor.

FRAMES FOR SECTION 5-3

1 To *factor* an integer is to express it as the product of at least two integers. The number 36, for example, can be factored into pairs of positive integers in five different ways (since multiplication is commutative, we regard $a \times b$ and $b \times a$ as the same pair of integers):

$36 = 1 \times 36$

$36 = 2 \times$ __

$36 =$ __ $\times 12$

$36 =$ __ $\times 9$

$36 =$ __ $\times$ __

---

18 3 4 6×6

---

2 The number 24 can be factored into pairs of positive integers in four ways:

$24 = 1 \times 24$

$24 =$ __ $\times$ __

$24 =$ __ $\times$ __

$24 =$ __ $\times$ __

2×12 3×8 4×6

---

3 The numbers 1, 2, and 3 can be factored into pairs of positive integers in only one way:

$1 = 1 \times 1$ $2 = 2 \times 1$ $3 = 3 \times 1$

In how many ways can the following numbers be factored into pairs of positive integers?

a. 4 *b.* 5 *c.* 6

---

a. two, 1×4 and 2×2 *b.* one, 1×5 *c.* two, 1×6 and 2×3

---

4 Numbers like 1, 2, 3, and 5 which can be factored into pairs of positive integers in only one way (as 1×1, 1×2, 1×3, 1×5, etc.) are called *prime numbers.* Which of the following are prime numbers: 1, 2, 3, 4, 5, 6, 7, 8, 9, 10, 11, 12, 13, 14, 15?

1, 2, 3, 5, 7, 11, 13 are prime numbers.

$4 = 1 \times 4$ or 2×2
$6 = 1 \times 6$ or 2×3
$8 = 1 \times 8$ or 2×4
$9 = 1 \times 9$ or 3×3
$10 = 1 \times 10$ or 2×5
$12 = 1 \times 12$ or 2×6 or 3×4
$14 = 1 \times 14$ or 2×7
$15 = 1 \times 15$ or 3×5

---

5 A positive integer is said to be *factored completely* if all its factors are prime numbers. Thus $36 = 2 \times 2 \times 3 \times 3$ is factored completely because 2 and 3 are both prime numbers (1 is usually omitted). $36 = 3 \times 12$ is not factored completely because 12 is not a prime number.

a. Is $24 = 2 \times 2 \times 6$ factored completely?

b. Is $18 = 2 \times 3 \times 3$ factored completely?

---

a. no, because 6 is not a prime number *b.* yes

---

6 Which of the following are factored completely?

a. $72 = 2 \times 2 \times 2 \times 9$

b. $45 = 3 \times 3 \times 5$

c. $16 = 4 \times 4$

d. $54 = 2 \times 3 \times 3 \times 3$

---

b and *d*

---

7 Factor the following numbers completely:

a. 16 *b.* 72 *c.* 99 *d.* 105

---

a. $2 \times 2 \times 2 \times 2$ *b.* $2 \times 2 \times 2 \times 3 \times 3$ *c.* $3 \times 3 \times 11$
d. $3 \times 5 \times 7$

---

8 Negative numbers can be factored by taking -1 as the first factor. Thus $-105 = -1 \times 105 = -1 \times 3 \times 5 \times 7$. Factor the following numbers completely:

a. 75 *b.* -108 *c.* 112 *d.* -84

a. $3 \times 5 \times 5$ *b.* $-1 \times 2 \times 2 \times 3 \times 3 \times 3$ *c.* $2 \times 2 \times 2 \times 2 \times 7$
d. $-1 \times 2 \times 2 \times 3 \times 7$

9 We shall return to factoring later in this section, but now we must consider the different forms of a rational number. A single point on the number line corresponds to a single rational number, and the set of equivalent fractions which graph onto a point are different forms of that rational number. Because equivalent fractions are forms of the same rational number, we use the equals sign to express the relationship between them. Thus

$$\frac{3}{7} = \frac{6}{?} = \frac{?}{21} = \frac{?}{-21}$$

14 9 −9

10 Since rational numbers can be represented by an infinite number of fractions, it is convenient to decide on one particular form for common use. We define the simplest possible form of an integral rational number as n or $-n$, where n is a positive integer. So the simplest form of $\frac{8}{2}$ is ______, and the simplest form of $\frac{-6}{3}$ is ______, and the simplest form of $\frac{20}{-4}$ is ______.

4 −2 −5

11 The simplest form of a nonintegral rational number is defined as $\frac{a}{b}$ or $-\frac{a}{b}$, where a and b are the *smallest* possible positive integers. So the simplest form of $\frac{2}{6}$ is ______, and the simplest form of $\frac{-5}{10}$ is ______, and the simplest form of $\frac{6}{-8}$ is ______.

$\frac{1}{3}$ $-\frac{1}{2}$ $-\frac{3}{4}$

12 Give the simplest forms of the following rational numbers:

a. $\frac{14}{-7}$ *b.* $\frac{-10}{18}$ *c.* $\frac{-8}{-20}$

a. -2 *b.* $-\frac{5}{9}$ *c.* $\frac{2}{5}$

13 Give the simplest forms of the following rational numbers:

a. $\frac{-21}{14}$ *b.* $\frac{15}{3}$ *c.* $\frac{9}{-15}$

---

a. $-\frac{3}{2}$ *b.* 5 *c.* $-\frac{3}{5}$

---

14 Notice that simplifying fractions involves dividing top and bottom by a common factor, using the rule that

$$\frac{a \times n}{b \times n} = \frac{a}{b}$$

We can find this common factor n by factoring the numerator and the denominator. Thus in Frame 13*a*

$$\frac{-21}{14} = \frac{-1 \times 3 \times 7}{2 \times 7} = \frac{-3}{2} = -\frac{3}{2}$$

after dividing top and bottom by the common factor ______, and in Frame 13*b*

$$\frac{15}{3} = \frac{5 \times 3}{1 \times 3} = \frac{5}{1} = 5$$

after dividing top and bottom by the common factor ______.

---

7 3

---

15 Completely factor the numerators and denominators of the following fractions and then convert the fractions to their simplest forms:

a. $\frac{30}{-36} = \frac{\not{2} \times \not{3} \times 5}{-1 \times \not{2} \times 2 \times \not{3} \times 3} = \frac{5}{-6} = -\frac{5}{6}$

after dividing by common factors ______ and ______.

b. $\frac{75}{105} = \frac{\quad \times \quad \times \quad}{\quad \times \quad \times \quad} =$ ______

after dividing by common factors ______ and ______.

---

a. 2; 3 *b.* $\frac{\not{5} \times \not{3} \times 5}{\not{5} \times \not{3} \times 7} = \frac{5}{7}$; 5; 3

---

16 Completely factor the numerators and denominators of the following fractions and then convert the fractions to their simplest forms:

a. $\dfrac{-45}{72} = \dfrac{\quad\times\quad\times\quad\times\quad}{\quad\times\quad\times\quad\times\quad\times\quad} = \_\_\_\_$

after dividing by common factors ____ and ____.

b. $\dfrac{48}{88} = \dfrac{\quad\times\quad\times\quad\times\quad\times\quad}{\quad\times\quad\times\quad\times\quad} = \_\_\_\_$

after dividing by common factors ____, ____, and ____.

---

a. $\dfrac{-1 \times 5 \times \cancel{3} \times \cancel{3}}{2 \times 2 \times 2 \times \cancel{3} \times \cancel{3}} = -\dfrac{5}{8}$; 3; 3 *b.* $\dfrac{\cancel{2} \times \cancel{2} \times \cancel{2} \times 2 \times 3}{\cancel{2} \times \cancel{2} \times \cancel{2} \times 11} = \dfrac{6}{11}$; 2; 2; 2

---

17 Often it is quicker to spot a common factor (not necessarily a prime factor) and divide by it as shown:

$\dfrac{\overset{6}{\cancel{48}}}{\underset{11}{\cancel{88}}} = \dfrac{6}{11}$ Dividing by 8

Simplify $\dfrac{18}{21}$ by this method.

$\dfrac{\overset{6}{\cancel{18}}}{\underset{7}{\cancel{21}}} = \dfrac{6}{7}$ Dividing by 3

---

18 Simplify $\dfrac{33}{55}$, $\dfrac{16}{28}$, and $\dfrac{-15}{10}$ by the same method.

$\dfrac{\overset{3}{\cancel{33}}}{\underset{5}{\cancel{55}}} = \dfrac{3}{5}$ Dividing by 11

$\dfrac{\overset{4}{\cancel{16}}}{\underset{7}{\cancel{28}}} = \dfrac{4}{7}$ Dividing by 4

$\dfrac{\overset{-3}{\cancel{-15}}}{\underset{2}{\cancel{10}}} = \dfrac{-3}{2} = -\dfrac{3}{2}$ Dividing by 5

---

19 This method of simplification can be continued until it is no longer possible to find a common factor. Thus, to simplify $\frac{72}{18}$:

$$\frac{\overset{8}{\cancel{72}}}{\underset{2}{\cancel{18}}} = \frac{\overset{4}{\cancel{8}}}{\underset{1}{\cancel{2}}} = 4 \qquad \text{Dividing by 9 and then by 2}$$

or

$$\frac{\overset{12}{\cancel{72}}}{\underset{3}{\cancel{18}}} = \frac{\overset{4}{\cancel{12}}}{\underset{1}{\cancel{3}}} = 4 \qquad \text{Dividing by 6 and then by 3}$$

It does not matter which common factor you eliminate first, but it is worth beginning with larger factors such as 6 and 9 rather than small prime factors like 2 and 3 whenever possible. Simplify $\frac{72}{108}$ by this same method.

$$\frac{\overset{8}{\cancel{72}}}{\underset{12}{\cancel{108}}} = \frac{\overset{2}{\cancel{8}}}{\underset{3}{\cancel{12}}} = \frac{2}{3} \qquad \text{Dividing by 9 and then by 4}$$

or

$$\frac{72}{108} = \frac{\overset{2}{\cancel{6}}}{\underset{3}{\cancel{9}}} = \frac{2}{3} \qquad \text{Dividing by 12 and then by 3}$$

Either route can be used.

---

20 Simplify $\frac{45}{75}$ by the same method.

$$\frac{\overset{9}{\cancel{45}}}{\underset{15}{\cancel{75}}} = \frac{\overset{3}{\cancel{9}}}{\underset{5}{\cancel{15}}} = \frac{3}{5} \qquad \text{Dividing by 5 and 3}$$

or

$$\frac{\overset{15}{\cancel{45}}}{\underset{25}{\cancel{75}}} = \frac{\overset{3}{\cancel{15}}}{\underset{5}{\cancel{25}}} = \frac{3}{5} \qquad \text{Dividing by 3 and 5}$$

Either route can be used.

---

21 Simplify $\dfrac{168}{-48}$ by the same method.

$$\dfrac{\overset{42}{\cancel{168}}}{\underset{-12}{\cancel{-48}}} = \dfrac{\overset{7}{\cancel{42}}}{\underset{-2}{\cancel{-12}}} = \dfrac{7}{-2} = -\dfrac{7}{2}$$ Dividing by 4 and 6

Any other route which eliminated factors of 24 was probably correct.

---

22 With large negative numbers it is more convenient to convert to the form $\dfrac{a}{b}$ or $-\dfrac{a}{b}$ before simplification. Thus

$$\dfrac{154}{-110} = -\dfrac{\overset{77}{\cancel{154}}}{\underset{55}{\cancel{110}}} = \underline{\qquad} \quad \text{and} \quad \dfrac{-160}{24} = \underline{\qquad}$$

$$-\dfrac{\overset{7}{\cancel{77}}}{\underset{5}{\cancel{55}}} = -\dfrac{7}{5} \qquad -\dfrac{\overset{20}{\cancel{160}}}{\underset{3}{\cancel{24}}} = -\dfrac{20}{3}$$ Other routes which gave the same answers were probably correct.

---

23 Simplify $\dfrac{210}{-84}$ and $\dfrac{-175}{-70}$

$$-\dfrac{\overset{30}{\cancel{210}}}{\underset{12}{\cancel{84}}} = -\dfrac{\overset{5}{\cancel{30}}}{\underset{2}{\cancel{12}}} = -\dfrac{5}{2}$$ Dividing by 7 and 5

$$\dfrac{\overset{35}{\cancel{175}}}{\underset{14}{\cancel{70}}} = \dfrac{\overset{5}{\cancel{35}}}{\underset{2}{\cancel{14}}} = \dfrac{5}{2}$$ Dividing by 5 and 7

Other routes which eliminated factors of 35 were probably correct.

---

24 Simplify $\dfrac{112}{-21}$ and $\dfrac{-45}{54}$

$$-\dfrac{\overset{16}{\cancel{112}}}{\underset{3}{\cancel{21}}} = -\dfrac{16}{3}$$ Dividing by 7

$$-\dfrac{\overset{5}{\cancel{45}}}{\underset{6}{\cancel{54}}} = -\dfrac{5}{6}$$ Dividing by 9

---

25 Simplify $\frac{110}{-66}$ and $\frac{-288}{-96}$

$$-\frac{\overset{10}{\cancel{110}}}{\underset{6}{\cancel{66}}} = -\frac{\overset{5}{\cancel{10}}}{\underset{3}{\cancel{6}}} = -\frac{5}{3}$$ Dividing by 11 and 2

$$\frac{\overset{72}{\cancel{288}}}{\underset{24}{\cancel{96}}} = \frac{\overset{6}{\cancel{72}}}{\underset{2}{\cancel{24}}} = \frac{\overset{3}{\cancel{6}}}{\underset{1}{\cancel{2}}} = 3$$ Dividing by 4, 12, and 2

You have finished the program for Sec. 5-3. Now take the posttest to see how much you have learned.

POSTTEST 5-3

Completely factor the following numbers:

1. 112
2. −90

Convert the following rational numbers to their simplest forms:

3. $\frac{105}{70}$
4. $\frac{-144}{96}$
5. $\frac{54}{-30}$
6. $\frac{-80}{-36}$

When you have finished the posttest, correct it yourself using the answers on page 241. Then check all the questions you got wrong on the list below.

1. ______ 4. ______

2. ______ 5. ______

3. ______ 6. ______

If you compare your performance with that on Pretest 5-3, you will see how much you have improved. If you made more than one mistake on this posttest, review the material and take the pretest again. Don't move on to the next section until you can get all the questions right.

Now proceed to the next section whose number you circled on page 202. If no further numbers were circled, take Final Test 5 (page 238).

Section 5-4
Mixed-numeral Forms

Do this section if you made a mistake in Questions 8 to 10 of Diagnostic Test 5. Begin by taking the pretest below, which will show you how much of Sec. 5-4 you already know and what you will be able to do by the time you have finished it.

PRETEST 5-4

Add the following numbers:

1. $\frac{5}{8} + \frac{-6}{8} =$ ______ 2. $\frac{-3}{7} + \frac{-9}{7} =$ ______

Convert the following mixed numerals into fractions:

3. $-2\frac{4}{7}$ 4. $3\frac{5}{8}$ 5. $-4\frac{2}{11}$

Convert the following fractions into mixed numerals:

6. $\frac{-33}{-10}$ 7. $\frac{23}{-11}$ 8. $\frac{-37}{9}$

When you have finished the pretest, correct it yourself using the answers on page 242. Then check all the questions you got wrong on the list below.

1.	______	5.	______
2.	______	6.	______
3.	______	7.	______
4.	______	8.	______

If you made any mistakes (even if it was only *one* mistake), read the introduction and go through the program.

If you made no mistakes at all, skip Sec. 5-4, because you already know its material, and go on to take Final Test 5 (page 238).

Introduction

The distributive law of division over addition can be used to define the addition of fractions which have the same denominator. Here this definition is used to derive the mixed-numeral form of a nonintegral rational number, which consists of an integer and a fraction less than 1 written together, for example, $3\frac{4}{7}$. You then practice converting mixed-numeral and simplest-fraction forms.

FRAMES FOR SECTION 5-4

1 In Chap. 2 we used the distributive law of division over addition for the set of positive integers. This law states that $(a+b) \div c = (\_\_\_\_) + (\_\_\_\_)$, where a, b, and c are positive integers.

$(a \div c) + (b \div c)$

---

2 If we assume that the distributive law is true for all integers (and not just positive integers), we state that $\frac{a+b}{c} = \frac{?}{c} + \frac{?}{c}$, where a, b, and c are integers and c is not equal to 0.

$\frac{a}{c} + \frac{b}{c}$

---

3 Written in reverse order as

$$\frac{a}{c} + \frac{b}{c} = \frac{a+b}{c}$$

this defines the addition of rational numbers with a common denominator. Use this definition of addition to add the following rational numbers:

a. $\frac{3}{5} + \frac{4}{5} = \frac{?}{5}$ *b.* $\frac{-4}{6} + \frac{-3}{6} = \frac{?}{6}$ *c.* $\frac{5}{7} + \frac{-8}{7} = \_\_\_\_$

---

a. 7 *b.* −7 *c.* $\frac{-3}{7}$

---

4 Add the following rational numbers:

a. $\frac{7}{5} + \frac{2}{5} = \_\_\_\_$ *b.* $\frac{-3}{8} + \frac{-11}{8} = \_\_\_\_$ *c.* $\frac{-16}{9} + \frac{14}{9} = \_\_\_\_$

---

a. $\frac{9}{5}$ *b.* $\frac{-14}{8}$ *c.* $\frac{-2}{9}$

---

5 Add the following rational numbers:

a. $\frac{-6}{7} + \frac{-3}{7} = \_\_\_\_$ *b.* $\frac{-10}{11} + \frac{4}{11} = \_\_\_\_$ *c.* $\frac{15}{8} + \frac{-12}{8} = \_\_\_\_$

---

a. $\frac{-9}{7}$ *b.* $\frac{-6}{11}$ *c.* $\frac{3}{8}$

---

6 Give the simplest forms of the following rational numbers:

a. $\frac{24}{6}$ *b.* $\frac{10}{5}$ *c.* $\frac{-28}{4}$ *d.* $\frac{-18}{9}$

---

a. 4 *b.* 2 *c.* −7 *d.* −2

---

7 Using the distributive law once more, we see that

$$\frac{25}{6} = \frac{24 + ?}{6} = \frac{24}{6} + \_\_\_\_ = \underset{\text{what integer?}}{\_\_\_\_\_\_\_\_} + \_\_\_\_\_\_\_\_$$

and

$$\frac{13}{5} = \frac{10 + ?}{5} = \frac{10}{5} + \_\_\_\_ = \underset{\text{what integer?}}{\_\_\_\_\_\_\_\_} + \_\_\_\_\_\_\_\_$$

---

1 $\frac{1}{6}$ $4 + \frac{1}{6}$ 3 $\frac{3}{5}$ $2 + \frac{3}{5}$

---

8 $4 + \frac{1}{6}$ can also be written as $4\frac{1}{6}$, which is called the *mixed-numeral* form of the rational number $\frac{25}{6}$. Give the mixed-numeral form of the rational number $\frac{13}{5}$.

$2\frac{3}{5}$

---

9 *a.* Write $6 + \frac{5}{8}$ in mixed-numeral form.

b. Write $5\frac{2}{3}$ in integer + fraction form.

---

a. $6\frac{5}{8}$ *b.* $5 + \frac{2}{3}$

---

10 The negative mixed numeral $-5\frac{2}{3}$ is defined as the negative of $5\frac{2}{3}$. So $-5\frac{2}{3}$ is not equal to $-5 + \frac{2}{3}$, but it is equal to $5\frac{2}{3} \times -1$, which is $\left(5 + \frac{2}{3}\right) \times -1 =$ ____ $\times -1 +$ ____ $\times -1 =$ ____ $+$ ____. Write $-7\frac{1}{4}$ in integer + fraction form.

---

5 $\frac{2}{3}$ $-5 + \left(-\frac{2}{3}\right)$ $-7 + \left(-\frac{1}{4}\right)$

---

11 Write the following mixed numerals in integer + fraction form:

a. $6\frac{3}{8}$ *b.* $-8\frac{1}{5}$ *c.* $-3\frac{4}{9}$

---

a. $6+\dfrac{3}{8}$ *b.* $-8+\left(-\dfrac{1}{5}\right)$ *c.* $-3+\left(-\dfrac{4}{9}\right)$

---

12 A mixed numeral can be converted into a fraction by replacing the integer part by a fraction with the same denominator. Thus

$$1\tfrac{3}{7}=1+\frac{3}{7}=\frac{7}{7}+\frac{3}{7}=\frac{10}{7}$$

and

$$-2\tfrac{1}{4}=-\left(2+\frac{1}{4}\right)=-\left(\frac{8}{4}+\frac{1}{4}\right)=-\frac{9}{4}$$

Similarly

$$4\tfrac{2}{9}=4+\frac{2}{9}=\frac{36}{9}+\frac{2}{9}=\frac{?}{?}$$

and

$$-5\tfrac{3}{8}=-\left(5+\frac{3}{8}\right)=-\left(\frac{40}{8}+\frac{3}{8}\right)=-\frac{?}{?}$$

$\dfrac{38}{9}$ $-\dfrac{43}{8}$

---

13 Convert the following mixed numerals into fractions:

a. $3\frac{3}{7}=3+\dfrac{3}{7}=\dfrac{21}{7}+\dfrac{3}{7}=$ _____ *b.* $6\frac{7}{8}=6+\dfrac{7}{8}=\dfrac{?}{8}+\dfrac{7}{8}=$ _____

c. $(-5\frac{2}{3})=-\left(5+\dfrac{2}{3}\right)=-\left(\dfrac{?}{3}+\dfrac{2}{3}\right)=$ _____

---

a. $\dfrac{24}{7}$ *b.* $48;\ \dfrac{55}{8}$ *c.* $15;\ -\dfrac{17}{3}$

---

14 Convert the following mixed numerals into fractions:

a. $-4\frac{3}{4}=-\left(4+\dfrac{3}{4}\right)=-\left(\dfrac{?}{4}+\dfrac{3}{4}\right)=$ _____

b. $-3\frac{2}{5}=-\left(\dfrac{?}{5}+\dfrac{2}{5}\right)=$ _____

c. $6\frac{1}{2}=\dfrac{?}{2}+\dfrac{1}{2}=$ _____

a. 16; $-\frac{19}{4}$ *b.* 15; $-\frac{17}{5}$ *c.* 12; $\frac{13}{2}$

15 Convert the following mixed numerals into fractions:

a. $3\frac{7}{9} = \frac{?}{9} + \frac{7}{9} =$ ______

b. $4\frac{5}{6} =$ ______

c. $-8\frac{2}{3} =$ ______

a. 27; $\frac{34}{9}$ *b.* $\frac{29}{6}$ *c.* $-\frac{26}{3}$

16 Convert the following mixed numerals into fractions:

a. $-9\frac{3}{7}$ *b.* $4\frac{1}{6}$ *c.* $-3\frac{7}{10}$

a. $-\frac{66}{7}$ *b.* $\frac{25}{6}$ *c.* $-\frac{37}{10}$

17 Convert the following mixed numerals into fractions:

a. $-3\frac{2}{11}$ *b.* $2\frac{5}{8}$ *c.* $-4\frac{4}{5}$

a. $-\frac{35}{11}$ *b.* $\frac{21}{8}$ *c.* $-\frac{24}{5}$

18 To convert fractions to mixed numerals we use the reverse process. Thus

$$\frac{23}{4} = \underbrace{\frac{20}{4}}_{\text{integer}} + \underbrace{\frac{3}{4}}_{\text{fraction less than 1}} = 5\frac{3}{4}$$

a. $$\frac{17}{5} = \underbrace{\frac{15}{5}}_{\text{integer}} + \underbrace{\frac{?}{5}}_{\text{fraction less than 1}} = \underbrace{\qquad\qquad}_{\text{mixed numeral}}$$

b. $$-\frac{32}{7} = -\left(\underbrace{\frac{28}{7}}_{\text{integer}} + \underbrace{\frac{?}{7}}_{\text{fraction less than 1}}\right) = \underbrace{\qquad\qquad}_{\text{mixed numeral}}$$

a. 2; $3\frac{2}{5}$ *b.* 4; $-4\frac{4}{7}$

19 Convert the following to mixed numerals:

a. $\frac{74}{9} = \frac{72}{9} + \frac{?}{9} =$ ______

b. $\frac{-33}{5} = -\frac{33}{5} = -\left(\frac{30}{5} + \frac{?}{5}\right) =$ ______

---

a. 2; $8\frac{2}{9}$ *b.* 3; $-6\frac{3}{5}$

---

20 Convert the following to mixed numerals:

a. $\frac{-59}{7} = -\frac{59}{7} = -\left(\frac{56}{7} + \frac{?}{7}\right) =$ ______

b. $\frac{41}{-9} = -\frac{41}{9} =$ ______

c. $\frac{-61}{-8} = \frac{61}{8} =$ ______

---

a. 3; $-8\frac{3}{7}$ *b.* $-4\frac{5}{9}$ *c.* $7\frac{5}{8}$

---

21 Convert the following to mixed numerals:

a. $\frac{-33}{-7}$ *b.* $\frac{-32}{5}$ *c.* $\frac{34}{-9}$

---

a. $4\frac{5}{7}$ *b.* $-6\frac{2}{5}$ *c.* $-3\frac{7}{9}$

---

22 Convert the following to mixed numerals:

a. $\frac{43}{-6}$ *b.* $\frac{-39}{-7}$ *c.* $\frac{13}{-4}$

---

a. $-7\frac{1}{6}$ *b.* $5\frac{4}{7}$ *c.* $-3\frac{1}{4}$

You have finished the program for Sec. 5-4. Now take the posttest to see how much you have learned.

POSTTEST 5-4

Add the following numbers:

1. $\frac{-3}{9} + \frac{4}{9} =$ ______

2. $\frac{-5}{6} + \frac{-8}{6} =$ ______

Convert the following mixed numerals into fractions:

3. $-3\frac{2}{9}$ 4. $4\frac{7}{8}$ 5. $-3\frac{6}{11}$

Convert the following fractions into mixed numerals:

6. $\frac{-67}{7}$ 7. $\frac{34}{-9}$ 8. $\frac{-47}{-11}$

When you have finished the posttest, correct it yourself using the answers on page 242. Then check all the questions you got wrong on the list below.

1.	______	5.	______
2.	______	6.	______
3.	______	7.	______
4.	______	8.	______

If you compare your performance with that on Pretest 5-4, you will see how much you have improved. If you made more than one mistake on this posttest, review the material and take the pretest again. Don't move on to the next section until you can get all the questions right.

Now proceed with Final Test 5.

FINAL TEST 5

Draw sections of the number line and graph the following fractions:

1. $\frac{-11}{4}$ 2. $-\frac{14}{3}$

Select values for m and n which make the second fraction in each pair equivalent to the first. Fractions which represent the same rational number are equivalent.

3. $-\frac{5}{6}$ and $\frac{m}{-48}$ 4. $\frac{9}{8}$ and $\frac{-63}{n}$

5. Completely factor the integer -168.

Express the following rational numbers in their simplest forms:

6. $\frac{128}{-80}$ 7. $\frac{-126}{-72}$ 8. $\frac{-11}{7} + \frac{6}{7}$

9. Express the mixed numeral $-9\frac{5}{9}$ in fractional form.

10. Express the fraction $\dfrac{-53}{-8}$ in mixed-numeral form.

The answers to this test are on page 242. Check the questions you got wrong on the list below and compare your results with those on Diagnostic Test 5 to see how much you have improved.

1. _____	3. _____	5. _____	8. _____
2. _____	4. _____	6. _____	9. _____
		7. _____	10. _____

If you made more than one mistake, review the material and then take the diagnostic test again. Make sure that you can get all the questions right before you finish the chapter.

Now go on to Chap. 6.

ANSWERS TO TESTS FOR CHAPTER 5

DIAGNOSTIC TEST 5

1. 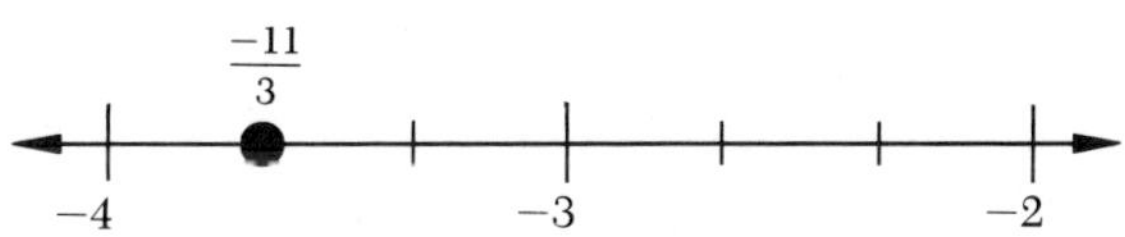

2. 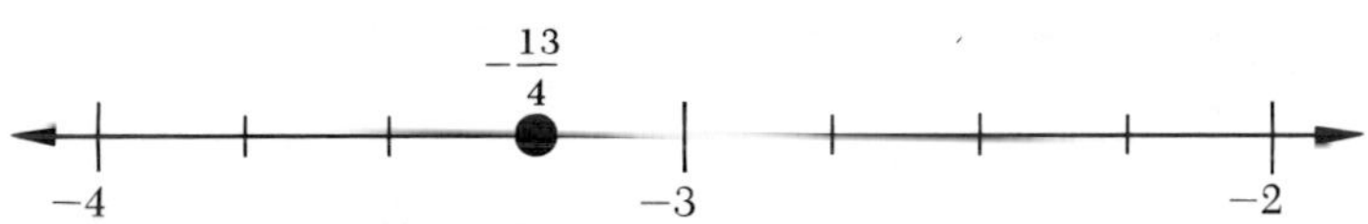

3. $m = 32$

4. $n = -49$

5. $-1 \times 2 \times 2 \times 2 \times 3 \times 5$

6. $-\dfrac{4}{3}$

7. $\dfrac{7}{6}$

8. $\dfrac{-5}{9}$ or $-\dfrac{5}{9}$

9. $-\dfrac{53}{7}$ or $\dfrac{-53}{7}$

10. $4\frac{8}{11}$

PRETEST 5-1

1. $\frac{3}{5}, 7, \frac{-7}{-9}, -5, \frac{0}{8}, \frac{4}{-6}$ $\left(\text{but not } \frac{-6}{0}\right)$

2. $\frac{10}{-5}, \frac{-8}{2}$, and $\frac{-12}{-3}$

3. 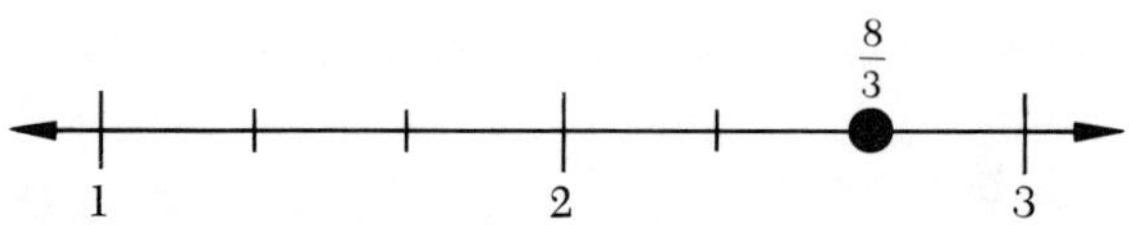

4. 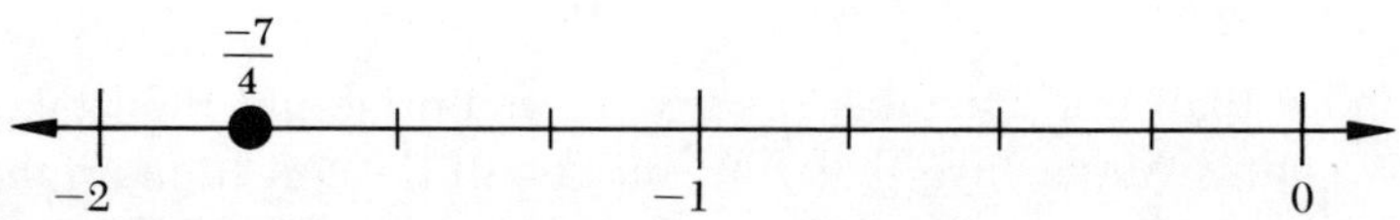

5. 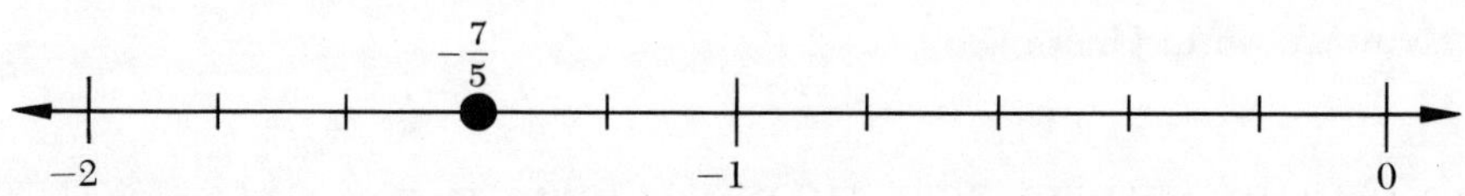

6. $-\frac{6}{7}$ or $\frac{-6}{7}$ 7. $\frac{9}{8}$ 8. $\frac{4}{11}$

POSTTEST 5-1

1. $\frac{8}{9}, 5, \frac{-3}{-8}, -6, \frac{0}{5}, \frac{3}{-9}$ $\left(\text{but not } \frac{-7}{0}\right)$

2. $\frac{16}{-4}, \frac{-12}{2}$, and $\frac{-15}{-5}$

3. 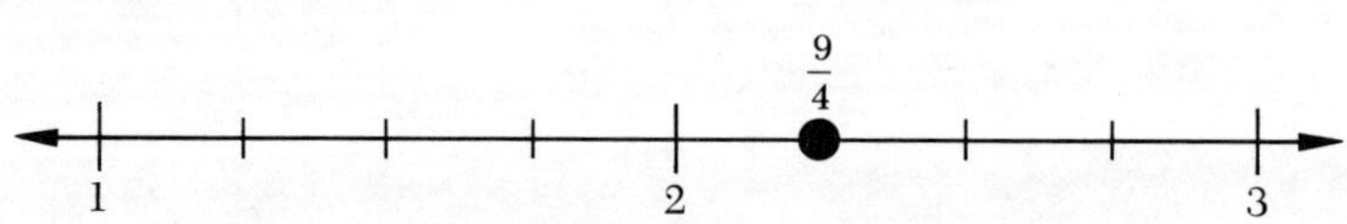

4. 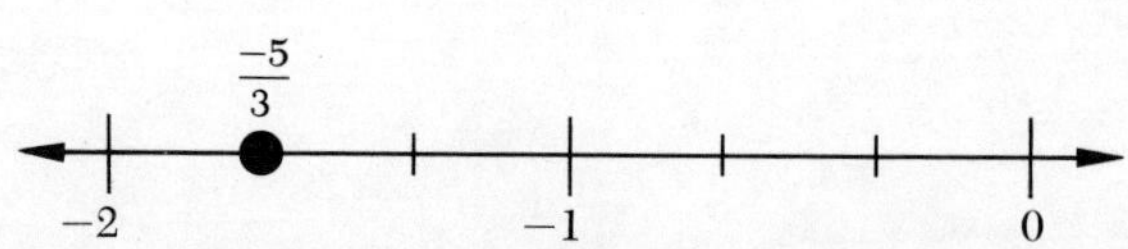

5. 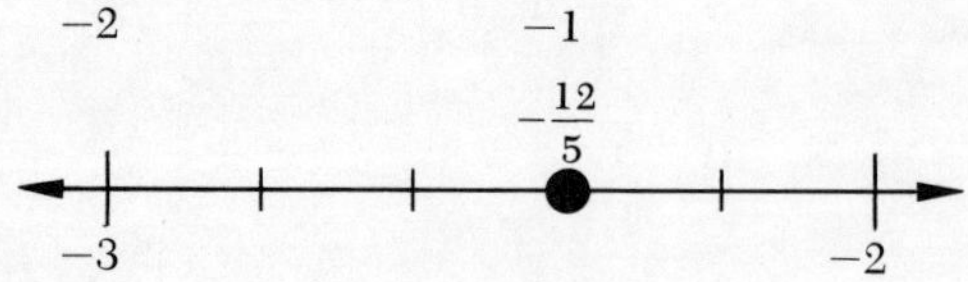

6. $\frac{-8}{9}$ or $-\frac{8}{9}$ 7. $\frac{7}{11}$ 8. $\frac{6}{13}$

PRETEST 5-2

1. yes
2. no
3. yes
4. no
5. $a = 9$
6. $b = -21$
7. $c = -40$
8. $d = 9$
9. $e = -2$
10. $f = -12$

POSTTEST 5-2

1. yes
2. no
3. yes
4. no
5. $a = 28$
6. $b = -36$
7. $c = -81$
8. $d = 12$
9. $e = -2$
10. $f = -16$

PRETEST 5-3

1. $2 \times 2 \times 2 \times 2 \times 2 \times 3$
2. $-1 \times 2 \times 5 \times 7$
3. $\frac{3}{2}$
4. $-\frac{12}{7}$
5. $-\frac{14}{5}$
6. $\frac{35}{11}$

POSTTEST 5-3

1. $2 \times 2 \times 2 \times 2 \times 7$
2. $-1 \times 2 \times 3 \times 3 \times 5$
3. $\frac{3}{2}$
4. $-\frac{3}{2}$
5. $-\frac{9}{5}$
6. $\frac{20}{9}$

PRETEST 5-4

1. $\frac{-1}{8}$ or $-\frac{1}{8}$
2. $\frac{-12}{7}$ or $-\frac{12}{7}$
3. $-\frac{18}{7}$
4. $\frac{29}{8}$
5. $-\frac{46}{11}$
6. $3\frac{3}{10}$
7. $-2\frac{1}{11}$
8. $-4\frac{1}{9}$

POSTTEST 5-4

1. $\frac{1}{9}$
2. $\frac{-13}{6}$ or $-\frac{13}{6}$
3. $-\frac{29}{9}$
4. $\frac{39}{8}$
5. $-\frac{39}{11}$
6. $-9\frac{4}{7}$
7. $-3\frac{7}{9}$
8. $4\frac{3}{11}$

FINAL TEST 5

1. 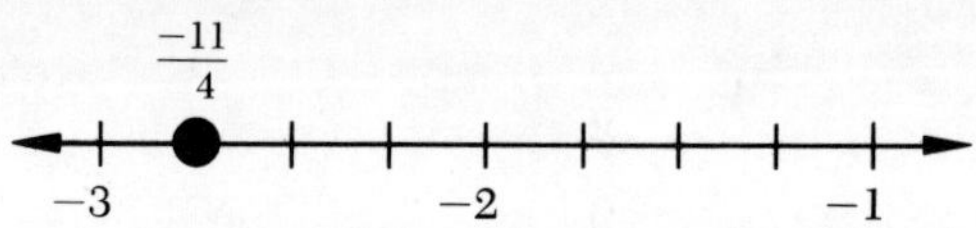

2. 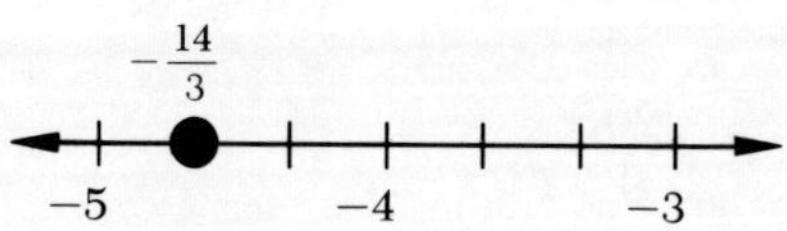

3. $m = 40$
4. $n = -56$
5. $-1 \times 2 \times 2 \times 2 \times 3 \times 7$
6. $-\frac{8}{5}$
7. $\frac{7}{4}$
8. $\frac{-5}{7}$ or $-\frac{5}{7}$
9. $-\frac{86}{9}$ or $\frac{-86}{9}$
10. $6\frac{5}{8}$

Chapter 6 Operations with Rational Numbers

This chapter introduces the multiplication, division, addition, and subtraction of rational numbers in both fractional and mixed-numeral forms. You practice these operations and learn how to extend them to the simplification of expressions which include combinations of operations.

Section 6-1 derives a definition of the multiplication of two rational numbers, and Sec. 6-2 introduces division as equivalent to multiplication by the reciprocal. In both sections you practice these operations with both fractional and mixed-numeral forms. Section 6-3 shows how fractions can be added and subtracted by conversion to forms which have a common denominator, and Sec. 6-4 extends the discussion to operations with mixed numerals. Section 6-5 introduces the concept of lowest common denominator (LCD) and uses it to evaluate multiple additions and subtractions, which Sec. 6-6 then extends to the evaluation of expressions containing combinations of additions and subtractions with multiplications and divisions.

This chapter requires completion of Chaps. 2, 4, and 5.

DIAGNOSTIC TEST 6

Simplify the following expressions:

1. $\frac{-3}{-4} \times \frac{5}{-6}$
2. $\frac{8}{-11} \div \frac{-4}{7}$
3. $3\frac{5}{6} \div (-4\frac{7}{12})$
4. $\frac{-9}{10} - \frac{11}{-7}$
5. $\frac{7}{-9} - \frac{-5}{-8}$
6. $4\frac{2}{7} - 1\frac{5}{8}$
7. $2\frac{3}{4} - 5\frac{4}{9}$

8. $\frac{8}{9}+\left(-\frac{7}{12}\right)-\left(-\frac{11}{8}\right)$

9. $\frac{3}{5}\left[-\frac{7}{10}-\left(-\frac{5}{6}\right)\right]$

10. $\dfrac{\frac{4}{9}-\left(-\frac{2}{7}\right)}{-\frac{5}{6}+\frac{9}{14}}$

When you have finished the diagnostic test, correct it yourself using the answers on page 296. Then check all the questions you got wrong on the list below:

1. ______	4. ______	8. ______
2. ______	5. ______	9. ______
3. ______	6. ______	10. ______
	7. ______	

If you made no mistakes at all, you need not do any work in this chapter and should go straight on to Chap. 7. But if you made any mistakes (even if it was only *one* mistake), follow these instructions:

	section	page
If you made any mistakes in Questions 1 to 3, circle Secs. 6-1 and 6-2.	6-1 6-2	245 255
If you made any mistakes in Questions 4 to 7, circle Secs. 6-3 and 6-4.	6-3 6-4	263 273
If you made any mistakes in Questions 8 to 10, circle Secs. 6-5 and 6-6.	6-5 6-6	279 289

Now work through the sections whose numbers you have circled, taking them in order. The sections begin on the pages indicated.

Section 6-1 Multiplication of Rational Numbers

Do this section if you made a mistake in Questions 1 to 3 of Diagnostic Test 6. Begin by taking the pretest below, which will show you how much of Sec. 6-1 you already know and what you will be able to do by the time you have finished it.

PRETEST 6-1

1. $\frac{-6}{11} \times \frac{1}{5} =$ ______
2. $\frac{2}{-5} \times \left(-\frac{3}{7}\right) =$ ______
3. $\frac{-2}{9} \times \frac{4}{9} =$ ______
4. $\frac{-16}{-33} \times \frac{-54}{24} =$ ______
5. $-3\frac{3}{9} \times (-2\frac{7}{10}) =$ ______
6. $4\frac{4}{11} \times (-3\frac{5}{16}) =$ ______

When you have finished the pretest, correct it yourself using the answers on page 296. Then check all the questions you got wrong on the list below.

1.	______	4.	______
2.	______	5.	______
3.	______	6.	______

If you made any mistakes (even if it was only *one* mistake), read the introduction and go through the program.

If you made no mistakes at all, skip Sec. 6-1, because you already know its material, and go on to Sec. 6-2.

Introduction

This section uses the definition of multiplication as multiple addition to derive a rule for multiplying rational numbers by integers. Combining this rule with the fact that $(a \times n)/(b \times n)$ is equivalent to a/b leads to the general rule for multiplying rational numbers by rational numbers:

$$\frac{a}{b} \times \frac{c}{d} = \frac{a \times c}{b \times d}$$

You practice this rule with fractions and then apply it to the multiplication of rational numbers in mixed-numeral form.

FRAMES FOR SECTION 6-1

1 Multiplication of rational numbers by positive integers is defined by the concept of multiplication as multiple addition. Thus

$\frac{3}{5}+\frac{3}{5}=$ _____ implies that $\frac{3}{5}\times 2=$ _____

$\frac{3}{5}+\frac{3}{5}+\frac{3}{5}=$ _____ implies that $\frac{3}{5}\times 3=$ _____

$\frac{3}{5}+\frac{3}{5}+\frac{3}{5}+\frac{3}{5}=$ _____ implies that $\frac{3}{5}\times 4=$ _____

$\frac{6}{5}$ $\frac{6}{5}$ $\frac{9}{5}$ $\frac{9}{5}$ $\frac{12}{5}$ $\frac{12}{5}$

---

2 From continuing this sequence one can deduce that $\frac{3}{5}\times n=\frac{?}{5}$, where n is any positive integer.

$3 \times n$

---

3 Similarly $\frac{-4}{5}+\frac{-4}{5}+\frac{-4}{5}=$ _____ implies that $\frac{-4}{5}\times 3=$ _____, and $\frac{6}{-7}+\frac{6}{-7}+\frac{6}{-7}+\frac{6}{-7}=$ _____ implies that $\frac{6}{-7}\times 4=$ _____.

$\frac{-12}{5}$ $\frac{-12}{5}$ $\frac{24}{-7}$ $\frac{24}{-7}$

---

4 In general it can be shown that if n is a positive integer and a and b are integers with b not equal to 0, $\frac{a}{b}\times n=\frac{?}{b}$

$a \times n$

---

5 Use this rule to multiply the following:

a. $\frac{4}{19}\times 6$ *b.* $\frac{-5}{11}\times 8$ *c.* $\frac{7}{-4}\times 5$

---

a. $\frac{24}{19}$ *b.* $\frac{-40}{11}$ or $-\frac{40}{11}$ *c.* $\frac{35}{-4}$ or $-\frac{35}{4}$

---

6 Multiply the following rational numbers:

a. $\frac{3}{-5} \times 7$ *b.* $\frac{-6}{-13} \times 3$ *c.* $-\frac{4}{9} \times 8$

---

a. $\frac{21}{-5}$ or $-\frac{21}{5}$ *b.* $\frac{-18}{-13}$ or $\frac{18}{13}$ *c.* $-\frac{32}{9}$ or $\frac{-32}{9}$

---

7 Multiply the following rational numbers:

a. $-\frac{3}{7} \times 5$ *b.* $\frac{-6}{-11} \times 4$ *c.* $-\frac{2}{9} \times 7$

---

a. $-\frac{15}{7}$ or $\frac{-15}{7}$ *b.* $\frac{-24}{-11}$ or $\frac{24}{11}$ *c.* $-\frac{14}{9}$ or $\frac{-14}{9}$

---

8 For integers we found that the operations "negative of" and "multiplying by −1" were equivalent. If we assume that this is also true for rational numbers, we see that

$$\frac{2}{5} \times (-3) = \frac{2}{5} \times (-1) \times 3 = \_\_\_\_\_\_ \times 3 = \_\_\_\_\_\_ = \frac{? \times (-3)}{5}$$

$-\frac{2}{5}$ $-\frac{6}{5}$ 2

---

9 In general

$$\frac{a}{b} \times (-k) = \frac{a}{b} \times (-1) \times k = \left(-\frac{a}{b}\right) \times k = \frac{? \times k}{b} = \frac{? \times (-k)}{b}$$

since if a and k are integers, we know that $(-a) \times k =$ ______ $\times (-k)$.

$-a$ a a

---

10 Therefore if a, b, and k are integers and b is not equal to 0,

$$\frac{a}{b} \times (-k) = \frac{a \times ?}{b}$$

and the statement that $\frac{a}{b} \times n =$ ______ is true for all integral values of n (not just positive values).

$(-k)$ $\frac{a \times n}{b}$

---

11 Use this rule to multiply the following:

a. $\frac{3}{8} \times (-5)$ *b.* $\frac{-2}{9} \times (-7)$ *c.* $\frac{-5}{-7} \times (-6)$

---

a. $\frac{-15}{8}$ or $-\frac{15}{8}$ *b.* $\frac{14}{9}$ *c.* $\frac{30}{-7}$ or $-\frac{30}{7}$

---

12 If we assume that the multiplication of rational numbers is commutative, we can multiply integers by rational numbers:

a. $6 \times \frac{5}{-13} = \frac{5}{-13} \times 6 =$ _____

b. $8 \times \frac{-2}{-7} =$ _____

c. $-3 \times \frac{6}{-11} =$ _____

---

a. $\frac{30}{-13}$ or $-\frac{30}{13}$ *b.* $\frac{-16}{-7}$ or $\frac{16}{7}$ *c.* $\frac{-18}{-11}$ or $\frac{18}{11}$

---

13 We now need to find how to multiply $\frac{a}{b}$ by a single rational number such as $\frac{1}{4}$. We can use the fact that $4 \times \frac{1}{4} = 1 = \frac{?}{4}$ to show that

$$\frac{\boldsymbol{a \times 4}}{\boldsymbol{b}} \times \frac{\mathbf{1}}{\mathbf{4}} = \frac{a}{b} \times \_\_\_\_\_ \times \frac{1}{4} = \frac{\boldsymbol{a}}{\boldsymbol{b}} \times \frac{\mathbf{?}}{\mathbf{4}}$$

---

4 4 4

---

14 In Chap. 5 we showed that $\frac{?}{?} = \frac{a \times n}{b \times n}$, which means that

$$\frac{\boldsymbol{a}}{\boldsymbol{b}} \times \frac{\mathbf{4}}{\mathbf{4}} = \frac{a}{b} \times 1 = \frac{a}{b} = \frac{\boldsymbol{a \times ?}}{\boldsymbol{b \times 4}}$$

Combining this with the last line of Frame 13, we see that

$$\frac{\boldsymbol{a \times 4}}{\boldsymbol{b}} \times \frac{\mathbf{1}}{\mathbf{4}} = \frac{a}{b} \times \frac{4}{4} = \frac{\boldsymbol{a \times ?}}{\boldsymbol{b \times 4}}$$

$\frac{a}{b}$ 4 4

---

15 If we then substitute k for $a \times 4$ in the last line of Frame 14, we get

$$\frac{k}{b} \times \frac{1}{4} = \frac{?}{? \times ?}$$

which can be used as a definition of multiplying by $\frac{1}{4}$.

$\frac{k}{b \times 4}$

---

16 Similarly it can be shown that $\frac{k}{b} \times \frac{1}{6} =$ ______ and in general that $\frac{k}{b} \times \frac{1}{n} =$ ______, where b, k, and n are integers with b and n not equal to 0.

$\frac{k}{b \times 6}$ $\frac{k}{b \times n}$

---

17 Apply this rule to the following:

a. $\frac{-5}{2} \times \frac{1}{3}$ *b.* $\frac{7}{-3} \times \frac{1}{4}$ *c.* $\frac{-8}{5} \times \frac{1}{-7}$

---

a. $\frac{-5}{6}$ or $-\frac{5}{6}$ *b.* $\frac{7}{-12}$ or $-\frac{7}{12}$ *c.* $\frac{-8}{-35}$ or $\frac{8}{35}$

---

18 Multiply the following:

a. $\frac{-7}{-8} \times \frac{1}{4}$ *b.* $\frac{-6}{5} \times \frac{1}{-5}$ *c.* $\frac{-11}{-6} \times \frac{1}{-8}$

---

a. $\frac{-7}{-32}$ or $\frac{7}{32}$ *b.* $\frac{-6}{-25}$ or $\frac{6}{25}$ *c.* $\frac{-11}{48}$ or $-\frac{11}{48}$

---

19 We can now multiply by $\frac{3}{5}$ because $\frac{3}{5} =$ ______ $\times \frac{1}{5}$. Thus

$$\frac{a}{b} \times \frac{3}{5} = \frac{a}{b} \times \_\_\_\_\_\_ \times \frac{1}{5} = \frac{a \times ?}{b} \times \frac{1}{5} = \frac{? \times ?}{? \times ?}$$

---

3 3 3 $\frac{a \times 3}{b \times 5}$

---

20 Similarly we can derive a general definition of the multiplication of rational numbers. If a, b, c, d are integers and b and d are not equal to 0,

$$\frac{a}{b} \times \frac{c}{d} = \frac{a}{b} \times \_ \times \frac{1}{?} = \frac{? \times ?}{?} \times \frac{1}{?} = \_\_\_\_\_\_$$

$c \times \frac{1}{d}$ $\quad \frac{a \times c}{b} \times \frac{1}{d}$ $\quad \frac{a \times c}{b \times d}$

---

21 Applying this definition, we find that:

a. $\frac{3}{2} \times \frac{7}{8} = \frac{3 \times 7}{2 \times 8} =$ _____

b. $\frac{4}{5} \times \frac{-6}{7} = \frac{4 \times (-6)}{5 \times 7} =$ _____

c. $\frac{3}{-7} \times \frac{-11}{-5} = \frac{? \times ?}{? \times ?} =$ _____

---

a. $\frac{21}{16}$ *b.* $\frac{-24}{35}$ or $-\frac{24}{35}$ *c.* $\frac{3 \times (-11)}{-7 \times (-5)} = \frac{-33}{35} = -\frac{33}{35}$

---

22 *a.* $\frac{-7}{9} \times \frac{2}{-5} = \frac{? \times ?}{? \times ?} =$ _____

b. $\frac{3}{8} \times \frac{-5}{8} =$ _____

c. $\frac{-4}{9} \times \frac{-2}{-11} =$ _____

---

a. $\frac{-7 \times 2}{9 \times (-5)} = \frac{-14}{-45} = \frac{14}{45}$ *b.* $\frac{-15}{64}$ or $-\frac{15}{64}$ *c.* $\frac{8}{-99}$ or $-\frac{8}{99}$

---

23 Time can be saved by canceling out factors which are common to both the top and the bottom. Instead of

$$\frac{5}{4} \times \frac{11}{10} = \frac{5 \times 11}{4 \times 10} = \frac{\overset{11}{\cancel{55}}}{\underset{8}{\cancel{40}}} = \frac{11}{8}$$

we write

$$\frac{5}{4} \times \frac{11}{10} = \frac{\overset{1}{\cancel{5}} \times 11}{4 \times \underset{2}{\cancel{10}}} = \frac{11}{8}$$

Similarly, by dividing the top and the bottom by 8 and 9 we see that

$$\frac{8}{27} \times \frac{63}{16} = \frac{8 \times 63}{27 \times 16} = \frac{?}{?}$$

$$\frac{\overset{1}{\cancel{8}} \times \overset{7}{\cancel{63}}}{\underset{3}{\cancel{27}} \times \underset{2}{\cancel{16}}} = \frac{7}{6}$$

---

24 By dividing the top and bottom by 5, 7, and 3,

$$\frac{42}{55} \times \frac{25}{21} = \frac{42 \times 25}{55 \times 21} = \frac{?}{?}$$

By canceling out factors common to both top and bottom,

$$\frac{27}{35} \times \frac{56}{45} = \frac{27 \times 56}{35 \times 45} = \frac{?}{?}$$

$\frac{10}{11}$ $\left[\frac{\overset{\overset{2}{\cancel{6}}}{\cancel{42}} \times \overset{5}{\cancel{25}}}{\underset{11}{\cancel{55}} \times \underset{\underset{1}{\cancel{3}}}{\cancel{21}}}\right]$ $\frac{24}{25}$ (after dividing through by 9 and by 7) $\left[\frac{\overset{3}{\cancel{27}} \times \overset{8}{\cancel{56}}}{\underset{5}{\cancel{35}} \times \underset{5}{\cancel{45}}}\right]$

---

25 With negative numbers it can be helpful to use the $-\frac{a}{b}$ form, so that just as $5 \times (-4) =$ _____, $\frac{3}{7} \times \left(-\frac{5}{8}\right) =$ _____, and just as $-3 \times (-6) =$ _____, $-\frac{3}{4} \times \left(-\frac{3}{5}\right) =$ _____.

-20 $\quad -\frac{15}{56}$ $\quad 18$ $\quad \frac{9}{20}$

---

26 Converting fractions to the form $\frac{a}{b}$ or $-\frac{a}{b}$ before multiplying, we get:

a. $\frac{5}{-7} \times \frac{-3}{-4} = -\frac{5}{7} \times \frac{3}{4} =$ ✓__ $\frac{5 \times 3}{7 \times 4} =$ _____

b. $\frac{-3}{5} \times \frac{-4}{11} = -\frac{3}{5} \times \left(-\frac{4}{11}\right) =$ ✓__ $\frac{3 \times 4}{5 \times 11} =$ _____

c. $\frac{-11}{-8} \times \frac{3}{-7} =$ _____ $\times$ _____ $=$ _____

Note: The check mark indicates that a plus or minus sign should be inserted in the blank which follows.

---

a. $-;\ -\dfrac{15}{28}$ *b.* $+;\ \dfrac{12}{55}$ *c.* $\dfrac{11}{8}\times\left(-\dfrac{3}{7}\right)=-\dfrac{33}{56}$

---

27 This method has the advantage of combining negative signs before canceling, so that

$$\frac{3}{-10}\times\frac{-25}{-9}=-\frac{3}{10}\times\frac{25}{9}=-\frac{3\times 25}{10\times 9}$$

which by canceling out factors common to both top and bottom, that is, 3 and 5, gives

$$-\frac{\overset{1}{\cancel{3}}\times\overset{5}{\cancel{25}}}{\underset{2}{\cancel{10}}\times\underset{3}{\cancel{9}}}=-\frac{?}{?}$$

and

$$\frac{-22}{45}\times\frac{54}{-55}=-\frac{22}{45}\times\left(-\frac{54}{55}\right)=+\frac{22\times 54}{45\times 55}$$

which by canceling out factors common to both top and bottom gives ______.

$-\dfrac{5}{6}$ $\dfrac{12}{25}$ $\left[\dfrac{\overset{2}{\cancel{22}}\times\overset{6}{\cancel{54}}}{\underset{5}{\cancel{45}}\times\underset{5}{\cancel{55}}}\quad\text{Dividing by 11 and 9}\right]$

---

28 *a.* $\dfrac{-18}{-77}\times\dfrac{-55}{63}=$ ______ $\times$ ______ $=\checkmark$__ $\left(\dfrac{18\times 55}{77\times 63}\right)$

which, by canceling factors common to both top and bottom, gives ______.

b. $\dfrac{64}{-21}\times\dfrac{-49}{40}=$ ______ $\times$ ______ $=\checkmark$__ $\dfrac{64\times 49}{21\times 40}$

which, by canceling factors common to both top and bottom gives ______.

---

a. $\dfrac{18}{77}\times\left(-\dfrac{55}{63}\right);\ -;\ -\dfrac{10}{49}$ $\left[\dfrac{\overset{2}{\cancel{18}}\times\overset{5}{\cancel{55}}}{\underset{7}{\cancel{77}}\times\underset{7}{\cancel{63}}}\quad\text{Dividing by 9 and 5}\right]$

b. $-\dfrac{64}{21}\times\left(-\dfrac{49}{40}\right);\ +;\ \dfrac{56}{15}$ $\left[\dfrac{\overset{8}{\cancel{64}}\times\overset{7}{\cancel{49}}}{\underset{3}{\cancel{21}}\times\underset{5}{\cancel{40}}}\quad\text{Dividing by 8 and 7}\right]$

---

29 Simplify the following:

a. $\frac{-32}{35} \times \frac{-28}{-36} =$ ______ *b.* $\frac{96}{-35} \times \frac{-45}{8} =$ ______

---

a. $-\frac{32}{35} \times \frac{28}{36} = -\frac{32 \times 28}{35 \times 36} = -\frac{32}{45}$ [Dividing by 4 and 7]

b. $-\frac{96}{35} \times \left(-\frac{45}{8}\right) = +\frac{96 \times 45}{35 \times 8} = \frac{108}{7}$ $\left[\frac{\overset{12}{\cancel{96}} \times \overset{9}{\cancel{45}}}{\underset{7}{\cancel{35}} \times \underset{1}{\cancel{8}}} \quad \text{Dividing by 8 and 5}\right]$

---

30 Mixed numerals should be converted to fractions before multiplication.

a. $2\frac{1}{4} \times (-1\frac{1}{2}) = \frac{9}{4} \times \left(-\frac{3}{2}\right) =$ ______

b. $-2\frac{1}{4} \times (-3\frac{1}{4}) =$ ______ $\times$ ______ $=$ ______

---

a. $-\frac{27}{8}$ *b.* $-\frac{9}{4} \times \left(-\frac{13}{4}\right) = \frac{117}{16}$

---

31 Again, canceling before multiplication can save time.

a. $-3\frac{1}{2} \times 2\frac{3}{7} =$ ______ $\times$ ______ $=$ ✔ $-\frac{? \times ?}{? \times ?} =$ ______

b. $-2\frac{2}{5} \times (-3\frac{1}{6}) =$ ______ $\times$ ______ $=$ ______

c. $2\frac{1}{3} \times (-3\frac{3}{14}) =$ ______ $\times$ ______ $=$ ______

---

a. $-\frac{7}{2} \times \frac{17}{7} = -\frac{7 \times 17}{2 \times 7} = -\frac{17}{2}$

b. $-\frac{12}{5} \times \left(-\frac{19}{6}\right) = \frac{12 \times 19}{5 \times 6} = \frac{38}{5}$ $\left[\frac{\overset{2}{\cancel{12}} \times 19}{5 \times \underset{1}{\cancel{6}}} \quad \text{Dividing by 6}\right]$

c. $\frac{7}{3} \times \left(-\frac{45}{14}\right) = -\frac{7 \times 45}{3 \times 14} = -\frac{15}{2}$ $\left[\frac{\overset{1}{\cancel{7}} \times \overset{15}{\cancel{45}}}{\underset{1}{\cancel{3}} \times \underset{2}{\cancel{14}}} \quad \text{Dividing by 7 and 3}\right]$

---

32 Simplify the following:

a. $-1\frac{8}{27} \times 12\frac{3}{5} =$ ______ *b.* $-6\frac{6}{7} \times (-7\frac{7}{8}) =$ ______

a. $-\frac{35}{27}\times\frac{63}{5}=-\frac{35\times 63}{27\times 5}=-\frac{49}{3}$ $\left[\frac{\overset{7}{\cancel{35}}\times\overset{7}{\cancel{63}}}{\underset{3}{\cancel{27}}\times\underset{1}{\cancel{5}}}\right]$ Dividing by 5 and 9

b. $-\frac{48}{7}\times\left(-\frac{63}{8}\right)=+\frac{48\times 63}{7\times 8}=54$ $\left[\frac{\overset{6}{\cancel{48}}\times\overset{9}{\cancel{63}}}{\underset{1}{\cancel{7}}\times\underset{1}{\cancel{8}}}\right]$ Dividing by 7 and 8

You have finished the program for Sec. 6-1. Now take the posttest to see how much you have learned.

POSTTEST 6-1

1. $\frac{-7}{9}\times\frac{1}{8}=$ ______
2. $\frac{3}{-8}\times\left(-\frac{5}{6}\right)=$ ______
3. $\frac{-4}{7}\times\frac{3}{7}=$ ______
4. $\frac{-24}{-35}\times\frac{-56}{9}=$ ______
5. $-3\frac{4}{12}\times(-2\frac{3}{10})=$ ______
6. $7\frac{5}{7}\times(-5\frac{2}{9})=$ ______

When you have finished the posttest, correct it yourself using the answers on page 296. Then check all the questions you got wrong on the list below.

1. ______ 4. ______

2. ______ 5. ______

3. ______ 6. ______

If you compare your performance with that on Pretest 6-1, you will see how much you have improved. If you made more than one mistake on this posttest, review the material and take the pretest again. Don't move on to the next section until you can get all the questions right.

Now proceed with Sec. 6-2.

Section 6-2
Division of Rational Numbers

Do this section if you made a mistake in Questions 1 to 3 of Diagnostic Test 6. Begin by taking the pretest below, which will show you how much of Sec. 6-2 you already know and what you will be able to do by the time you have finished it.

PRETEST 6-2

1. What is the reciprocal of $\frac{6}{-7}$?

2. What is the reciprocal of $-\frac{7}{3}$?

3. $\frac{-3}{-4} \div \frac{5}{-6} =$ ______

4. $\frac{7}{-10} \div \frac{-14}{17} =$ ______

5. $3\frac{3}{5} \div (-1\frac{2}{7}) =$ ______

6. $-2\frac{2}{9} \div (-3\frac{5}{18}) =$ ______

When you have finished the pretest, correct it yourself using the answers on page 297. Then check all the questions you got wrong on the list below.

1.	______	3.	______	5.	______
2.	______	4.	______	6.	______

If you made any mistakes (even if it was only *one* mistake), read the introduction and go through the program.

If you made no mistakes at all, skip Sec. 6-2, because you already know its material, and go on to the next section whose number you circled on page 244. If no further numbers were circled, take Final Test 6 (page 295).

Introduction

After the concept of a reciprocal is introduced, you are shown that division by a rational number is equivalent to multiplication by its reciprocal. You then practice dividing rational numbers with both fraction and mixed-numeral forms.

FRAMES FOR SECTION 6-2

1 We can define the negative of a by the equation $a + (-a) =$ ______.

0

---

2 Another useful definition is the *reciprocal* of *a*, defined by the equation $a \times (\text{reciprocal of } a) = 1$.

a. $2 \times \frac{1}{2} =$ ______.

b. What is the reciprocal of 2?

c. $\frac{1}{2} \times 2 =$ ______.

d. What is the reciprocal of $\frac{1}{2}$?

---

a. 1 *b.* $\frac{1}{2}$ *c.* 1 *d.* 2

---

3 *a.* $3 \times$ ______ $= 1$.

b. What is the reciprocal of 3?

c. $\frac{1}{4} \times$ ______ $= 1$.

d. What is the reciprocal of $\frac{1}{4}$?

---

a. $\frac{1}{3}$ *b.* $\frac{1}{3}$ *c.* 4 *d.* 4

---

4 *a.* $\frac{1}{7} \times$ ______ $= 1$.

b. What is the reciprocal of $\frac{1}{7}$?

c. $6 \times$ ______ $= 1$.

d. What is the reciprocal of $\frac{1}{6}$?

---

a. 7 *b.* 7 *c.* $\frac{1}{6}$ *d.* 6

---

5 $$\frac{3}{2} \times \frac{2}{3} = \frac{\overset{1}{\cancel{3}} \times \overset{1}{\cancel{2}}}{\underset{1}{\cancel{2}} \times \underset{1}{\cancel{3}}} = \frac{1}{1} = 1$$

and

$$\frac{5}{6} \times \frac{6}{5} = \frac{\overset{1}{\cancel{5}} \times \overset{1}{\cancel{6}}}{\underset{1}{\cancel{6}} \times \underset{1}{\cancel{5}}} = \frac{1}{1} = 1$$

So $\frac{2}{3}$ is the reciprocal of $\frac{3}{2}$, and ______ is the reciprocal of $\frac{5}{6}$. Since $\frac{3}{4} \times$ ______ $= 1$, ______ is the reciprocal of $\frac{3}{4}$.

$\frac{6}{5}$ $\frac{4}{3}$ $\frac{4}{3}$

---

6 *a.* $\frac{4}{7} \times$ ______ $= 1.$

b. What is the reciprocal of $\frac{4}{7}$?

c. $\frac{a}{b} \times$ ______ $= 1.$

d. What is the reciprocal of $\frac{a}{b}$?

---

a. $\frac{7}{4}$ *b.* $\frac{7}{4}$ *c.* $\frac{b}{a}$ *d.* $\frac{b}{a}$

---

7 $6 \div 3 = \frac{6}{3} = 6 \times \frac{1}{3}$

$a \div b = \frac{?}{?} =$ __ $\times \frac{1}{?}$

$a \div b = a \times$ ________ of b

$\frac{a}{b} = a \times \frac{1}{b}$ reciprocal

---

8 $z \div y = z \times$ ______ $= z \times$ ________ of y

$2 \div 4 = 2 \times$ ______

$2 \div \frac{1}{4} = 2 \times$ ______

$\frac{1}{y}$; reciprocal $\frac{1}{4}$ 4

---

9 $2 \div 7 = 2 \times$ ______ $2 \div \frac{1}{7} = 2 \times$ ______

$2 \div \frac{3}{4} = 2 \times$ ______ $2 \div \frac{4}{3} = 2 \times$ ______

$\frac{1}{7}$ 7 $\frac{4}{3}$ $\frac{3}{4}$

---

10 *a.* $2 \div \frac{5}{6} = 2 \times$ ______ $=$ ______ *b.* $7 \div \frac{3}{8} = 7 \times$ ______ $=$ ______

c. $4 \div \frac{3}{11} =$ ______ $\times$ ______ $=$ ______

a. $\frac{6}{5}; \frac{12}{5}$ *b.* $\frac{8}{3}; \frac{56}{3}$ *c.* $4 \times \frac{11}{3} = \frac{44}{3}$

11 *a.* $\frac{4}{9} \div \frac{3}{8} = \frac{4}{9} \times$ _____ $=$ _____

b. $\frac{3}{4} \div \frac{5}{6} = \frac{3}{4} \times$ _____ $=$ _____

c. $\frac{3}{8} \div \frac{4}{5} =$ _____ $\times$ _____ $=$ _____

a. $\frac{8}{3}; \frac{32}{27}$ *b.* $\frac{3}{\cancel{4}_2} \times \frac{\cancel{6}^3}{5} = \frac{9}{10}$ *c.* $\frac{3}{8} \times \frac{5}{4} = \frac{15}{32}$

12 *a.* $\frac{5}{11} \div \frac{15}{16} =$ _____ $\times$ _____ $=$ _____

b. $\frac{4}{9} \div \frac{17}{3} =$ _____

c. $\frac{3}{16} \div \frac{13}{8} =$ _____

a. $\frac{\cancel{5}^1}{11} \times \frac{16}{\cancel{15}_3} = \frac{16}{33}$ *b.* $\frac{4}{\cancel{9}_3} \times \frac{\cancel{3}^1}{17} = \frac{4}{51}$ *c.* $\frac{3}{\cancel{16}_2} \times \frac{\cancel{8}^1}{13} = \frac{3}{26}$

13 Which of the following statements are true?

a. $\frac{3}{7} \times \left(-\frac{7}{3}\right) = 1$

b. $-\frac{3}{7} \times \left(-\frac{7}{3}\right) = 1$

c. $-\frac{5}{8} \times \left(-\frac{8}{5}\right) = 1$

d. $-\frac{5}{8} \times \frac{8}{5} = 1$

b and *c*

14 Give answers of the form $\frac{a}{b}$ or $-\frac{a}{b}$. What is the reciprocal of:

a. $-\frac{3}{7}$ *b.* $\frac{-3}{7}$ *c.* $\frac{-3}{-7}$ *d.* $\frac{3}{-7}$

---

a. $-\frac{7}{3}$ *b.* $-\frac{7}{3}$ *c.* $\frac{7}{3}$ *d.* $-\frac{7}{3}$

---

15 *a.* Is the reciprocal of $\frac{-5}{8}$ equal to $\frac{8}{-5}$, that is, does $\frac{-5}{8} \times \frac{8}{-5} = 1$?

b. Is the reciprocal of $\frac{-7}{-9}$ equal to $\frac{-9}{-7}$?

c. Is the reciprocal of $\frac{2}{-3}$ equal to $\frac{-3}{2}$?

---

a. yes *b.* yes *c.* yes

---

16 Give the reciprocals of the following rational numbers:

a. $\frac{3}{-5}$ *b.* $\frac{-5}{-11}$ *c.* $-\frac{9}{8}$ *d.* $\frac{-6}{5}$

---

a. $\frac{-5}{3}$ or $-\frac{5}{3}$ *b.* $\frac{-11}{-5}$ or $\frac{11}{5}$ *c.* $-\frac{8}{9}$ *d.* $\frac{5}{-6}$ or $-\frac{5}{6}$

---

17 The following method is useful for divisions involving negative numbers:

a. $\frac{(-5)}{8} \div \frac{3}{-7} = \frac{-5}{8} \times \frac{-7}{3} = -\frac{5}{8} \times \left(-\frac{7}{3}\right) =$ ______

b. $\frac{3}{-8} \div \frac{-7}{-5} = \frac{3}{-8} \times \frac{-5}{-7} = -\frac{3}{8} \times \frac{5}{7} =$ ______

---

a. $\frac{35}{24}$ *b.* $-\frac{15}{56}$

---

18 *a.* $\frac{3}{-10} \div \frac{-5}{7} = -\frac{3}{10} \times \left(-\frac{7}{5}\right) =$ ______

b. $\frac{-5}{11} \div \frac{-3}{-8} = -\frac{5}{11} \times$ ______ $=$ ______

c. $\frac{8}{9} \div \frac{5}{-8} = \frac{8}{9} \times$ ______ $=$ ______

---

a. $\frac{21}{50}$ *b.* $\frac{8}{3}; -\frac{40}{33}$ *c.* $-\frac{8}{5}; -\frac{64}{45}$

---

19 *a.* $\frac{-4}{5} \div \frac{3}{-8} = \frac{-4}{5} \times$ _____ = _____

b. $\frac{7}{-8} \div \frac{-2}{-9} =$ _____ × _____ = _____

---

a. $-\frac{8}{3}; \frac{32}{15}$ *b.* $-\frac{7}{8} \times \frac{9}{2} = -\frac{63}{16}$

---

20 *a.* $\frac{3}{-7} \div \frac{-8}{3} =$ _____ × _____ = _____

b. $\frac{11}{-4} \div \frac{-10}{-9} =$ _____ × _____ = _____

---

a. $-\frac{3}{7} \times \left(-\frac{3}{8}\right) = \frac{9}{56}$ *b.* $-\frac{11}{4} \times \frac{9}{10} = -\frac{99}{40}$

---

21 Canceling out must be done at the multiplication stage (*not* at the division stage). Thus

$$\frac{6}{-7} \div \frac{-24}{-49} = -\frac{6}{7} \times \frac{49}{24} = -\frac{\overset{1}{\cancel{6}} \times \overset{7}{\cancel{49}}}{\underset{1}{\cancel{7}} \times \underset{4}{\cancel{24}}} = -\frac{7}{4}$$

dividing the top and bottom by 7 and 6.

$$\frac{8}{63} \div \frac{-4}{35} = \frac{8}{63} \times \left(-\frac{35}{4}\right) = -\frac{\overset{2}{\cancel{8}} \times \overset{5}{\cancel{35}}}{\underset{9}{\cancel{63}} \times \underset{1}{\cancel{4}}} = \_\_\_\_\_$$

dividing the top and bottom by _____ and by _____.

$-\frac{10}{9}$ 4 7

---

22 *a.* $\frac{22}{-9} \div \frac{-55}{36} = -\frac{22}{9} \times \left(-\frac{36}{55}\right) = \frac{\overset{2}{\cancel{22}} \times \overset{4}{\cancel{36}}}{\underset{1}{\cancel{9}} \times \underset{5}{\cancel{55}}} = \_\_\_\_\_$

dividing the top and bottom by _____ and by _____.

b. $\frac{52}{-9} \div \frac{-13}{-27} = -\frac{52}{9} \times \frac{27}{13} = -\frac{\overset{4}{\cancel{52}} \times \overset{3}{\cancel{27}}}{\underset{1}{\cancel{9}} \times \underset{1}{\cancel{13}}} = \_\_\_\_\_$

---

a. $\frac{8}{5}$; 9; 11 *b.* -12

---

23 *a.* $\frac{-20}{63} \div \frac{45}{28} = \_\_\_\_\_ \times \_\_\_\_\_ = \checkmark\_\ \frac{? \times ?}{? \times ?} = \_\_\_\_\_$

b. $\frac{-15}{64} \div \frac{55}{-32} = \_\_\_\_\_ \times \_\_\_\_\_ = \checkmark\_\ \frac{? \times ?}{? \times ?} = \_\_\_\_\_$

Note: The check mark indicates that a plus or minus sign should be inserted in the blank which follows.

---

a. $-\frac{20}{63} \times \frac{28}{45} = -\frac{\overset{4}{\cancel{20}} \times \overset{4}{\cancel{28}}}{\underset{9}{\cancel{63}} \times \underset{9}{\cancel{45}}} = -\frac{16}{81}$ *b.* $-\frac{15}{64} \times \left(-\frac{32}{55}\right) = \frac{\overset{3}{\cancel{15}} \times \overset{\overset{1}{\cancel{4}}}{\cancel{32}}}{\underset{\underset{2}{\cancel{8}}}{\cancel{64}} \times \underset{11}{\cancel{55}}} = \frac{3}{22}$

You could have canceled by a different route and still have been correct.

---

24 *a.* $\frac{5}{-64} \div \frac{-65}{16} = \_\_\_\_\_$

b. $\frac{-21}{32} \div \frac{-63}{28} = \_\_\_\_\_$

---

a. $-\frac{5}{64} \times \left(-\frac{16}{65}\right) = \frac{\overset{1}{\cancel{5}} \times \overset{1}{\cancel{16}}}{\underset{4}{\cancel{64}} \times \underset{13}{\cancel{65}}} = \frac{1}{52}$ *b.* $-\frac{21}{32} \times \left(-\frac{28}{63}\right) = \frac{\overset{\overset{1}{\cancel{3}}}{\cancel{21}} \times \overset{7}{\cancel{28}}}{\underset{8}{\cancel{32}} \times \underset{\underset{3}{\cancel{9}}}{\cancel{63}}} = \frac{7}{24}$

---

25 Mixed numerals should be converted to fractions at the *division* stage.

a. $1\frac{3}{4} \div (-2\frac{1}{5}) = \frac{?}{?} \div \left(-\frac{?}{?}\right) = \frac{?}{?} \times \left(-\frac{?}{?}\right) =$ ______

b. $-2\frac{5}{8} \div (-3\frac{3}{7}) = -\frac{?}{?} \div \left(-\frac{?}{?}\right) =$ ______

c. $-1\frac{5}{6} \div 2\frac{1}{5} =$ ______

---

a. $\frac{7}{4} \div \left(-\frac{11}{5}\right) = \frac{7}{4} \times \left(-\frac{5}{11}\right) = -\frac{35}{44}$

b. $-\frac{21}{8} \div \left(-\frac{24}{7}\right) = -\frac{21}{8} \times \left(-\frac{7}{24}\right) = \frac{\overset{7}{\cancel{21}} \times 7}{8 \times \underset{8}{\cancel{24}}} = \frac{49}{64}$

c. $-\frac{11}{6} \div \frac{11}{5} = -\frac{11}{16} \times \frac{5}{11} = -\frac{\cancel{11} \times 5}{6 \times \cancel{11}} = -\frac{5}{6}$

---

26 *a.* $-2\frac{4}{9} \div (-6\frac{2}{7}) =$ ______

b. $3\frac{3}{8} \div (-1\frac{17}{28}) =$ ______

---

a. $-\frac{22}{9} \div \left(-\frac{44}{7}\right) = -\frac{22}{9} \times \left(-\frac{7}{44}\right) = \frac{\overset{1}{\cancel{22}} \times 7}{9 \times \underset{2}{\cancel{44}}} = \frac{7}{18}$

b. $\frac{27}{8} \div \left(-\frac{45}{28}\right) = \frac{27}{8} \times \left(-\frac{28}{45}\right) = -\frac{\overset{3}{\cancel{27}} \times \overset{7}{\cancel{28}}}{\underset{2}{\cancel{8}} \times \underset{5}{\cancel{45}}} = -\frac{21}{10}$

You have finished the program for Sec. 6-2. Now take the posttest to see how much you have learned.

POSTTEST 6-2

What is the reciprocal of:

1. $\frac{4}{-5}$

2. $\left(-\frac{9}{7}\right)$

3. $\frac{-2}{-5} \div \frac{7}{-8} =$ ______

4. $\frac{9}{-11} \div \frac{7}{-8} =$ ______

5. $4\frac{2}{5} \div (-1\frac{4}{7}) =$ ______

6. $-3\frac{3}{4} \div (-2\frac{5}{16}) =$ ______

When you have finished the posttest, correct it yourself using the answers on page 297. Then check all the questions you got wrong on the list below.

1. ______ 3. ______ 5. ______

2. ______ 4. ______ 6. ______

If you compare your performance with that on Pretest 6-2, you will see how much you have improved. If you made more than one mistake on this posttest, review the material and take the pretest again. Don't move on to the next section until you can get all the questions right.

Now proceed to the next section whose number you circled on page 244. If no other sections were circled, take Final Test 6 (page 295).

Section 6-3 Addition and Subtraction of Fractions

Do this section if you made a mistake in Questions 4 to 7 of Diagnostic Test 6. Begin by taking the pretest below, which will show you how much of Sec. 6-3 you already know and what you will be able to do by the time you have finished it.

PRETEST 6-3

1. $-\frac{6}{7}-\left(\frac{-15}{7}\right)=$ ______
2. $\frac{7}{3}+\left(-\frac{5}{12}\right)=$ ______
3. $\frac{4}{5}-\frac{7}{8}=$ ______
4. $-\frac{5}{6}-\frac{4}{7}=$ ______
5. $-\frac{1}{9}+\left(-\frac{3}{4}\right)=$ ______
6. $\frac{-5}{11}-\frac{-1}{-3}=$ ______
7. $\frac{-3}{-10}-\frac{8}{-7}=$ ______
8. $\frac{4}{-9}+\frac{-7}{-8}=$ ______

When you have finished the pretest, correct it yourself using the answers on page 297. Then check all the questions you got wrong on the list below.

1. ______ 4. ______ 7. ______

2. ______ 5. ______ 8. ______

3. ______ 6. ______

If you made any mistakes (even if it was only *one* mistake), read the introduction and go through the program.

If you made no mistakes at all, skip Sec. 6-3, because you already know its material, and go on to Sec. 6-4 (page 273).

Introduction

Chapter 5 used the distributive law to derive a rule for the addition of rational numbers with a common denominator. This section extends the rule to the subtraction of rational numbers with a common denominator and then shows how fractions of different denominators can be added or subtracted by prior conversion to equivalent forms with the same denominator.

FRAMES FOR SECTION 6-3

1 In Chap. 5 we showed that the distributive law could be used to define the addition of rational numbers of common denominator. Thus

$$\frac{a}{c}+\frac{b}{c}=\frac{?}{c}$$

$a+b$

---

2 For example,

$$\frac{7}{5}+\frac{4}{5}= \_\_\_\_\_ \qquad \frac{-13}{4}+\frac{7}{4}= \_\_\_\_\_ \qquad \frac{-5}{6}+\frac{-8}{6}= \_\_\_\_\_$$

$\frac{11}{5}$ $\quad \frac{-6}{4}$ or $-\frac{3}{2}$ $\quad \frac{-13}{6}$ or $-\frac{13}{6}$

---

3 Rational numbers of the form $-\frac{a}{b}$ can then be added through conversion to the equivalent form $\frac{-a}{b}$.

$$\frac{2}{3}+\left(-\frac{1}{3}\right)=\frac{2}{3}+\frac{?}{3}=\frac{?+?}{3}= \_\_\_\_\_$$

and

$$-\frac{4}{5}+\left(-\frac{3}{5}\right)=\frac{?}{5}+\frac{?}{5}=\frac{?+?}{5}= \_\_\_\_\_$$

-1 $\quad 2+(-1)$ $\quad \frac{1}{3}$ $\quad -4$ $\quad -3$ $\quad -4+(-3)$ $\quad \frac{-7}{5}$ or $-\frac{7}{5}$

---

4 Similarly

$$-\frac{3}{7}+\frac{5}{7}=\frac{?+?}{7}=\_\_\_\_\_\quad\text{and}\quad -\frac{5}{9}+\left(-\frac{2}{9}\right)=\frac{?+?}{9}=\_\_\_\_\_$$

$-3+5$ $\quad\frac{2}{7}\quad$ $-5+(-2)$ $\quad\frac{-7}{9}$ or $-\frac{7}{9}$

---

5 Subtraction of rational numbers is defined as the addition of the negative. So

$$\frac{7}{11}-\frac{4}{11}=\frac{7}{11}+\left(-\frac{4}{11}\right)=\frac{?+?}{11}=\_\_\_\_\_$$

and

$$\frac{9}{11}-\left(-\frac{3}{11}\right)=\frac{9}{11}+\frac{3}{11}=\frac{?+?}{11}=\_\_\_\_\_$$

$7+(-4)$ $\quad\frac{3}{11}\quad$ $9+3$ $\quad\frac{12}{11}$

---

6 Thus

$$\frac{11}{5}-\frac{7}{5}=\frac{11+(-7)}{5}=\_\_\_\_\_$$

and

$$\frac{8}{9}-\left(-\frac{3}{9}\right)=\frac{8+?}{9}=\_\_\_\_\_$$

$\frac{-4}{5}$ or $-\frac{4}{5}$ $\quad 3\quad$ $\frac{11}{9}$

---

7 $-\frac{5}{8}-\frac{2}{8}=\frac{-5+?}{?}=\_\_\_\_\_$

$$-\frac{6}{7}-\frac{3}{7}=\_\_\_\_\_$$

$\frac{-5+(-2)}{8}=\frac{-7}{8}$ or $-\frac{7}{8}$ $\quad\frac{-6+(-3)}{7}=\frac{-9}{7}$ or $-\frac{9}{7}$

---

8 Give negative answers in the simplest possible form, i.e., as $-\frac{a}{b}$ with a and b the smallest possible integers. Thus

$$-\frac{3}{4}-\frac{12}{4}=\frac{?+?}{4}=-\frac{?}{?}$$

and

$$-\frac{27}{6}-\left(-\frac{14}{6}\right)=\frac{?+?}{6}=\_\_\_\_\_$$

$-3+(-12)$ $\quad -\frac{15}{4}$ $\quad -27+14$ $\quad -\frac{13}{6}$

---

9 *a.* $\frac{8}{9}-\frac{16}{9}=\_\_\_\_\_$

b. $-\frac{15}{8}-\frac{10}{8}=\_\_\_\_\_$

c. $-\frac{9}{7}-\left(-\frac{13}{7}\right)=\_\_\_\_\_$

---

a. $-\frac{8}{9}$ *b.* $-\frac{25}{8}$ *c.* $\frac{4}{7}$

---

10 So far we have defined the addition and subtraction of rational numbers only when they are represented by fractions with a common denominator. When rational numbers are represented by fractions with different denominators, our definition no longer applies. But we can replace these fractions by other equivalent fractions which do have the same denominator. Thus

$$\frac{1}{2}+\frac{1}{4}=\frac{2}{4}+\frac{1}{4}=\frac{?}{4}$$

since $\frac{1}{2}$ and $\frac{?}{4}$ are equivalent; and

$$\frac{2}{3}-\frac{1}{6}=\frac{4}{6}-\frac{1}{6}=\frac{4}{6}+\left(-\frac{1}{6}\right)=\_\_\_\_\_$$

since $\frac{2}{3}$ and $\frac{?}{6}$ are equivalent.

---

3 $\quad$ 2 $\quad$ $\frac{3}{6}\left(\text{or } \frac{1}{2}\right)$ $\quad$ 4

---

11 *a.* $\frac{1}{5}+\frac{1}{10}=\frac{2}{10}+\frac{1}{10}=$ ______

since $\frac{1}{5}$ and $\frac{?}{10}$ are equivalent fractions.

b. $\frac{4}{15}-\frac{1}{5}=\frac{4}{15}+\left(-\frac{3}{15}\right)=$ ______

since $\frac{1}{5}$ and $\frac{?}{15}$ are equivalent fractions.

c. $\frac{1}{20}+\left(-\frac{2}{5}\right)=\frac{1}{20}+\left(-\frac{8}{20}\right)=$ ______

since $-\frac{2}{5}$ and $-\frac{?}{20}$ are equivalent fractions.

---

a. $\frac{3}{10}$; 2 *b.* $\frac{1}{15}$; 3 *c.* $-\frac{7}{20}$; 8

---

12 Notice that we use the rule $\frac{a}{b}=\frac{a}{b}\times\frac{c}{c}=\frac{a\times c}{b\times c}$.

a. To get a fraction $\frac{k}{20}$ equivalent to $\frac{2}{5}$

$\frac{2}{5}=\frac{2}{5}\times\frac{4}{4}=\frac{?}{20}$

b. To get a fraction $\frac{k}{24}$ equivalent to $\frac{5}{8}$

$\frac{5}{8}=\frac{5}{8}\times\frac{?}{?}=\frac{?}{24}$

c. To get a fraction $\frac{k}{42}$ equivalent to $\frac{3}{7}$

$\frac{3}{7}=\frac{3}{7}\times\frac{?}{?}=\frac{?}{42}$

---

a. 8 *b.* $\frac{3}{3}$; 15 *c.* $\frac{6}{6}$; 18

---

13 *a.* $\frac{3}{7}+\frac{5}{42}=\frac{?}{42}+\frac{5}{42}=$ _____

since $\frac{3}{7}$ and $\frac{?}{42}$ are equivalent fractions.

b. $\frac{5}{8}-\frac{7}{24}=\frac{?}{24}+\left(-\frac{7}{24}\right)=$ _____

since $\frac{5}{8}$ and $\frac{?}{24}$ are equivalent fractions.

c. $-\frac{3}{5}-\frac{17}{25}=-\frac{?}{25}+\left(-\frac{17}{25}\right)=$ _____

since $-\frac{3}{5}$ and $-\frac{?}{25}$ are equivalent fractions.

---

a. 18; $\frac{23}{42}$; 18 *b.* 15; $\frac{8}{24}$ $\left(\text{or } \frac{1}{3}\right)$; 15 *c.* 15; $-\frac{32}{25}$; 15

---

14 *a.* $\frac{5}{6}+\frac{5}{36}=\frac{?}{36}+\frac{5}{36}=$ _____

since $\frac{5}{6}$ and $\frac{?}{36}$ are equivalent fractions.

b. $\frac{7}{20}+\left(-\frac{3}{4}\right)=\frac{7}{20}+\left(-\frac{?}{20}\right)=$ _____

c. $\frac{37}{48}-\frac{5}{8}=\frac{37}{48}+\left(-\frac{?}{48}\right)=$ _____

---

a. 30; $\frac{35}{36}$; 30 *b.* 15; $-\frac{8}{20}$ or $-\frac{2}{5}$ *c.* 30; $\frac{7}{48}$

---

15 *a.* $\frac{1}{6}-\left(-\frac{11}{24}\right)=\frac{?}{24}+\frac{11}{24}=$ _____

b. $\frac{4}{11}-\left(-\frac{13}{55}\right)=$ _____

c. $-\frac{3}{7}+\frac{10}{84}=$ _____

---

a. 4; $\frac{15}{24}$ or $\frac{5}{8}$ *b.* $\frac{20}{55}+\frac{13}{55}=\frac{33}{55}$ or $\frac{3}{5}$ *c.* $-\frac{36}{84}+\frac{10}{84}=-\frac{26}{84}$ or $-\frac{13}{42}$

---

16 *a.* $-\frac{17}{72}+\left(-\frac{5}{12}\right)=$ _____ *b.* $-\frac{4}{9}-\left(-\frac{33}{45}\right)=$ _____

a. $-\frac{17}{72}+\left(-\frac{30}{72}\right)=-\frac{47}{72}$ *b.* $-\frac{20}{45}+\frac{33}{45}=\frac{13}{45}$

17 In a case like $\frac{2}{5}+\frac{3}{4}$, both fractions have to be replaced by equivalent fractions. This can be done by multiplying the top and the bottom of each fraction by the denominator of the other fraction.

a. $\frac{2}{5}+\frac{3}{4}=\frac{2}{5}\cdot\frac{4}{4}+\frac{3}{4}\cdot\frac{5}{5}=\frac{?}{20}+\frac{?}{20}=$ _____

b. $\frac{2}{3}-\frac{3}{7}=\frac{2}{3}\cdot\frac{7}{7}+\left(-\frac{3}{7}\cdot\frac{3}{3}\right)=\frac{?}{?}+\left(-\frac{?}{?}\right)=\frac{?+?}{21}=$ _____

a. $\frac{8}{20}+\frac{15}{20}=\frac{23}{20}$ *b.* $\frac{14}{21}+\left(-\frac{9}{21}\right)=\frac{14+(-9)}{21}=\frac{5}{21}$

18 *a.* $\frac{3}{5}+\frac{3}{8}=\left(\frac{3}{5}\cdot\frac{8}{8}\right)+\left(\frac{3}{8}\cdot\frac{5}{5}\right)=\frac{?}{?}+\frac{?}{?}=\frac{?+?}{?}=$ _____

b. $-\frac{6}{7}-\frac{1}{6}=\left(-\frac{6}{7}\cdot\frac{6}{6}\right)+\left(-\frac{1}{6}\cdot\frac{7}{7}\right)=\left(-\frac{?}{?}\right)+\left(-\frac{?}{?}\right)=\frac{?+?}{?}=$ _____

c. $\frac{4}{9}-\frac{2}{5}=\frac{4}{9}\cdot\frac{?}{?}+\left(-\frac{2}{5}\cdot\frac{?}{?}\right)=\frac{?+?}{45}=$ _____

a. $\frac{24}{40}+\frac{15}{40}=\frac{24+15}{40}=\frac{39}{40}$ *b.* $-\frac{36}{42}+\left(-\frac{7}{42}\right)=\frac{-36+(-7)}{42}=-\frac{43}{42}$

c. $\frac{4}{9}\cdot\frac{5}{5}-\frac{2}{5}\cdot\frac{9}{9}=\frac{20+(-18)}{45}=\frac{2}{45}$

19 *a.* $\frac{3}{7}+\frac{5}{8}=\frac{3}{7}\cdot\frac{?}{?}+\frac{5}{8}\cdot\frac{?}{?}=\frac{?+?}{56}=$ _____

b. $-\frac{5}{8}-\left(-\frac{3}{11}\right)=\left(-\frac{5}{8}\cdot\frac{?}{?}\right)+\frac{3}{11}\cdot\frac{?}{?}=\frac{?+?}{?}=$ _____

c. $\frac{2}{3}-\frac{7}{8}=\frac{2}{3}\cdot\frac{?}{?}+\left(-\frac{7}{8}\cdot\frac{?}{?}\right)=\frac{?+?}{?}=$ _____

a. $\frac{3}{7}\cdot\frac{8}{8}+\frac{5}{8}\cdot\frac{7}{7}=\frac{24+35}{56}=\frac{59}{56}$

b. $\left(-\frac{5}{8}\cdot\frac{11}{11}\right)+\frac{3}{11}\cdot\frac{8}{8}=\frac{-55+24}{88}=-\frac{31}{88}$

c. $\frac{2}{3}\cdot\frac{8}{8}+\left(-\frac{7}{8}\cdot\frac{3}{3}\right)=\frac{16+(-21)}{24}=-\frac{5}{24}$

20 *a.* $-\frac{4}{7}-\left(-\frac{5}{9}\right)=\left(-\frac{4}{7}\cdot\frac{?}{?}\right)+\frac{5}{9}\cdot\frac{?}{?}=\frac{?+?}{?}=$ ______

b. $\frac{1}{2}-\frac{7}{9}=\frac{?+?}{?}=$ ______ *c.* $\frac{3}{8}-\left(-\frac{4}{9}\right)=\frac{?+?}{?}=$ ______

---

a. $\left(-\frac{4}{7}\cdot\frac{9}{9}\right)+\frac{5}{9}\cdot\frac{7}{7}=\frac{-36+35}{63}=-\frac{1}{63}$ *b.* $\frac{9+(-14)}{18}=-\frac{5}{18}$

c. $\frac{27+32}{72}=\frac{59}{72}$

---

21 *a.* $\frac{5}{6}-\frac{6}{7}=\frac{?+?}{?}=$ ______ *b.* $\frac{5}{11}+\frac{4}{9}=$ ______

c. $\frac{4}{5}-\left(-\frac{1}{4}\right)=$ ______

---

a. $\frac{35+(-36)}{42}=-\frac{1}{42}$ *b.* $\frac{45+44}{99}=\frac{89}{99}$ *c.* $\frac{16+(-5)}{20}=\frac{11}{20}$

---

22 *a.* $-\frac{6}{7}-\frac{3}{4}=$ ______ *b.* $-\frac{3}{11}+\frac{4}{5}=$ ______

c. $\frac{5}{9}-\left(-\frac{7}{8}\right)=$ ______

---

a. $\frac{-24+(-21)}{28}=-\frac{45}{28}$ *b.* $\frac{-15+44}{55}=\frac{29}{55}$ *c.* $\frac{40+63}{72}=\frac{103}{72}$

---

23 *a.* $\frac{3}{11}+\frac{5}{6}=$ ______

b. $-\frac{4}{7}-\left(-\frac{7}{8}\right)=$ ______

c. $-\frac{9}{5}-\frac{5}{9}=$ ______

---

a. $\frac{18+55}{66}=\frac{73}{66}$ *b.* $\frac{-32+49}{56}=\frac{17}{56}$ *c.* $\frac{-81+(-25)}{45}=-\frac{106}{45}$

---

24 *a.* $\frac{2}{5}-\frac{9}{11}=$ ______ *b.* $-\frac{3}{8}-\left(-\frac{2}{7}\right)=$ ______

c. $\frac{3}{7}+\left(-\frac{1}{2}\right)=$ ______

a. $\frac{22+(-45)}{55} = -\frac{23}{55}$ *b.* $\frac{-21+16}{56} = -\frac{5}{56}$ *c.* $\frac{6+(-7)}{14} = -\frac{1}{14}$

---

25 Convert the following fractions to the form $\frac{a}{b}$ or $-\frac{a}{b}$, where a and b are positive integers:

$\frac{4}{-7}$ $\frac{-2}{-3}$ $\frac{3}{-5}$ $\frac{-7}{-6}$ $\frac{-8}{3}$

$-\frac{4}{7}$ $\frac{2}{3}$ $-\frac{3}{5}$ $\frac{7}{6}$ $-\frac{8}{3}$

---

26 Fractions containing negative integers can be converted to the form $\frac{a}{b}$ or $-\frac{a}{b}$ before adding or subtracting. Thus

$\frac{3}{-8} + \frac{5}{-3} = -\frac{3}{8} + \left(-\frac{5}{3}\right) =$ ______ and

$\frac{-5}{4} + \frac{-2}{5} =$ ______ $+$ ______ $=$ ______

$\frac{-9+(-40)}{24} = -\frac{49}{24}$ $-\frac{5}{4} + \left(-\frac{2}{5}\right) = \frac{-25+(-8)}{20} = -\frac{33}{20}$

---

27 Similarly

$\frac{-7}{-3} - \frac{-3}{2} = \frac{7}{3} - \frac{3}{2} =$ ______

$\frac{3}{-5} - \frac{4}{-7} =$ ______ $-$ ______ $=$ ______

$\frac{14+(-9)}{6} = \frac{5}{6}$ $-\frac{3}{5} - \left(-\frac{4}{7}\right) = \frac{-21+20}{35} = -\frac{1}{35}$

---

28 *a.* $\frac{-4}{7} - \frac{-5}{11} =$ ______ *b.* $\frac{-2}{3} + \frac{3}{-5} =$ ______

---

a. $-\frac{4}{7} - \left(-\frac{5}{11}\right) = \frac{-44+35}{77} = -\frac{9}{77}$

b. $-\frac{2}{3} + \left(-\frac{3}{5}\right) = \frac{-10+(-9)}{15} = -\frac{19}{15}$

---

29 *a.* $\frac{-5}{11} - \frac{-3}{-8} =$ ______

b. $\frac{-4}{-3} + \frac{3}{-5} =$ ______

---

a. $-\frac{5}{11} - \frac{3}{8} = \frac{-40 + (-33)}{88} = -\frac{73}{88}$

b. $\frac{4}{3} + \left(-\frac{3}{5}\right) = \frac{20 + (-9)}{15} = \frac{11}{15}$

---

30 *a.* $\frac{4}{-7} + \frac{-2}{-3} =$ ______

b. $\frac{-5}{9} - \frac{9}{-2} =$ ______

---

a. $-\frac{4}{7} + \frac{2}{3} = \frac{-12 + 14}{21} = \frac{2}{21}$ *b.* $-\frac{5}{9} - \left(-\frac{9}{2}\right) = \frac{-10 + 81}{18} = \frac{71}{18}$

You have finished the program for Sec. 6-3. Now take the posttest to see how much you have learned.

POSTTEST 6-3

1. $-\frac{4}{9} - \left(-\frac{14}{9}\right) =$ ______
2. $\frac{5}{4} + \left(-\frac{3}{16}\right) =$ ______
3. $\frac{5}{8} - \frac{6}{7} =$ ______
4. $-\frac{7}{10} - \frac{2}{3} =$ ______
5. $-\frac{2}{5} + \left(-\frac{5}{6}\right) =$ ______
6. $\frac{-7}{10} - \frac{-1}{-4} =$ ______
7. $\frac{-4}{-11} - \frac{8}{-5} =$ ______
8. $\frac{3}{-8} + \frac{-5}{-9} =$ ______

When you have finished the posttest, correct it yourself using the answers on page 297. Then check all the questions you got wrong on the list below:

1. ______	4. ______	7. ______
2. ______	5. ______	8. ______
3. ______	6. ______	

If you compare your performance with that on Pretest 6-3, you will see how much you have improved. If you made more than one mistake on this posttest, review the material and take the pretest again. Don't move on to the next section until you can get all the questions right.

Now proceed with Sec. 6-4.

Section 6-4 Addition and Subtraction of Mixed Numerals

Do this section if you made a mistake in Questions 4 to 7 of Diagnostic Test 6. Begin by taking the pretest below, which will show you how much of Sec. 6-4 you already know and what you will be able to do by the time you have finished it.

PRETEST 6-4

1. $5 - \dfrac{3}{7} =$ ______
2. $-4 - \dfrac{5}{6} =$ ______
3. $3 + \dfrac{2}{13} =$ ______
4. $3\frac{5}{9} + 6\frac{2}{5} =$ ______
5. $3\frac{1}{10} - 1\frac{7}{8} =$ ______
6. $2\frac{3}{4} - 3\frac{4}{7} =$ ______

When you have finished the pretest, correct it yourself using the answers on page 297. Then check all the questions you got wrong on the list below:

1. ______	3. ______	5. ______
2. ______	4. ______	6. ______

If you made any mistakes (even if it was only *one* mistake), read the introduction and go through the program.

If you made no mistakes at all, skip Sec. 6-4, because you already know its material, and go on to the next section whose number you circled on page 244. If no further numbers were circled, take Final Test 6 (page 295).

Introduction

After introducing methods of adding and subtracting integers and fractions, this section shows you how mixed numerals can be added and subtracted by separating the integer and fraction components.

FRAMES FOR SECTION 6-4

1 $1 - \frac{4}{7} = \frac{?}{7} - \frac{4}{7} = \frac{? + (-4)}{7} =$ _____ $\quad 1 - \frac{5}{6} = \frac{?}{6} - \frac{5}{6} =$ _____

---

7 $\quad$ 7 $\quad \frac{3}{7} \quad$ 6 $\quad \frac{1}{6}$

---

2 $1 - \frac{5}{8} =$ _____ $\quad 1 - \frac{4}{9} =$ _____

$\frac{3}{8} \quad \frac{5}{9}$

---

3 $2 - \frac{5}{8} = 1 + 1 - \frac{5}{8} = 1 + \frac{?}{8} =$ _____ (mixed-numeral form)

$3 - \frac{4}{9} = 2 + 1 - \frac{4}{9} = 2 + \frac{?}{9} =$ _____ (mixed-numeral form)

---

3 $\quad 1\frac{3}{8} \quad$ 5 $\quad 2\frac{5}{9}$

---

4 Give the answers to the following in mixed-numeral form:

a. $5 - \frac{6}{7} =$ _____ $\quad$ *b.* $4 - \frac{3}{11} =$ _____ $\quad$ *c.* $3 - \frac{2}{9} =$ _____

---

a. $4\frac{1}{7}$ $\quad$ *b.* $3\frac{8}{11}$ $\quad$ *c.* $2\frac{7}{9}$

---

5 Remembering the definitions of positive and negative mixed numerals, we can write

$5\frac{1}{9} = 5 +$ _____ $\quad$ and $\quad -5\frac{1}{9} = -5 +$ _____

$\frac{1}{9} \quad \left(-\frac{1}{9}\right)$

---

6 Thus $-7 + \left(-\frac{5}{6}\right) =$ _____

$4 + \frac{7}{12} =$ _____

$-3 - \frac{2}{9} =$ _____

$-7\frac{5}{6} \quad 4\frac{7}{12} \quad -3\frac{2}{9}$

---

7 Give the answers to the following in mixed-numeral forms:

a. $3+\frac{5}{8}$ *b.* $3-\frac{5}{8}$ *c.* $(-3)-\frac{5}{8}$

---

a. $3\frac{5}{8}$ *b.* $2\frac{3}{8}$ *c.* $-3\frac{5}{8}$

---

8 We still have not considered situations like $-3+\frac{5}{8}$. If we multiply by -1 and assume that the distributive law of multiplication over addition is true for all rational numbers, we get

$$\left(-3+\frac{5}{8}\right)\times(-1)=[-3\times(-1)]+\left[\frac{5}{8}\times(-1)\right]=\_\_\_\_\_+\_\_\_\_\_=\_\_\_\_\_$$

Since multiplying by -1 is equivalent to "taking the negative of," we can deduce that $-3+\frac{5}{8}=$ _____.

$3+\left(-\frac{5}{8}\right)=2\frac{3}{8}$ $-2\frac{3}{8}$ (because its negative is $2\frac{3}{8}$)

---

9 We can write this as

$$-3+\frac{5}{8}=-\left[3+\left(-\frac{5}{8}\right)\right]=\_\_\_\_\_$$

Similarly

$$-4+\frac{6}{11}=-\left[4+\left(-\frac{6}{11}\right)\right]=\_\_\_\_\_$$

and

$$-5+\frac{3}{10}=-\left[5+\left(-\frac{3}{10}\right)\right]=\_\_\_\_\_$$

$-2\frac{3}{8}$ $-3\frac{5}{11}$ $-4\frac{7}{10}$

---

10 *a.* $-7+\frac{4}{9}=-\left[7+\left(-\frac{4}{9}\right)\right]=$ _____

b. $-8+\frac{2}{3}=$ _____

c. $-5+\frac{1}{4}=$ _____

---

a. $-6\frac{5}{9}$ *b.* $-7\frac{1}{3}$ *c.* $-4\frac{3}{4}$

---

11 Give the answers to the following in mixed-numeral form:

a. $2 + \frac{7}{9}$ *b.* $3 - \frac{5}{6}$

c. $(-5) + \frac{3}{8}$ *d.* $(-6) - \frac{7}{10}$

---

a. $2\frac{7}{9}$ *b.* $2\frac{1}{6}$ *c.* $-4\frac{5}{8}$ *d.* $-6\frac{7}{10}$

---

12 Give the answers to the following in mixed-numeral form:

a. $-7 + \frac{5}{9}$ *b.* $4 + \frac{3}{7}$

c. $-2 - \frac{7}{8}$ *d.* $3 - \frac{3}{4}$

---

a. $-6\frac{4}{9}$ *b.* $4\frac{3}{7}$ *c.* $-2\frac{7}{8}$ *d.* $2\frac{1}{4}$

---

13 Addition and subtraction of mixed numerals is probably easiest if you separate integer and fraction components as shown below:

a. $2\frac{1}{5} + 3\frac{1}{4} = (2 + \frac{1}{5}) + (3 + \frac{1}{4}) = (2 + 3) + (\frac{1}{5} + \frac{1}{4}) = \_\_\_\_ + \frac{? + ?}{20} = 5\,\frac{?}{20}$

b. $5\frac{5}{8} + 2\frac{2}{7} = (5 + 2) + (\frac{5}{8} + \frac{2}{7}) = \_\_\_\_ + \frac{? + ?}{?} = \_\_\_\_$

c. $3\frac{2}{5} + 5\frac{1}{3} = \_\_\_\_ + \frac{? + ?}{?} = \_\_\_\_$

---

a. $5 + \frac{4 + 5}{20} = 5\frac{9}{20}$ *b.* $7 + \frac{35 + 16}{56} = 7\frac{51}{56}$ *c.* $8 + \frac{6 + 5}{15} = 8\frac{11}{15}$

---

14 *a.* $3\frac{3}{4} - 1\frac{1}{3} = (3 - 1) + (\frac{3}{4} - \frac{1}{3}) = \_\_\_\_ + \frac{? + ?}{12} = \_\_\_\_ + \frac{?}{12} = \_\_\_\_$

b. $3\frac{1}{3} - 1\frac{3}{4} = (3 - 1) + (\frac{1}{3} - \frac{3}{4}) = \_\_\_\_ + \frac{? + ?}{12} = \_\_\_\_ + \frac{?}{12} = \_\_\_\_$

---

a. $2 + \frac{9 + (-4)}{12} = 2 + \frac{5}{12} = 2\frac{5}{12}$ *b.* $2 + \frac{4 + (-9)}{12} = 2 + \left(-\frac{5}{12}\right) = 1\frac{7}{12}$

---

15 *a.* $2\frac{6}{7} - 4\frac{5}{8} = (2 - 4) + (\frac{6}{7} - \frac{5}{8}) = \_\_\_\_ + \frac{? + ?}{?} = \_\_\_\_ + \frac{?}{?} = \_\_\_\_$

b. $1\frac{1}{5} - 4\frac{2}{7} = \_\_\_\_ + \frac{? + ?}{?} = \_\_\_\_$

a. $-2+\dfrac{48+(-35)}{56}=-2+\dfrac{13}{56}=-1\tfrac{43}{56}$ *b.* $-3+\dfrac{7+(-10)}{35}=-3\tfrac{3}{35}$

16 *a.* $5\tfrac{2}{3}-1\tfrac{4}{5}=$ ______ $+\dfrac{?+?}{?}=$ ______

b. $2\tfrac{1}{4}-6\tfrac{1}{5}=$ ______

a. $4+\dfrac{10+(-12)}{15}=4+\left(-\dfrac{2}{15}\right)=3\tfrac{13}{15}$

b. $-4+\dfrac{5+(-4)}{20}=(-4)+\dfrac{1}{20}=-3\tfrac{19}{20}$

17 *a.* $3\tfrac{1}{8}-5\tfrac{2}{5}=$ ______

b. $5\tfrac{1}{8}-3\tfrac{2}{9}=$ ______

a. $-2+\dfrac{5+(-16)}{40}=-2+\left(-\dfrac{11}{40}\right)=-2\tfrac{11}{40}$

b. $2+\dfrac{9+(-16)}{72}=2+\left(-\dfrac{7}{72}\right)=1\tfrac{65}{72}$

18 When the fractional component is greater than 1, it is customary to transfer 1 to the integer component. Thus

$$3\,\frac{5}{6}+4\,\frac{3}{5}=7+\frac{25+18}{30}=7+\frac{43}{30}=8+\text{\_\_\_\_\_\_}=\text{\_\_\_\_\_\_}$$

$1\tfrac{3}{4}+5\tfrac{4}{11}=$ ______ and $3\tfrac{5}{8}+2\tfrac{5}{7}=$ ______

$\dfrac{13}{30}$ $8\tfrac{13}{30}$ $7\tfrac{5}{44}$ $\left[6+\dfrac{33+16}{44}=6+\dfrac{49}{44}\right]$ $6\tfrac{19}{56}$ $\left[5+\dfrac{35+40}{56}=5+\dfrac{75}{56}\right]$

19 *a.* $1\tfrac{4}{11}-4\tfrac{2}{7}=$ ______

b. $5\tfrac{5}{6}+2\tfrac{3}{11}=$ ______

c. $7\tfrac{1}{4}-3\tfrac{5}{9}=$ ______

a. $-3+\dfrac{28+(-22)}{77}=-3+\dfrac{6}{77}=-2\tfrac{71}{77}$ *b.* $7+\dfrac{55+18}{66}=7+\dfrac{73}{66}=8\tfrac{7}{66}$

c. $4+\dfrac{9+(-20)}{36}=4+\left(-\dfrac{11}{36}\right)=3\tfrac{25}{36}$

20 *a.* $3\frac{1}{8} - 2\frac{5}{9} =$ ______

b. $1\frac{2}{5} + 6\frac{3}{4} =$ ______

c. $2\frac{7}{10} - 3\frac{6}{7} =$ ______

---

a. $1 + \frac{9 + (-40)}{72} = 1 + \left(-\frac{31}{72}\right) = \frac{41}{72}$ *b.* $7 + \frac{8 + 15}{20} = 7 + \frac{23}{20} = 8\frac{3}{20}$

c. $-1 + \frac{49 + (-60)}{70} = -1 + \left(-\frac{11}{70}\right) = -1\frac{11}{70}$

You have finished the program for this section. Now take the posttest to see how much you have learned.

POSTTEST 6-4

1. $6 - \frac{3}{8} =$ ______
2. $-5 - \frac{6}{7} =$ ______
3. $-4 + \frac{5}{11} =$ ______
4. $4\frac{7}{8} + 5\frac{3}{4} =$ ______
5. $5\frac{2}{9} - 1\frac{5}{7} =$ ______
6. $1\frac{2}{3} - 3\frac{1}{6} =$ ______

When you have finished the posttest, correct it yourself using the answers on page 297. Then check all the questions you got wrong on the list below:

1. ______ 4. ______

2. ______ 5. ______

3. ______ 6. ______

If you compare your performance with that on Pretest 6-4, you will see how much you have improved. If you made more than one mistake on this posttest, review the material and then take the pretest again. Don't move on to the next section until you can get all the questions right.

Now proceed to the next section whose number you circled on page 244. If no further sections were circled, take Final Test 6 (page 295).

Section 6-5
The Lowest Common Denominator and Multiple Additions

Do this section if you made a mistake in Questions 8 to 10 of Diagnostic Test 6. Begin by taking the pretest below, which will show you how much of Sec. 6-5 you already know and what you will be able to do by the time you have finished it.

PRETEST 6-5

Give the lowest common denominators of the following expressions:

1. $\frac{17}{25}+\frac{19}{15}+\frac{11}{6}$
2. $\frac{3}{8}-\frac{5}{18}+\frac{7}{3}$

Simplify the following expressions:

3. $\frac{7}{10}-\frac{12}{25}-\frac{3}{4}$
4. $-\frac{7}{12}+\frac{4}{9}+\frac{7}{30}$
5. $-\frac{5}{6}+\left(-\frac{7}{8}\right)-\left(-\frac{9}{10}\right)$
6. $\frac{8}{15}-\left(-\frac{5}{21}\right)+\left(-\frac{16}{35}\right)$

When you have finished the pretest, correct it yourself using the answers on page 298. Then check all the questions you got wrong on the list below.

1. ______ 3. ______ 5. ______

2. ______ 4. ______ 6. ______

If you made any mistakes (even if it was only *one* mistake), read the introduction and go through the program.

If you made no mistakes at all, skip Sec. 6-5, because you already know its material, and go on to Sec. 6-6.

Introduction

After revising and developing the concept of factoring, this section shows how three rational numbers can be added together by converting them to forms with the same denominator. It introduces the concept of the lowest common denominator (LCD) and shows how factoring gives a quick method for finding the LCD. You learn how to use the LCD to evaluate expressions containing additions and subtractions of positive and negative fractions.

FRAMES FOR SECTION 6-5

1 Simplify the following fractions:

a. $\frac{88}{16}$ *b.* $\frac{96}{16}$ *c.* $\frac{144}{96}$

---

a. $\frac{11}{2}$ *b.* 6 *c.* $\frac{3}{2}$

---

2 Completely factor the following numbers:

a. 16 *b.* 32 *c.* 27 *d.* 96 *e.* 88 *f.* 144

---

a. $2 \times 2 \times 2 \times 2$ *b.* $2 \times 2 \times 2 \times 2 \times 2$ *c.* $3 \times 3 \times 3$
d. $2 \times 2 \times 2 \times 2 \times 2 \times 3$ *e.* $2 \times 2 \times 2 \times 11$ *f.* $2 \times 2 \times 2 \times 2 \times 3 \times 3$

---

3 A longer way of simplifying $\frac{88}{16}$ would have been the conversion of 88 and 16 to completely factored forms and cancelation of their common factors:

$$\frac{88}{16} = \frac{2 \times 2 \times 2 \times 11}{2 \times 2 \times 2 \times 2} = \frac{11}{2}$$

You will notice that since the completely factored form of 88 contains three 2s, $8 = 2 \times 2 \times 2$ (is/is not) ________ a factor of 88, but $16 = 2 \times 2 \times 2 \times 2$ (is/is not) ________ a factor of 88 and $32 = 2 \times 2 \times 2 \times 2 \times 2$ (is/is not) ________ a factor of 88.

---

is; is not; is not

---

4 *a.* The completely factored form of 96 contains ______ 2s.

b. Is 8 a factor of 96?

c. Is 16 a factor of 96?

d. Is 32 a factor of 96?

---

a. five *b.* yes *c.* yes *d.* yes

---

5 *a.* The completely factored form of 144 contains ______ 2s and ______ 3s.

b. Is 16 a factor of 144?

c. Is 32 a factor of 144?

d. Is 9 a factor of 144?

e. Is 27 a factor of 144?

---

a. four; two *b.* yes *c.* no *d.* yes *e.* no

---

6 *a.* Completely factor 24. *b.* Is 24 a factor of 88?

c. Is 24 a factor of 96? *d.* Is 24 a factor of 144?

---

a. $2 \times 2 \times 2 \times 3$ *b.* no *c.* yes *d.* yes

---

7 If n is an integer which is not divisible by 2 or 3, which of the following numbers have 24 as a factor?

a. $2 \times 2 \times 2 \times 2 \times n$ *b.* $2 \times 2 \times 2 \times 2 \times 3 \times 3 \times n$

c. $2 \times 2 \times 3 \times 3 \times n$ *d.* $2 \times 2 \times 2 \times 3 \times n$

---

b and *d*

---

8 If m is an integer which is not divisible by 2 or 3 or 7, which of the following numbers have 42 as a factor?

a. $2 \times 2 \times 7 \times 7 \times m$ *b.* $2 \times 2 \times 3 \times 3 \times 7 \times 7 \times m$

c. $2 \times 2 \times 2 \times 3 \times 3 \times m$ *d.* $2 \times 3 \times 7 \times m$

---

b and *d*

---

9 We add rational numbers by converting them to forms which have the same denominator. For example, to convert $\frac{1}{4} - \frac{1}{3} + \frac{1}{6}$ to a form with a common denominator of 12, we can multiply each fraction by $\frac{12}{12}$ (equivalent to 1).

a. $$\frac{1}{\cancel{4}_1} \cdot \frac{\cancel{12}^3}{12} - \frac{1}{\cancel{3}_1} \cdot \frac{\cancel{12}^4}{12} + \frac{1}{\cancel{6}_1} \cdot \frac{\cancel{12}^2}{12} = \frac{3}{12} - \frac{4}{12} + \frac{2}{12} = \frac{? + ? + ?}{12} = \_\_\_\_\_\_$$

b. Is $\frac{1}{4}$ equivalent to $\frac{3}{12}$?

c. Is $\frac{1}{3}$ equivalent to $\frac{4}{12}$?

d. Is $\frac{1}{6}$ equivalent to $\frac{2}{12}$?

---

a. $3 + (-4) + 2$; $\frac{1}{12}$ *b.* yes *c.* yes *d.* yes

---

10 *a.* To convert $\frac{5}{8}+\frac{1}{6}-\frac{7}{12}$ to a form with a common denominator of 24, we multiply each fraction by $\frac{24}{24}$ (= 1):

$$\frac{5}{\underset{1}{\cancel{8}}}\cdot\frac{\overset{3}{\cancel{24}}}{24}+\frac{1}{\underset{1}{\cancel{6}}}\cdot\frac{\overset{4}{\cancel{24}}}{24}-\frac{7}{\underset{1}{\cancel{12}}}\cdot\frac{\overset{2}{\cancel{24}}}{24}=\frac{?}{24}+\frac{?}{24}-\frac{?}{24}=\frac{?+?+?}{24}=\_\_\_\_$$

b. Convert $\frac{7}{16}-\frac{5}{12}-\frac{3}{8}$ to a form with a common denominator of 48 and simplify:

$$\frac{7}{16}\cdot\frac{48}{48}-\frac{5}{12}\cdot\frac{48}{48}-\frac{3}{8}\cdot\frac{48}{48}=\frac{?+?+?}{48}=-\frac{?}{48}$$

---

a. $\frac{15}{24}+\frac{4}{24}-\frac{14}{24}=\frac{15+4+(-14)}{24}=\frac{5}{24}$

b. $\frac{7}{\underset{1}{\cancel{16}}}\cdot\frac{\overset{3}{\cancel{48}}}{48}-\frac{5}{\underset{1}{\cancel{12}}}\cdot\frac{\overset{4}{\cancel{48}}}{48}-\frac{3}{\underset{1}{\cancel{8}}}\cdot\frac{\overset{6}{\cancel{48}}}{48}=\frac{21+(-20)+(-18)}{48}=-\frac{17}{48}$

---

11 *a.* Convert $-\frac{4}{9}+\frac{5}{6}+\frac{3}{8}$ to a form with a common denominator of 72 and simplify:

$$-\frac{4}{9}\cdot\frac{72}{72}+\frac{5}{6}\cdot\frac{72}{72}+\frac{3}{8}\cdot\frac{72}{72}=\frac{?+?+?}{72}=\frac{?}{72}$$

b. Convert $-\frac{3}{7}+\frac{5}{3}-\frac{1}{6}$ to a form with a common denominator of 42 and simplify.

---

a. $-\frac{4}{\underset{1}{\cancel{9}}}\cdot\frac{\overset{8}{\cancel{72}}}{72}+\frac{5}{\underset{1}{\cancel{6}}}\cdot\frac{\overset{12}{\cancel{72}}}{72}+\frac{3}{\underset{1}{\cancel{8}}}\cdot\frac{\overset{9}{\cancel{72}}}{72}=\frac{-32+60+27}{72}=\frac{55}{72}$

b. $-\frac{3}{\underset{1}{\cancel{7}}}\cdot\frac{\overset{6}{\cancel{42}}}{42}+\frac{5}{\underset{1}{\cancel{3}}}\cdot\frac{\overset{14}{\cancel{42}}}{42}-\frac{1}{\underset{1}{\cancel{6}}}\cdot\frac{\overset{7}{\cancel{42}}}{42}=\frac{-18+70+(-7)}{42}=\frac{\overset{15}{\cancel{45}}}{\underset{14}{\cancel{42}}}=\frac{15}{14}$

---

12 Convert $\frac{5}{3}+\frac{11}{25}-\frac{7}{15}$ to a form with a common denominator of 75.

$$\frac{5}{\not{3}_1} \cdot \frac{\overset{25}{\not{75}}}{75} + \frac{11}{\not{25}_1} \cdot \frac{\overset{3}{\not{75}}}{75} - \frac{7}{\not{15}_1} \cdot \frac{\overset{5}{\not{75}}}{75} = \frac{125 + 33 + (-35)}{75} = \frac{123}{75}$$

13 If Frame 12 had used the expression $\frac{5}{3} + \frac{11}{25} - \frac{7}{10}$ and we had tried 75 as common denominator, we would have had trouble because $\frac{7}{\not{10}_2} \cdot \frac{\overset{15}{\not{75}}}{75} = \frac{?}{150}$ and $\frac{7}{10}$ (can/cannot) ________ be expressed as a fraction of the form $\frac{n}{75}$ because 10 (does/does not) ________ divide into 75.

105 cannot; does not

14 In order for a number to be a common denominator it must be divisible by each of the original denominators.

In Frame 10*a*, for example, the original denominators were 8, 6, and 12 and the common denominator was 24.

a. Do 8, 6, and 12 all divide into 24?

In Frame 11*a* the original denominators were 9, 6, and 8 and the common denominator was 72.

b. Do 9, 6, and 8 all divide into 72?

a. yes *b*. yes

15 In order for a number to be a common denominator it must be divisible by each of the original denominators. Which of the following numbers are common denominators of $-\frac{3}{10} + \frac{5}{25} + \frac{1}{6} - \frac{4}{15}$?

a. 75 *b*. 60 *c*. 150 *d*. 300

c and *d* (75 is not divisible by 10; 60 is not divisible by 25)

16 Which of the following numbers are common denominators of $\frac{16}{21} - \frac{9}{14} + \frac{5}{8} - \frac{1}{6}$?

a. 42 *b.* 48 *c.* 84 *d.* 168

---

d (42 is not divisible by 8; 48 is not divisible by 14 or by 21; 84 is not divisible by 8)

---

17 It is usually best to use the lowest possible number as a common denominator. This number is called the lowest common denominator (LCD).

a. What is the LCD of the expressions in Frame 15?

b. What is the LCD of the expressions in Frame 16?

---

a. 150 *b.* 168

---

18 A systematic method of finding the LCD involves completely factoring each denominator. Suppose we have $\frac{7}{16} + \frac{9}{28} + \frac{5}{42}$.

$16 = 2 \times 2 \times 2 \times 2$

$28 = 2 \times 2 \times 7$

$42 = 2 \times 3 \times 7$

In order to be divisible by 16, the LCD must be of the form $2 \times 2 \times 2 \times 2 \times n$. In order to be divisible by 16 and by 28, the LCD must be of the form

$$\rlap{$\overbrace{}^{16}$}2 \times 2 \times \underbrace{2 \times 2 \times 7}_{28} \times m \quad \text{(the 2s are already included)}$$

In order to be divisible by 16, 28, and by 42, the LCD must be of the form

$$\rlap{$\overbrace{}^{16}$}2 \times 2 \times \rlap{$\underbrace{}_{28}$}2 \times \overbrace{2 \times 7 \times \_\_\_\_}^{42} \quad \text{(a 2 and a 7 are already included)}$$

The LCD is $16 \times$ _____ = _____.

---

3 21 336

---

19 Suppose we have $\frac{5}{8}+\frac{2}{30}+\frac{11}{25}$. $8 =$ ______ $\times$ ______ $\times$ ______; $30 =$ ______ $\times$ ______ $\times$ ______, and $25 =$ ______ $\times$ ______. In order to be divisible by 8, the LCD must be of the form ______ $\times$ ______ $\times$ ______ $\times\, n$. In order to be divisible by 8 and by 30, the LCD must be of the form ______ $\times$ ______ $\times$ ______ $\times$ ______ $\times$ ______ $\times\, m$. In order to be divisible by 8, 30, and 25, the LCD must be

______ $\times$ ______ $\times$ ______ $\times$ ______ $\times$ ______ $\times$ ______ $= 8 \times$ ______ $=$ ______

(Refer to Frame 18 if you wish.)

$2\times2\times2$ $\quad 2\times3\times5$ $\quad 5\times5$ $\quad 2\times2\times2$ $\quad 2\times2\times2\times3\times5$

$$\underbrace{2\times2\times2}_{8}\times\overbrace{3\times5}^{30}\times 5 = 8\times75 = 600$$

(30 brackets $2\times3\times5$; 25 brackets 5×5)

---

20 Suppose we have $\frac{4}{9}+\frac{5}{21}+\frac{3}{49}$.

$9 =$ ______ $\times$ ______ $\qquad 21 =$ ______ $\times$ ______ $\qquad 49 =$ ______ $\times$ ______

LCD $=$ ______ $\times$ ______ $\times$ ______ $\times$ ______ $=$ ______

(Refer back to Frames 18 and 19 if you have to.)

3×3 $\quad 3\times7$ $\quad 7\times7$ $\quad 3\times3\times7\times7 = 441$

(9 brackets 3×3; 21 brackets 3×7; 49 brackets 7×7)

---

21 What is the LCD of $\frac{7}{12}+\frac{3}{8}+\frac{7}{20}$?

$2\times2\times3\times2\times5 = 120$

(8 brackets $2\times2\times2$; 12 brackets $2\times2\times3$; 20 brackets $2\times2\times5$)

---

22 What is the LCD of $\frac{5}{6}+\frac{7}{4}+\frac{13}{14}$?

$2\times3\times2\times7 = 84$

(6 brackets 2×3; 14 brackets 2×7; 4 brackets 2×2)

---

23 Using the LCDs we found in Frames 21 and 22:

a. $\frac{7}{12}+\frac{3}{8}-\frac{7}{20}=\frac{7}{\underset{1}{\cancel{12}}}\cdot\frac{\overset{10}{\cancel{120}}}{120}+\frac{3}{\underset{1}{\cancel{8}}}\cdot\frac{\overset{15}{\cancel{120}}}{120}-\frac{7}{\underset{1}{\cancel{20}}}\cdot\frac{\overset{6}{\cancel{120}}}{120}=\frac{?+?+?}{120}=$ _____

b. $\frac{5}{6}-\frac{7}{4}+\frac{13}{14}=\frac{5}{6}\cdot\frac{?}{?}-\frac{7}{4}\cdot\frac{?}{?}+\frac{13}{14}\cdot\frac{?}{?}=\frac{?+?+?}{?}=$ _____

---

a. $\frac{70+45+(-42)}{120}=\frac{73}{120}$

b. $\frac{5}{\underset{1}{\cancel{6}}}\cdot\frac{\overset{14}{\cancel{84}}}{84}-\frac{7}{\underset{1}{\cancel{4}}}\cdot\frac{\overset{21}{\cancel{84}}}{84}+\frac{13}{\underset{1}{\cancel{14}}}\cdot\frac{\overset{6}{\cancel{84}}}{84}=\frac{70+(-147)+78}{84}=\frac{1}{84}$

---

24 *a.* What is the LCD of $-\frac{3}{16}+\frac{5}{12}-\frac{2}{9}$?

b. $-\frac{3}{16}+\frac{5}{12}-\frac{2}{9}=-\frac{3}{16}\cdot\frac{?}{?}+\frac{5}{12}\cdot\frac{?}{?}-\frac{2}{9}\cdot\frac{?}{?}=\frac{?+?+?}{?}=$ _____

---

a.
12
$2\times2\times2\times2\times3\times3=144$
16 9

b. $-\frac{3}{\underset{1}{\cancel{16}}}\cdot\frac{\overset{9}{\cancel{144}}}{144}+\frac{5}{\underset{1}{\cancel{12}}}\cdot\frac{\overset{12}{\cancel{144}}}{144}-\frac{2}{\underset{1}{\cancel{9}}}\cdot\frac{\overset{16}{\cancel{144}}}{144}=\frac{-27+60+(-32)}{144}=\frac{1}{144}$

---

25 *a.* What is the LCD of $-\frac{3}{5}-\frac{11}{30}+\frac{33}{50}$? *b.* $-\frac{3}{5}-\frac{11}{30}+\frac{33}{50}=$ _____

---

a.
30
$5\times2\times3\times5=150$
50

b. $-\frac{3}{\underset{1}{\cancel{5}}}\cdot\frac{\overset{30}{\cancel{150}}}{150}-\frac{11}{\underset{1}{\cancel{30}}}\cdot\frac{\overset{5}{\cancel{150}}}{150}+\frac{33}{\underset{1}{\cancel{50}}}\cdot\frac{\overset{3}{\cancel{150}}}{150}=\frac{-90+(-55)+99}{150}=-\frac{46}{150}\left(\text{or }-\frac{23}{75}\right)$

26 $\dfrac{13}{27} - \dfrac{3}{4} + \dfrac{11}{12} =$ ______

$$\text{LCD} = \overbrace{3 \times 3 \times 3}^{27} \times \overbrace{2 \times 2}^{4} = 108$$

$$\text{LCD} = 3 \times 3 \times \underbrace{3 \times 2 \times 2}_{12} = 108$$

$$\frac{13}{\cancel{27}_{\cancel{3}_{1}}} \cdot \frac{\overset{\overset{4}{\cancel{12}}}{\cancel{108}}}{108} - \frac{3}{\cancel{4}_{1}} \cdot \frac{\overset{27}{\cancel{108}}}{108} + \frac{11}{\cancel{12}_{1}} \cdot \frac{\overset{9}{\cancel{108}}}{108} = \frac{52 + (-81) + 99}{108} = \frac{70}{108} \left(\text{or } \frac{35}{54}\right)$$

---

27 $\dfrac{11}{24} + \dfrac{5}{6} - \dfrac{9}{16} =$ ______

$$\text{LCD} = \overbrace{2 \times 2 \times 2 \times 3}^{24} \times 2 = 48 \quad (6 = 3 \times 2)$$

$$\text{LCD} = \underbrace{2 \times 2 \times 2 \times 3 \times 2}_{16} = 48$$

$$\frac{11}{\cancel{24}_{1}} \cdot \frac{\overset{2}{\cancel{48}}}{48} + \frac{5}{\cancel{6}_{1}} \cdot \frac{\overset{8}{\cancel{48}}}{48} - \frac{9}{\cancel{16}_{1}} \times \frac{\overset{3}{\cancel{48}}}{48} = \frac{22 + 40 + (-27)}{48} = \frac{35}{48}$$

---

28 $\dfrac{3}{7} - \left(-\dfrac{9}{10}\right) + \left(-\dfrac{3}{14}\right) = \dfrac{3}{7} \_\_ \dfrac{9}{10} \_\_ \dfrac{3}{14}$ (insert correct signs) $=$ ______

$$\text{LCD} = 7 \times 2 \times 5 = 70$$

$$\frac{3}{7} + \frac{9}{10} - \frac{3}{14} = \frac{3}{\cancel{7}_{1}} \cdot \frac{\overset{10}{\cancel{70}}}{70} + \frac{9}{\cancel{10}_{1}} \cdot \frac{\overset{7}{\cancel{70}}}{70} - \frac{3}{\cancel{14}_{1}} \cdot \frac{\overset{5}{\cancel{70}}}{70} = \frac{30 + 63 + (-15)}{70} = \frac{78}{70} \text{ or } \frac{39}{35}$$

---

29 $-\dfrac{3}{10} + \left(-\dfrac{11}{15}\right) - \left(-\dfrac{7}{9}\right) = \dfrac{3}{10} \_\_ \dfrac{11}{15} \_\_ \dfrac{7}{9}$ (insert correct signs) $=$ ______

$$\text{LCD} = 2 \times 5 \times 3 \times 3 = 90$$

$$-\frac{3}{10} - \frac{11}{15} + \frac{7}{9} = -\frac{3}{\cancel{10}_{1}} \cdot \frac{\overset{9}{\cancel{90}}}{90} - \frac{11}{\cancel{15}_{1}} \cdot \frac{\overset{6}{\cancel{90}}}{90} + \frac{7}{\cancel{9}_{1}} \cdot \frac{\overset{10}{\cancel{90}}}{90} = \frac{-27 + (-66) + 70}{90} = -\frac{23}{90}$$

---

30 $-\frac{5}{6}+\left(-\frac{7}{9}\right)-\frac{7}{12}=$ ______

LCD $= 2 \times 3 \times 3 \times 2 = 36$

$$-\frac{5}{6}-\frac{7}{9}-\frac{7}{12}=-\frac{5}{\not{6}_1}\cdot\frac{\not{36}^{6}}{36}-\frac{7}{\not{9}_1}\cdot\frac{\not{36}^{4}}{36}-\frac{7}{\not{12}_1}\cdot\frac{\not{36}^{3}}{36}=\frac{-30+(-28)+(-21)}{36}=-\frac{79}{36}$$

---

31 $\frac{5}{12}-\left(-\frac{3}{10}\right)+\left(-\frac{8}{15}\right)=$ ______

LCD $= 2 \times 2 \times 3 \times 5 = 60$

$$\frac{5}{12}+\frac{3}{10}-\frac{8}{15}=\frac{5}{\not{12}_1}\cdot\frac{\not{60}^{5}}{60}+\frac{3}{\not{10}_1}\cdot\frac{\not{60}^{6}}{60}-\frac{8}{\not{15}_1}\cdot\frac{\not{60}^{4}}{60}=\frac{25+18+(-32)}{60}=\frac{11}{60}$$

You have finished the program for Sec. 6-5. Now take the posttest to see how much you have learned.

POSTTEST 6-5

Give the lowest common denominator of the following expressions:

1. $\frac{11}{7}+\frac{19}{3}+\frac{13}{14}$
2. $\frac{19}{10}-\frac{2}{25}-\frac{13}{4}$

Simplify the following expressions:

3. $\frac{3}{8}-\frac{7}{18}+\frac{2}{3}$
4. $-\frac{11}{12}+\frac{7}{15}+\frac{3}{20}$
5. $-\frac{13}{9}+\left(-\frac{11}{12}\right)-\left(-\frac{3}{8}\right)$
6. $\frac{4}{15}-\left(-\frac{7}{33}\right)+\left(-\frac{9}{55}\right)$

When you have finished the posttest, correct it yourself using the answers on page 298. Then check all the questions you got wrong on the list below.

1. ______	3. ______	5. ______
2. ______	4. ______	6. ______

If you compare your performance with that on Pretest 6-5, you will see how much you have improved. If you made more than one mistake on this posttest, review the material and take the pretest again. Don't move on to the next section until you can get all the questions right.

Now proceed with Sec. 6-6.

Section 6-6 Multiple Operations

Do this section if you made a mistake in Questions 8 to 10 of Diagnostic Test 6. Begin by taking the pretest below, which will show you how much of Sec. 6-6 you already know and what you will be able to do by the time you have finished it.

PRETEST 6-6

1. $-\frac{2}{5}\left(\frac{3}{4}-\frac{2}{3}\right)=$ ______
2. $\frac{5}{6}\left[-\frac{6}{7}-\left(-\frac{3}{8}\right)\right]=$ ______
3. $\left(\frac{6}{7}-\frac{2}{5}\right)\times\left(\frac{1}{6}+\frac{7}{8}\right)=$ ______
4. $\dfrac{\frac{3}{8}+\frac{2}{7}}{\frac{3}{7}-\frac{1}{4}}=$ ______
5. $\dfrac{-\frac{5}{12}+\left(-\frac{2}{5}\right)}{-\frac{7}{6}-\frac{7}{8}}=$ ______
6. $\dfrac{\frac{4}{9}-\left(-\frac{2}{7}\right)}{-\frac{5}{6}+\frac{11}{4}}=$ ______

When you have finished the pretest, correct it yourself using the answers on page 298. Then check all the questions you got wrong on the list below:

1. ______ 3. ______ 5. ______

2. ______ 4. ______ 6. ______

If you made any mistakes (even if it was only *one* mistake), read the introduction and go through the program.

If you made no mistakes at all, skip Sec. 6-6, because you already know its material, and take Final Test 6 (page 295).

Introduction

In this section you practice evaluating expressions that involve the combination of additions or subtractions with multiplications or divisions.

FRAMES FOR SECTION 6-6

1 The distributive law of multiplication over addition or subtraction can be shown to be valid for all rational numbers and not just for integers, and it can be used to simplify expressions:

$$\frac{3}{4}\left(\frac{2}{7}-\frac{4}{5}\right)=\left(\frac{3}{4}\times\frac{2}{7}\right)-\left(\frac{3}{4}\times\frac{4}{5}\right)=\frac{?}{?}-\frac{?}{?}=\_\_\_\_\_$$

$$\frac{3}{14}-\frac{3}{5}=\frac{15+(-42)}{70}=-\frac{27}{70}$$

---

2 An alternative method of simplifying the same expression is to complete the operations within parentheses before proceeding further. Thus

$$\frac{3}{4}\left(\frac{2}{7}-\frac{4}{5}\right)=\frac{3}{4}\left(\_\_\_\_\_\right)=\_\_\_\_\_$$

$$\frac{3}{\cancel{4}_{2}}\left(-\frac{\cancel{18}^{9}}{35}\right)=-\frac{27}{70}$$

---

3 In the first method (Frame 1) we carried out _____ (how many?) multiplication(s) of fractions and _____ subtraction(s); whereas in the second method (Frame 2) we carried out _____ multiplication(s) of fractions and _____ subtraction(s).

---

two; one; one; one

---

4 In general, therefore, one might expect the (first/second) _____ method to be the quicker.

---

second

---

5 Use this method to simplify the expressions below:

a. $\frac{2}{5}\left(\frac{1}{2}-\frac{1}{3}\right)=\frac{2}{5}\left(\_\_\_\_\_\right)=\_\_\_\_\_$

b. $\frac{5}{6}\left(\frac{3}{4}-\frac{5}{7}\right)=\_\_\_\_\_$

---

a. $\frac{2}{5}\left(\frac{1}{6}\right)=\frac{1}{15}$ *b.* $\frac{5}{6}\left(\frac{1}{28}\right)=\frac{5}{168}$

---

6 *a.* $\frac{3}{7}\left(\frac{3}{4}+\frac{1}{6}\right)=$ ______ *b.* $\frac{4}{9}\left(\frac{13}{4}-\frac{11}{5}\right)=$ ______

---

a. $\frac{\cancel{3}^{1}}{7}\left(\frac{11}{\cancel{12}_{4}}\right)=\frac{11}{28}$ *b.* $\frac{\cancel{4}^{1}}{\cancel{9}_{3}}\left(\frac{\cancel{21}^{7}}{\cancel{20}_{5}}\right)=\frac{7}{15}$

---

7 *a.* $-\frac{1}{3}\left(\frac{6}{11}-\frac{5}{2}\right)=$ ______ *b.* $\frac{1}{4}\left[\frac{4}{5}+\left(-\frac{7}{3}\right)\right]=$ ______

---

a. $-\frac{1}{3}\left(-\frac{43}{22}\right)=\frac{43}{66}$ *b.* $\frac{1}{4}\left(-\frac{23}{15}\right)=-\frac{23}{60}$

---

8 *a.* $-\frac{2}{5}\left[-\frac{4}{9}+\left(-\frac{3}{4}\right)\right]=$ ______ *b.* $\frac{2}{7}\left[\frac{2}{3}-\left(-\frac{14}{5}\right)\right]=$ ______

---

a. $\left(-\frac{\cancel{2}^{1}}{5}\right)\left(-\frac{43}{\cancel{36}_{18}}\right)=\frac{43}{90}$ *b.* $\frac{2}{7}\left(\frac{52}{15}\right)=\frac{104}{105}$

---

9 *a.* $\frac{3}{5}\left[-\frac{10}{7}+\left(-\frac{11}{6}\right)\right]=$ ______ *b.* $-\frac{5}{8}\left[-\frac{9}{4}-\left(-\frac{14}{3}\right)\right]=$ ______

---

a. $\frac{\cancel{3}^{1}}{5}\left(-\frac{137}{\cancel{42}_{14}}\right)=-\frac{137}{70}$ *b.* $-\frac{5}{8}\left(\frac{29}{12}\right)=-\frac{145}{96}$

---

10 When there are more terms, it remains quicker to perform the operations within parentheses first. Thus

$$\left(\frac{1}{4}+\frac{1}{5}\right)\times\left(\frac{3}{10}-\frac{1}{9}\right)=\frac{?}{20}\times\frac{?}{90}=\frac{?\times ?}{20\times 90}=\text{\_\_\_\_\_\_}$$

and

$$\left(\frac{2}{3}+\frac{3}{5}\right)\times\left(\frac{2}{7}-\frac{1}{6}\right)=\text{\_\_\_\_\_\_}\times\text{\_\_\_\_\_\_}=\text{\_\_\_\_\_\_}$$

(Don't forget to simplify by canceling common factors before completing the multiplication.)

$$\frac{9}{20} \times \frac{17}{90} = \frac{\overset{1}{\cancel{9}} \times 17}{20 \times \underset{10}{\cancel{90}}} = \frac{17}{200} \qquad \frac{19}{15} \times \frac{5}{42} = \frac{19 \times \overset{1}{\cancel{5}}}{\underset{3}{\cancel{15}} \times 42} = \frac{19}{126}$$

11 *a.* $\left(\frac{7}{4} - \frac{11}{3}\right) \times \left(\frac{5}{6} - \frac{7}{2}\right) =$ ______ $\times$ ______ $=$ ______

b. $\left[\frac{6}{5} + \left(-\frac{5}{3}\right)\right] \times \left[-\frac{4}{7} - \left(-\frac{5}{6}\right)\right] =$ ______ $\times$ ______ $=$ ______

a. $-\frac{23}{12} \times \left(-\frac{32}{12}\right) = \frac{23 \times \overset{2}{\overset{\cancel{8}}{\cancel{32}}}}{\underset{3}{\cancel{12}} \times \underset{3}{\cancel{12}}} = \frac{46}{9}$ *b.* $-\frac{7}{15} \times \frac{11}{42} = -\frac{\overset{1}{\cancel{7}} \times 11}{15 \times \underset{6}{\cancel{42}}} = -\frac{11}{90}$

12 *a.* $\left[-\frac{2}{9} + \left(-\frac{3}{4}\right)\right] \times \left[\frac{7}{8} - \left(-\frac{4}{7}\right)\right] =$ ______

b. $\left[\frac{7}{3} + \left(-\frac{3}{7}\right)\right] \times \left[-\frac{3}{5} - \frac{5}{8}\right] =$ ______

a. $-\frac{35}{36} \times \frac{81}{56} = -\frac{\overset{5}{\cancel{35}} \times \overset{9}{\cancel{81}}}{\underset{4}{\cancel{36}} \times \underset{8}{\cancel{56}}} = -\frac{45}{32}$ *b.* $\frac{40}{21} \times \left(-\frac{49}{40}\right) = -\frac{\overset{1}{\cancel{40}} \times \overset{7}{\cancel{49}}}{\underset{3}{\cancel{21}} \times \underset{1}{\cancel{40}}} = -\frac{7}{3}$

13 Since division is equivalent to multiplication by the reciprocal, the same method can be used for dividing expressions. Thus

$\left(\frac{3}{5} - \frac{3}{8}\right) \div \left(\frac{4}{7} - \frac{7}{8}\right) =$ ______ $\div$ ______ $=$ ______ $\times$ ______ $=$ ______

and

$\frac{\frac{2}{9} + \frac{5}{6}}{\frac{4}{9} - \frac{7}{11}} = \left(\frac{2}{9} + \frac{5}{6}\right) \div \left(\frac{4}{9} - \frac{7}{11}\right) =$ ______ $\div$ ______ $=$ ______ $\times$ ______ $=$ ______

$$\frac{9}{40} \div \left(-\frac{17}{56}\right) = \frac{9}{40} \times \left(-\frac{56}{17}\right) = -\frac{9 \times \overset{7}{\cancel{56}}}{\underset{5}{\cancel{40}} \times 17} = -\frac{63}{85}$$

$$\frac{57}{54} \div \left(-\frac{19}{99}\right) = \frac{57}{54} \times \left(-\frac{99}{19}\right) = -\frac{\overset{\overset{1}{\cancel{19}}}{\cancel{57}} \times \overset{11}{\cancel{99}}}{\underset{\underset{2}{\cancel{18}}}{\cancel{54}} \times \underset{1}{\cancel{19}}} = -\frac{11}{2}$$

---

14 It is usually quickest to keep the denominator in factor form, as this shows you where to cancel.

a. $\dfrac{\frac{11}{6} - \frac{8}{7}}{\frac{5}{14} + \frac{2}{9}} = \dfrac{? + ?}{6 \times 7} \div \dfrac{? + ?}{9 \times 14} = \dfrac{?}{6 \times 7} \times \dfrac{9 \times 14}{?} = $ ______

b. $\dfrac{\frac{3}{8} - \frac{2}{5}}{\frac{7}{16} - \frac{2}{3}} = \dfrac{? + ?}{8 \times 5} \div \dfrac{? + ?}{16 \times 3} = $ ______

---

a. $\dfrac{77 + (-48)}{6 \times 7} \div \dfrac{45 + 28}{14 \times 9} = \dfrac{29}{\underset{\underset{1}{\cancel{3}}}{\cancel{6}} \times \underset{1}{\cancel{7}}} \times \dfrac{\overset{\overset{1}{\cancel{2}}}{\cancel{14}} \times \overset{3}{\cancel{9}}}{73} = \dfrac{87}{73}$

b. $\dfrac{15 + (-16)}{8 \times 5} \div \dfrac{21 + (-32)}{16 \times 3} = -\dfrac{1}{\underset{1}{\cancel{8}} \times 5} \times \left(-\dfrac{\overset{2}{\cancel{16}} \times 3}{11}\right) = \dfrac{6}{55}$

---

15 A safe procedure when negative numbers are involved is to insert the sign of the answer before canceling. (The check mark reminds you to do this.) Thus:

a. $\dfrac{\frac{11}{8} - \frac{9}{7}}{\frac{1}{6} - \frac{9}{14}} = \dfrac{77 + (-72)}{8 \times 7} \div \dfrac{14 + (-54)}{6 \times 14} = \dfrac{5}{8 \times 7} \times \dfrac{6 \times 14}{-40} = $ ✓__ ______

b. $\dfrac{\frac{3}{5} - \frac{5}{12}}{\frac{8}{11} - \frac{5}{6}} = \dfrac{? + ?}{5 \times 12} \div \dfrac{? + ?}{11 \times 6} = \dfrac{?}{5 \times 12} \times \dfrac{11 \times 6}{?} = $ ✓__ ______

---

a. $$\frac{\overset{1}{\cancel{5}}}{8 \times \underset{1}{\cancel{7}}} \times \frac{\overset{3}{\cancel{6}} \times \overset{\overset{1}{\cancel{2}}}{\cancel{14}}}{\underset{\underset{\underset{2}{\cancel{4}}}{\cancel{8}}}{\cancel{40}}} = -\frac{3}{16}$$

b. $$\frac{36 + (-25)}{5 \times 12} \div \frac{48 + (-55)}{11 \times 6} = \frac{11}{5 \times \underset{2}{\cancel{12}}} \times \frac{11 \times \overset{1}{\cancel{6}}}{-7} = -\frac{121}{70}$$

---

16 *a.* $$\frac{\frac{7}{16} - \frac{5}{9}}{\frac{7}{12} - \frac{3}{5}} = \frac{? + ?}{16 \times 9} \div \frac{? + ?}{12 \times 5} = \_\_\_\_$$

b. $$\frac{\frac{3}{7} - (-\frac{1}{6})}{\frac{-5}{14} + (-\frac{7}{9})} = \frac{? + ?}{7 \times 6} \div \frac{? + ?}{14 \times 9} = \_\_\_\_$$

---

a. $$\frac{63 + (-80)}{16 \times 9} \div \frac{35 + (-36)}{12 \times 5} = \frac{-17}{16 \times 9} \times \frac{12 \times 5}{-1} = \frac{85}{12}$$

b. $$\frac{18 + 7}{7 \times 6} \div \frac{(-45) + (-98)}{14 \times 9} = \frac{25}{7 \times 6} \times \frac{14 \times 9}{-143} = -\frac{75}{143}$$

---

17 *a.* $$\frac{-\frac{2}{9} - (-\frac{3}{8})}{\frac{3}{7} + (-\frac{4}{9})} = \frac{? + ?}{9 \times 8} \div \frac{? + ?}{7 \times 9} = \_\_\_\_$$

b. $$\frac{\frac{8}{9} - \frac{7}{6}}{-\frac{3}{8} - (-\frac{7}{27})} = \_\_\_\_$$

---

a. $$\frac{-16 + 27}{9 \times 8} \div \frac{27 + (-28)}{7 \times 9} = \frac{11}{9 \times 8} \times \frac{7 \times 9}{-1} = -\frac{77}{8}$$

b. $$\frac{48 + (-63)}{9 \times 6} \div \frac{-81 + 56}{8 \times 27} = \frac{-15}{9 \times 6} \times \frac{8 \times 27}{-25} = \frac{12}{5}$$

---

18 $$\frac{\frac{5}{7} + (-\frac{3}{5})}{-\frac{7}{10} - (-\frac{3}{4})} = \_\_\_\_$$

$$\frac{25 + (-21)}{7 \times 5} \div \frac{-28 + 30}{10 \times 4} = \frac{4}{7 \times 5} \times \frac{10 \times 4}{2} = \frac{16}{7}$$

---

19 $$\frac{-\frac{8}{21} - (-\frac{1}{4})}{-\frac{2}{7} + (-\frac{2}{15})} = \_\_\_\_$$

$$\frac{-32+21}{21\times 4} \div \frac{-30+(-14)}{7\times 15} = \frac{-11}{21\times 4} \times \frac{7\times 15}{-44} = \frac{5}{16}$$

You have finished the program for Sec. 6-6. Now take the posttest to see how much you have learned.

POSTTEST 6-6

1. $-\frac{3}{7}\left(\frac{5}{8}-\frac{1}{6}\right) =$ ______

2. $\frac{2}{3}\left[-\frac{3}{4}-\left(-\frac{4}{7}\right)\right] =$ ______

3. $\left(\frac{3}{11}-\frac{1}{6}\right)\times\left(\frac{8}{7}-\frac{4}{9}\right) =$ ______

4. $\frac{\frac{5}{16}+\frac{3}{5}}{\frac{4}{5}-\frac{7}{8}} =$ ______

5. $\frac{-\frac{5}{9}+(-\frac{2}{5})}{-\frac{5}{6}-\frac{7}{10}} =$ ______

6. $\frac{\frac{2}{7}-(-\frac{5}{12})}{-\frac{7}{4}+\frac{11}{6}} =$ ______

When you have finished the posttest, correct it yourself using the answers on page 298. Then check all the questions you got wrong on the list below.

1. ______ 3. ______ 5. ______

2. ______ 4. ______ 6. ______

If you compare your performance with that on Pretest 6-6, you will see how much you have improved. If you made more than one mistake on this posttest, review the material and take the pretest again. Don't move on until you can get all the questions right.

Now proceed with Final Test 6.

FINAL TEST 6

Simplify the following expressions:

1. $\frac{-4}{-5}\times\frac{3}{-7}$

2. $\frac{16}{-9}\div\frac{-8}{5}$

3. $4\frac{2}{3}\div(-5\frac{5}{6})$

4. $\frac{-7}{10}-\frac{11}{-9}$

5. $\frac{9}{-7}-\frac{-4}{-3}$

6. $3\frac{3}{8}-1\frac{5}{7}$

7. $1\frac{5}{6}-7\frac{3}{11}$

8. $\frac{5}{18}+\left(-\frac{13}{24}\right)-\left(-\frac{9}{8}\right)$

9. $\frac{5}{8}\left[-\frac{3}{5}-\left(-\frac{7}{9}\right)\right]$

10. $\frac{\frac{4}{9}-\frac{11}{12}}{-\frac{5}{8}+(-\frac{7}{6})}$

The answers to this test are on page 298. Check all the questions you got wrong on the list below and compare your results with those on Diagnostic Test 6 to see how much you have improved.

1. ______	4. ______	8. ______
2. ______	5. ______	9. ______
3. ______	6. ______	10. ______
	7. ______	

If you made more than one mistake, review the material and then take the diagnostic test again. Make sure that you can get all the questions right before you finish the chapter.

Now proceed to Chapter 7.

ANSWERS TO TESTS FOR CHAPTER 6

DIAGNOSTIC TEST 6

1. $-\frac{5}{8}$
2. $\frac{14}{11}$
3. $-\frac{46}{55}$
4. $\frac{47}{70}$
5. $-\frac{101}{72}$
6. $2\frac{37}{56}$ or $\frac{149}{56}$
7. $-2\frac{25}{36}$ or $-\frac{97}{36}$
8. $\frac{121}{72}$ or $1\frac{49}{72}$
9. $\frac{2}{25}$
10. $-\frac{23}{6}$

PRETEST 6-1

1. $\frac{-6}{55}$ or $-\frac{6}{55}$
2. $\frac{6}{35}$
3. $-\frac{8}{81}$
4. $-\frac{12}{11}$
5. 9
6. $-\frac{159}{11}$ or $-14\frac{5}{11}$

POSTTEST 6-1

1. $\frac{-7}{72}$ or $-\frac{7}{72}$
2. $\frac{5}{16}$
3. $-\frac{12}{49}$
4. $-\frac{64}{15}$
5. $\frac{23}{3}$ or $7\frac{2}{3}$
6. $-\frac{282}{7}$ or $-40\frac{2}{7}$

PRETEST 6-2

1. $\frac{-7}{6}$ or $-\frac{7}{6}$
2. $-\frac{3}{7}$
3. $-\frac{9}{10}$
4. $\frac{17}{20}$
5. $-\frac{14}{5}$ or $-2\frac{4}{5}$
6. $\frac{40}{59}$

POSTTEST 6-2

1. $\frac{-5}{4}$ or $-\frac{5}{4}$
2. $-\frac{7}{9}$
3. $-\frac{16}{35}$
4. $\frac{72}{77}$
5. $-\frac{14}{5}$
6. $\frac{60}{37}$

PRETEST 6-3

1. $\frac{9}{7}$
2. $\frac{23}{12}$
3. $-\frac{3}{40}$
4. $-\frac{59}{42}$
5. $-\frac{31}{36}$
6. $-\frac{26}{33}$
7. $\frac{101}{70}$
8. $\frac{31}{72}$

POSTTEST 6-3

1. $\frac{10}{9}$
2. $\frac{17}{16}$
3. $-\frac{13}{56}$
4. $-\frac{41}{30}$
5. $-\frac{37}{30}$
6. $-\frac{19}{20}$
7. $\frac{108}{55}$
8. $\frac{13}{72}$

PRETEST 6-4

1. $4\frac{4}{7}$
2. $-4\frac{5}{6}$
3. $-2\frac{11}{13}$
4. $10\frac{2}{9}$
5. $1\frac{9}{40}$
6. $-\frac{23}{28}$

POSTTEST 6-4

1. $5\frac{5}{8}$
2. $-5\frac{6}{7}$
3. $-3\frac{6}{11}$
4. $10\frac{5}{8}$
5. $3\frac{32}{63}$
6. $-1\frac{1}{2}$

PRETEST 6-5

1. 150
2. 72
3. $-\frac{53}{100}$
4. $\frac{17}{180}$
5. $-\frac{97}{120}$
6. $\frac{11}{35}$

POSTTEST 6-5

1. 42
2. 100
3. $\frac{47}{72}$
4. $-\frac{3}{10}$ or $-\frac{18}{60}$
5. $-\frac{143}{72}$
6. $\frac{52}{165}$

PRETEST 6-6

1. $-\frac{1}{30}$
2. $-\frac{45}{112}$
3. $\frac{10}{21}$
4. $\frac{37}{10}$
5. $\frac{2}{5}$
6. $\frac{8}{21}$

POSTTEST 6-6

1. $-\frac{11}{56}$
2. $-\frac{5}{42}$
3. $\frac{2}{27}$
4. $-\frac{73}{6}$
5. $\frac{43}{69}$
6. $\frac{59}{7}$

FINAL TEST 6

1. $-\frac{12}{35}$
2. $\frac{10}{9}$
3. $-\frac{4}{5}$
4. $\frac{47}{90}$
5. $-\frac{55}{21}$
6. $1\frac{37}{56}$ or $\frac{93}{56}$
7. $-5\frac{29}{66}$ or $-\frac{359}{66}$
8. $\frac{31}{36}$
9. $\frac{1}{9}$
10. $\frac{34}{129}$

Chapter 7
Introduction to Decimals

This chapter introduces decimal notation and shows how most rational numbers can be represented in decimal form. The metric system of units and multiplication or division of decimals by multiples of 10 are also included.

Section 7-1 shows how decimals can be converted to mixed numerals or fractions. Section 7-2 then converts fractions whose denominators are multiples of 10 to decimals and introduces multiplication and division of decimals by multiples of 10. Section 7-3 uses this to introduce the metric system of units for length and weight and to practice conversion of measurements from one metric unit to another Section 7-4 then deals with general methods of converting fractions to decimals and introduces infinite-decimal notation.

Chapter 7 requires completion of Chaps. 3, 5, and 6.

DIAGNOSTIC TEST 7

1. Convert 39.073 to mixed-numeral form.
2. Convert 0.000651 to fractional form.
3. Convert $\frac{235}{100{,}000}$ to decimal form.
4. $0.516 \div 1000 =$ _____
5. $0.00053 \times 10{,}000 =$ _____
6. 13 centimeters = _____ kilometer(s)
7. 0.63 gram = _____ milligram(s)

Convert the following fractions to decimal forms (using infinite-decimal notation where necessary):

8. $\frac{321}{8}$
9. $\frac{25}{9{,}000}$
10. $\frac{234}{70}$

When you have finished the diagnostic test, correct it yourself using the answers on page 327. Then check all the questions you got wrong on the list below:

1. ______	3. ______	6. ______	8. ______
2. ______	4. ______	7. ______	9. ______
	5. ______		10. ______

If you made no mistakes at all, you need not do any work in this chapter and should go straight on to Chap. 8. But if you made any mistakes (even if it was only *one* mistake), follow these instructions:

	section	page
If you made any mistakes in Question 1 or 2, circle Sec. 7-1.	7-1	301
If you made any mistakes in Questions 3 to 5, circle Sec. 7-2.	7-2	308
If you made any mistakes in Question 6 or 7, circle Sec. 7-3.	7-3	316
If you made any mistakes in Questions 8 to 10, circle Sec. 7-4.	7-4	321

You should now work through the sections whose numbers you have circled, taking them in numerical order. The sections begin on the pages indicated.

Section 7-1
Definition of Decimals and Conversion to Fractions or Mixed-numeral Forms

Do this section if you made a mistake in Question 1 or 2 of Diagnostic Test 7. Begin by taking the pretest below, which will show you how much of Sec. 7-1 you already know and what you will be able to do by the time you have finished it.

PRETEST 7-1

Give the mixed-numeral form of the following decimals ($6\frac{1}{8}$ is the mixed-numeral form of the fraction $\frac{49}{8}$):

1. 30.53   2. 212.062   3. 51.0039

Give the fractional forms of the following decimals:

4. 1,800.7   5. 0.00096   6. 20.012795

When you have finished the pretest, correct it yourself using the answers on page 327. Then check all the questions you got wrong on the list below:

1. ______   3. ______   5. ______

2. ______   4. ______   6. ______

If you made any mistakes (even if it was only *one* mistake), read the introduction and go through the program.

If you made no mistakes at all, skip Sec. 7-1, because you already know its material, and go on to the next section whose number you circled on page 300. If no further numbers were circled, take Final Test 7 (page 326).

Introduction

In this section you learn decimal notation and how the value of each digit in a decimal is determined by its distance and direction from the decimal point. Then you practice converting decimal forms of rational numbers to mixed-numeral and fractional forms.

FRAMES FOR SECTION 7-1

1 Our number system has a base of 10. For example, 3,890 is our notation for writing

3 × ______ + 8 × ______ + 9 × ______ + 0 × 1

1,000   100   10

---

2 If we divide 3,890 by 10, we get ______, which can be written as 3 × ______ + 8 × ______ + 9 × ______.

---

389 100 10 1

---

3 If we divide 389 by 10, we get the fraction $\frac{389}{10}$, which can be written in mixed-numeral form as ______ or as 3 × ______ + 8 × ______ + 9 × ______.

$38\frac{9}{10}$ 10 1 $\frac{1}{10}$

---

4 Do you notice the pattern:

$$3{,}890 = 3 \times 1{,}000 + 8 \times 100 + 9 \times 10$$

$$389 = 3 \times 100 + 8 \times 10 + 9 \times 1$$

$$\frac{389}{10} = 3 \times 10 + 8 \times 1 + 9 \times \frac{1}{10}$$

$$\frac{389}{10} \div 10 = \frac{389}{100} = 3 \times \text{\_\_\_\_\_\_} + 8 \times \text{\_\_\_\_\_\_} + 9 \times \text{\_\_\_\_\_\_}$$

---

1 $\frac{1}{10}$ $\frac{1}{100}$

---

5 Similarly we can divide each term by 10 to show that

$$\frac{389}{100} \div 10 = \frac{389}{1{,}000} = 3 \times \text{\_\_\_\_\_\_} + 8 \times \text{\_\_\_\_\_\_} + 9 \times \text{\_\_\_\_\_\_}$$

$\frac{1}{10}$ $\frac{1}{100}$ $\frac{1}{1{,}000}$

---

6 This pattern is the basis of the *decimal notation* for representing rational numbers in which the value of each digit is determined by its distance and direction from the *decimal point* (.).

Thus in the number 123.45:

a. The digit 1 represents $1 \times 100 =$ ______.

b. The digit 2 represents $2 \times 10 =$ ______.

c. The digit 3 represents $3 \times 1 =$ ______.

d. The digit 4 represents $4 \times \frac{1}{10} =$ ______.

e. The digit 5 represents $5 \times \frac{1}{100} =$ ______.

100 20 3 $\frac{4}{10}$ $\frac{5}{100}$

7 *a.* $\frac{4}{10} + \frac{5}{100} = \frac{?}{100}$

so we can write 123.45 in mixed-numeral form as $123\,\frac{?}{100}$.

b. $123 = \frac{?}{100}$

so we can write 123.45 in fractional form as $\frac{?}{100}$.

a. 45; 45 *b.* 12,300; 12,345

8 In the number 345.678:

a. The digit 3 represents 3 × ______ = ______.

b. The digit 4 represents 4 × ______ = ______.

c. The digit 5 represents 5 × ______ = ______.

d. The digit 6 represents 6 × ______ = ______.

e. The digit 7 represents 7 × ______ = ______.

f. The digit 8 represents 8 × ______ = ______.

a. 100; 300 *b.* 10; 40 *c.* 1; 5 *d.* $\frac{1}{10}$; $\frac{6}{10}$ *e.* $\frac{1}{100}$; $\frac{7}{100}$

f. $\frac{1}{1,000}$; $\frac{8}{1,000}$

9 *a.* $\frac{6}{10} + \frac{7}{100} + \frac{8}{1,000} = \frac{?}{1,000}$

So the mixed-numeral form of 345.678 is ______.

b. Write 345.678 in fractional form.

a. 678; $345\,\frac{678}{1,000}$ *b.* $\frac{345,678}{1,000}$

10 In the number 23.75648:

a. The digit 7 represents 7 × ______ = ______.

b. The digit 5 represents 5 × ______ = ______.

c. The digit 6 represents 6 × ______ = ______.

d. The digit 4 represents 4 × ______ = ______.

e. The digit 8 represents 8 × ______ = ______.

---

a. $\frac{1}{10}$; $\frac{7}{10}$ *b.* $\frac{1}{100}$; $\frac{5}{100}$ *c.* $\frac{1}{1,000}$; $\frac{6}{1,000}$ *d.* $\frac{1}{10,000}$; $\frac{4}{10,000}$

e. $\frac{1}{100,000}$; $\frac{8}{100,000}$

---

11 *a.* Give the mixed-numeral form of 23.75648.

b. Give the fractional form of 23.75648.

---

a. $23\frac{75,648}{100,000}$ *b.* $\frac{2,375,648}{100,000}$

---

12 Give the mixed-numeral form of:

a. 3.7 *b.* 4.63 *c.* 5.731

---

a. $3\frac{7}{10}$ *b.* $4\frac{63}{100}$ *c.* $5\frac{731}{1,000}$

---

13 *a.* 3.7 has ______ digit(s) on the right of the decimal point, and $\frac{7}{10}$ has ______ zero(s) in its denominator.

b. 4.63 has ______ digit(s) on the right of the decimal point, and $\frac{63}{100}$ has ______ zero(s) in its denominator.

c. 5.731 has ______ digit(s) on the right of the decimal point, and $\frac{731}{1,000}$ has ______ zero(s) in its denominator.

---

a. one; one *b.* two; two *c.* three; three

---

14 *a.* How many digits are on the right of the decimal point in 5.632179?

b. How many zeros are in the denominator of its mixed-numeral form?

c. What is its mixed-numeral form?

a. six *b.* six *c.* $5\frac{632,179}{1,000,000}$

15 *a.* How many digits are on the right of the decimal point in 63.5293?

b. How many zeros are in the denominator of its mixed-numeral form?

c. What is its mixed-numeral form?

d. What is its fractional form?

a. four *b.* four *c.* $63\frac{5,293}{10,000}$ *d* $\frac{635,293}{10,000}$

16 *a.* How many digits are on the right of the decimal point in 729.305?

b. How many zeros are in the denominator of its fractional form?

c. What is its fractional form?

a. three *b.* three *c.* $\frac{729,305}{1,000}$

17 Give the fractional forms of the following numbers:

a. 23.5603 *b.* 24.07132 *c.* 25.002

a. $\frac{235,603}{10,000}$ *b.* $\frac{2,407,132}{100,000}$ *c.* $\frac{25,002}{1,000}$

18 In which of the following integers is the zero digit necessary?

a. 650 *b.* 608 *c.* 069

a and *b*

19 These numbers can be considered as shorter forms of the decimals 650., 608., and 069., which with zero digits removed would become 65., 68., and 69. In which of these numbers does removal of the zero digit change the position of other digits relative to the decimal point?

650. and 608.; in both cases removal of the zero digit brings the 6 one position closer to the decimal point and changes the value.

20 Removal of the zero digit from the decimals 15.032, 1.805, and 1.880 gives 15.32, 1.85 and 1.88.

a. In which of these numbers does removal of the zero digit change the position of other digits relative to the decimal point?

b. In which of these numbers does the zero digit change the value?

---

a. 15.032 and 1.805 *b.* 15.032 and 1.805

---

21 3.560; 4.079; .083

a. In which of the above numbers does removal of the zero digit change the position of other digits relative to the decimal point?

b. In which of the above numbers does the zero digit change the value?

---

a. 4.079 and .083 *b.* 4.079 and .083

---

22 Although the form .083 is often used, it is more usual to write 0.083. This prevents the decimal point from being mistaken for punctuation.

a. How many digits are on the right of the decimal point in 0.083?

b. What is the fractional form of 0.083?

---

a. three *b.* $\frac{83}{1,000}$

---

23 *a.* How many digits are on the right of the decimal point in 0.00095?

b. What is the fractional form of 0.00095?

---

a. five *b.* $\frac{95}{100,000}$

---

24 Give the fractional forms of the following decimals:

a. 0.063 *b.* 79.025 *c.* 1.00692

---

a. $\frac{63}{1,000}$ *b.* $\frac{79,025}{1,000}$ *c.* $\frac{100,692}{100,000}$

---

25 Give the mixed-numeral forms of the following decimals:

a. 79.025 *b.* 1.00692 *c.* 312.0352

a. $79\frac{25}{1,000}$ *b.* $1\frac{692}{100,000}$ *c.* $312\frac{352}{10,000}$

26 Give the mixed-numeral forms of the following decimals:

a. 35.007 *b.* 103.0695 *c.* 2.02305

a. $35\frac{7}{1,000}$ *b.* $103\frac{695}{10,000}$ *c.* $2\frac{2,305}{100,000}$

27 Give the fractional forms of the following decimals:

a. 2.5063 *b.* 300.68 *c.* 0.000053

a. $\frac{25,063}{10,000}$ *b.* $\frac{30,068}{100}$ *c.* $\frac{53}{1,000,000}$

You have finished the program for Sec. 7-1. Now take the posttest to see how much you have learned.

POSTTEST 7-1

Give the mixed-numeral forms of the following decimals:

1. 40.273 2. 329.05 3. 23.0078

Give the fractional forms of the following decimals:

4. 790.6 5. 0.000057 6. 30.02398

When you have finished the posttest, correct it yourself using the answers on page 327. Then check all the questions you got wrong on the list below:

1. _____ 3. _____ 5. _____

2. _____ 4. _____ 6. _____

If you compare your performance with that on Pretest 7-1, you will see how much you have improved. If you made more than one mistake on this posttest, review the material and take the pretest again. Don't move on to the next section until you can get all the questions right.

Now proceed to the next section whose number you circled on page 300. If no other sections were circled, take Final Test 7 (page 326).

Section 7-2 Multiplication or Division of Decimals by Multiples of 10

Do this section if you made a mistake in Questions 3 to 5 of Diagnostic Test 7. Begin by taking the pretest below, which will show you how much of Sec. 7-2 you already know and what you will be able to do by the time you have finished it.

PRETEST 7-2

Give the following fractions in decimal form:

1. $\dfrac{32{,}019}{1000}$ 2. $\dfrac{91}{10{,}000}$

Give the answers to the following in decimal form:

3. $0.428 \div 100$
4. 0.0073×1000
5. $1.08 \div 100$
6. $0.097 \times 10{,}000$
7. $0.00023 \times 1{,}000{,}000$
8. $42.39 \div 100{,}000$

When you have finished the pretest, correct it yourself using the answers on page 328. Then check all the questions you got wrong on the list below:

1. ______ 3. ______ 5. ______ 7. ______

2. ______ 4. ______ 6. ______ 8. ______

If you made any mistakes (even if it was only *one* mistake), read the introduction and go through the program.

If you made no mistakes at all, skip Sec. 7-2, because you already know its material, and go on to the next section whose number you circled on page 300. If no further numbers were circled, take Final Test 7 (page 326).

Introduction

You learn how fractions whose denominators are multiples of 10 can be converted to decimal forms. This leads to definitions of the division of a decimal by a multiple of 10 and the multiplication of a decimal by a multiple of 10.

FRAMES FOR SECTION 7-2

1 Give the fractional forms of the following decimals:

a. 3.2065 *b.* 420.23 *c.* 5.006795

---

a. $\frac{32{,}065}{10{,}000}$ *b.* $\frac{42{,}023}{100}$ *c.* $\frac{5{,}006{,}795}{1{,}000{,}000}$

---

2 *a.* How many digits are there to the right of the decimal point in 3.2065? How many zeros are there in the denominator of $\frac{32{,}065}{10{,}000}$?

b. How many digits are there to the right of the decimal point in 420.23? How many zeros are there in the denominator of $\frac{42{,}023}{100}$?

c. How many digits are there to the right of the decimal point in 5.006795? How many zeros are there in the denominator of $\frac{5{,}006{,}795}{1{,}000{,}000}$?

---

a. four; four *b.* two; two *c.* six; six

---

3 *a.* How many zeros are there in the denominator of $\frac{6{,}324}{100}$?

b. How many digits are there to the right of the decimal point in the decimal form of $\frac{6{,}324}{100}$?

c. What is the decimal form of $\frac{6{,}324}{100}$?

---

a. two *b.* two *c.* 63.24

---

4 *a.* How many zeros are there in the denominator of $\frac{620{,}653}{10{,}000}$?

b. How many digits are there to the right of the decimal point in the decimal form of $\frac{620{,}653}{10{,}000}$?

c. What is the decimal form of $\frac{620{,}653}{10{,}000}$?

---

a. four *b.* four *c.* 62.0653

---

5 *a.* How many zeros are there in the denominator of $\frac{53}{1,000}$?

b. How many digits are there to the right of the decimal point in the decimal form of $\frac{53}{1,000}$?

c. What is the decimal form of $\frac{53}{1,000}$?

---

a. three *b.* three *c.* 0.053

---

6 *a.* How many zeros are there in the denominator of $\frac{76}{100,000}$?

b. What is its decimal form?

---

a. five *b.* 0.00076

---

7 Give the decimal forms of the following fractions:

a. $\frac{325}{1,000,000}$ *b.* $\frac{7,329}{10}$ *c.* $\frac{63}{10,000}$

---

a. 0.000325 *b.* 732.9 *c.* 0.0063

---

8 Give the decimal forms of the following fractions:

a. $\frac{43,256}{1,000}$ *b.* $\frac{232}{100,000}$ *c.* $\frac{8}{10,000}$

---

a. 43.256 *b.* 0.00232 *c.* 0.0008

---

9 *a.* The fraction $\frac{43,256}{1,000} = 43,256 \div 1,000$, so the decimal form of $43,256 \div 1,000$ is ______.

b. $43,256 \div 100 = \frac{43,256}{100} =$ ______ (decimal form)

c. $43,256 \div 10 = \frac{43,256}{10} =$ ______ (decimal form)

---

a. 43.256 *b.* 432.56 *c.* 4,325.6

---

10 Notice that in dividing by a multiple of 10, none of the digits is altered. Only the position of the decimal point is changed. Writing 43,256 as 43,256., we can see:

a. That dividing by 10 to get 4325.6 is equivalent to moving the decimal point ______ place(s) to the (left/right) ______.

b. That dividing by 100 to get 432.56 is equivalent to moving the decimal point ______ place(s) to the ______.

c. That dividing by 1,000 to get 43.256 is equivalent to moving the decimal point ______ place(s) to the ______.

---

a. one; left *b.* two; left *c.* three; left

---

11 a. Dividing by 100 is equivalent to dividing by 10 ______ (how many?) times.

b. Dividing by 1,000 is equivalent to dividing by 10 ______ times.

c. Dividing by 10,000 is equivalent to dividing by 10 ______ times.

d. How many zeros in 10,000?

---

a. two *b.* three *c.* four *d.* four

---

12 *a.* How many zeros in 100,000?

b. Dividing by 100,000 is equivalent to dividing by 10 ______ times.

c. What is the decimal form of $\frac{6,324,798}{100,000}$?

d. How many digits are there to the right of the decimal point?

---

a. five *b.* five *c.* 63.24798 *d.* five

---

13 Dividing 6,324,798 by 100,000 to get 63.24798 is equivalent to moving the decimal point ______ places to the (left/right) ______.

---

five; left

---

14 *a.* What is the decimal form of $\frac{32,567}{10,000}$?

b. Dividing by 10,000 is equivalent to dividing by 10 ______ times.

c. Dividing 32,567 by 10,000 is equivalent to moving the decimal point ______ places to the (left/right) ______.

a. 3.2567 *b.* four *c.* four; left

15 *a.* Dividing by 1,000 is equivalent to dividing by 10 ______ times.

b. Dividing a number by 1,000 is equivalent to moving the decimal point ______ places to the (left/right) ______.

c. Use this method to divide 7,364 and 2,398.1 by 1,000.

a. three *b.* three; left *c.* 7.364; 2.3981

16 Divide the following numbers by 1,000:

a. 23,674 *b.* 283.62 *c.* 5,329.6

a. 23.674 *b.* 0.28362 *c.* 5.3296

17 *a.* How many zeros in 100,000?

b. Dividing a number by 100,000 is equivalent to moving the decimal point ______ places to the (left/right) ______.

c. Use this method to divide 53,267 and 76,543.21 by 100,000.

a. five *b.* five; left *c.* 0.53267; 0.7654321

18 It is helpful to use arrows to indicate movement of the decimal point, and it may also be necessary to add some zero digits. Thus $232 \div 100{,}000$ can be written as 000232 = 0.00232. Use the same method to divide:

a. 232 by 1,000,000 *b.* 11 by 10,000 *c.* 123 by 1,000

a. 000232: 0.000232 *b.* 000011: 0.0011 *c.* 123: 0.123

19 Use the same method to divide:

a. 11.4 by 100,000 *b.* 12.324 by 100

c. 729.36 by 1,000,000

a. 000011.4: 0.000114 *b.* 12.324: 0.12324

c. 000729.36: 0.00072936

20 *a.* $3.23 \div 10 =$ ______ *b.* $3.23 \times 10 =$ ______

c. $3.23 \div 100 =$ ______ *d.* $3.23 \times 100 =$ ______

---

a. 0.323 *b.* 32.3 *c.* 0.0323 *d.* 323

---

21 *a.* Just as dividing by 10 is equivalent to moving the decimal point ______ place(s) to the (left/right) ______, multiplying by 10 is equivalent to moving the decimal point ______ place(s) to the ______.

b. Just as dividing by 100 is equivalent to moving the decimal point ______ place(s) to the ______, multiplying by 100 is equivalent to moving the decimal point ______ place(s) to the ______.

---

a. one, left; one, right *b.* two, left; two, right

---

22 *a.* How many 0s in 10,000?

b. Multiplying a number by 10,000 is equivalent to moving the decimal point ______ place(s) to the ______.

c. $3.2675 \times 10{,}000 =$ ______

d. $0.0056793 \times 10{,}000 =$ ______

---

a. four *b.* four; right *c.* 3.2675; 32,675 *d.* 0.0056793; 56.793

---

23 Use this method to multiply the following numbers:

a. 0.00379×100

b. 36.7345×1000

c. $0.0008878 \times 100{,}000$

---

a. 0.00379; 0.379 *b.* 36.7345; 36,734.5 *c.* 0.0008878; 88.78

---

24 Extra zero digits may also be needed in multiplication by multiples of 10. Thus $83 \times 10{,}000 = (83.0000) = 830{,}000$. Use the same method to multiply the following numbers:

a. 23.21 by 100,000

b. 6.75 by 1,000

c. 123 by 100

---

a. 23.2100000; 2,321,000 *b.* 6.75000; 6,750 *c.* 123.000; 12,300

---

25 *a.* 463.2 × 100,000

b. 5.271 × 1,000

c. 0.0035 × 1,000,000

---

a. 463.20000: 46,320,000   *b.* 5.271: 5,271   *c.* 0.003500: 3,500

---

26 *a.* In multiplying by multiples of 10, you move the decimal point to the (left/right) ______ and end up with a (larger/smaller) ______ number.

b. In dividing by multiples of 10, you move the decimal point to the (left/right) ______ and end up with a (larger/smaller) ______ number.

---

a. right; larger   *b.* left; smaller

---

27 Do the following problems as shown in example *a.* First decide whether the answer will be larger or smaller than the number you start with, then decide how many places to move the decimal point, then work out the answer by the method you have been shown.

	Larger or smaller	*Number of places*	*Working*	*Answer*
a. 0.002 ÷ 100	smaller	two	000.002	0.00002
b. 0.653 × 1,000				
c. 0.98 ÷ 10				
d. 320 ÷ 10,000				
e. 0.073 × 100,000				
f. 47.3 ÷ 1,000,000				

Note: Make sure you show your arrows.

---

b. larger; three; 0.65300; 653.00   *c.* smaller; one; 000.98; 0.098

d. smaller; four; 000320.0; 0.032   *e.* larger; five; 0.0730000; 7,300

f. smaller; six; 00000047.3; 0.0000473

---

28 Do the following problems in a similar manner:

	Larger or smaller	*Number of places*	*Working*	*Answer*
a. 0.036 × 10				
b. 0.069 ÷ 1,000				
c. 0.72 × 100				
d. 0.000328 × 10,000				
e. 5.678 ÷ 100,000				
f. 0.0073 × 1,000,000				

Note: Make sure you show your arrows.

---

a. larger; one; 0.036; 0.36 *b.* smaller; three; 0000.069; 0.000069

c. larger; two; 0.7200; 72.0 *d.* larger; four; 0.000328; 3.28

e. smaller; five; 000005.678; 0.00005678

f. larger; six; 0.00730000; 7,300

---

29 Do the following problems by the method you have been shown. Show your arrows!

a. 0.356 ÷ 1,000 *b.* 0.013 × 100

c. 10.9 ÷ 100 *d.* 0.068 × 100,000

e. 0.00069 × 1,000,000 *f.* 36.98 ÷ 100,000

---

a. 0000.356; 0.000356 *b.* 0.013; 1.3 *c.* 010.9; 0.109

d. 0.068000; 6,800 *e.* 0.00069000; 690 *f.* 000036.98; 0.0003698

You have finished the program for Sec. 7-2. Now take the posttest.

POSTTEST 7-2

Give the following fractions in decimal forms:

1. $\dfrac{56{,}207}{100}$

2. $\dfrac{317}{10{,}000}$

Give the answers to the following in decimal form:

3. $0.327 \div 1{,}000$
4. 0.0019×100
5. $2.07 \div 10$
6. $0.079 \times 10{,}000$
7. $0.00061 \times 1{,}000{,}000$
8. $31.27 \div 100{,}000$

When you have finished the posttest, correct it yourself using the answers on page 328. Then check all the questions you got wrong on the list below:

1. ______	4. ______	7. ______
2. ______	5. ______	8. ______
3. ______	6. ______	

If you compare your performance with that on Pretest 7-2, you will see how much you have improved. If you made more than one mistake on the posttest, review the material and take the pretest again. Don't move on to the next section until you can get all the questions right.

Now proceed to the next section whose number you circled on page 300. If no other sections were circled, take Final Test 7 (page 326).

Section 7-3 Metric Units

Do this section if you made a mistake in Question 6 or 7 of Diagnostic Test 7. Begin by taking the pretest below, which will show you how much of Sec. 7-3 you already know and what you will be able to do by the time you have finished it.

PRETEST 7-3

Convert the following measurements to the units indicated:

1. 520 kilometers = ______ meter(s)
2. 33 centimeters = ______ kilometer(s)
3. 0.52 meter = ______ millimeter(s)
4. 0.035 millimeter = ______ centimeter(s)
5. 0.065 gram = ______ kilogram(s)
6. 125 milligrams = ______ gram(s)

When you have finished the pretest, correct it yourself using the answers on page 328. Then check all the questions you got wrong on the list below:

1. ______ 4. ______

2. ______ 5. ______

3. ______ 6. ______

If you made any mistakes (even if it was only *one* mistake), read the introduction and go through the program.

If you made no mistakes at all, skip Sec. 7-3, because you already know its material, and go on to the next section whose number you circled on page 300. If no further numbers were circled, take Final Test 7 (page 326).

Introduction

Section 7-3 introduces the metric system of units for length and weight and gives you practice in converting measurements from one metric unit to another.

FRAMES FOR SECTION 7-3

1 The basic unit of the metric system of length is the *meter,* which is just over a yard (1 meter = 1.09 yards). Only one larger unit, the *kilometer,* is in common use, and 1 kilometer = 1,000 meters. How many meters are the following distances?

a. 2 kilometers *b.* 5.7 kilometers *c.* 0.83 kilometer

---

a. 2,000 *b.* 5,700 *c.* 830

---

2 How many kilometers are the following distances?

a. 10,000 meters *b.* 6,342 meters *c.* 235 meters

---

a. 10 *b.* 6.342 *c.* 0.235

---

3 Meters are abbreviated as m and kilometers as km:

a. 23 m = ______ km *b.* 0.86 km = ______ m

---

a. 0.023 *b.* 860

---

4 Two metric units shorter than the meter are in common use. The smaller, the *millimeter,* is equal to one-thousandth of a meter. Thus 1 millimeter = $\frac{1}{1{,}000}$ meter = 0.001 meter. How many millimeters are the following distances?

a. 5 meters *b.* 0.73 meter *c.* 0.028 meter

---

a. 5,000 *b.* 730 *c.* 28

---

5 Millimeters are abbreviated as mm.

a. 5,231 mm = ______ m *b.* 29 mm = ______ m

c. 5.3 mm = ______ m

---

a. 5.231 *b.* 0.029 *c.* 0.0053

---

6 *a.* 0.56 km = ______ m = ______ mm

b. 732.1 mm = ______ m = ______ km

---

a. 560; 560,000 *b.* 0.7321; 0.0007321

---

7 The other common unit, the *centimeter,* is easily remembered, because just as a cent is one-hundredth of a dollar, a centimeter is one-hundredth of a meter. 1 centimeter = $\frac{1}{100}$ meter. How many centimeters are the following distances?

a. 3 meters *b.* 0.621 meter *c.* 0.0053 meter

---

a. 300 *b.* 62.1 *c.* 0.53

---

8 1 meter = ______ centimeters = ______ millimeters. How many millimeters in a centimeter?

---

100 1,000 10

---

9 The abbreviation for centimeter is cm.

a. 23 mm = ______ cm *b.* 7 mm = ______ cm

c. 5.6 mm = ______ cm

---

a. 2.3 *b.* 0.7 *c.* 0.56

---

10 *a.* 23 cm = ______ mm *b.* 112 cm = ______ m

c. 5.6 cm = ______ mm *d.* 5.6 cm = ______ m

---

a. 230 *b.* 1.12 *c.* 56 *d.* 0.056

---

11 *a.* 52 cm = ______ m = ______ km

b. 0.57 km = ______ m = ______ cm

---

a. 0.52; 0.00052 *b.* 570; 57,000

---

12 *a.* 112 cm = ______ km

b. 2.5 cm = ______ m

c. 13.65 cm = ______ mm

---

a. 0.00112 *b.* 0.025 *c.* 136.5

---

13 *a.* 5.3 km = ______ cm *b.* 0.132 km = ______ m

c. 0.067 km = ______ mm

---

a. 530,000 *b.* 132 *c.* 67,000

---

14 *a.* 2.5 mm = ______ m

b. 25.7 mm = ______ km

c. 325 mm = ______ cm

---

a. 0.0025 *b.* 0.0000257 *c.* 32.5

---

15 The metric system of weight is very similar. The basic unit is the *gram*, which is about 0.035 ounce, and there are two other common units:

a. 1 kilogram = ______ gram(s) *b.* 1 milligram = ______ gram(s)

---

a. 1,000 *b.* $\frac{1}{1,000}$

---

16 The abbreviation for gram is g, and the abbreviation for kilogram is kg.

a. 3.1 kg = ______ g *b.* 22 g = ______ kg

---

a. 3,100 *b.* 0.022

---

17 The abbreviation for milligram is mg.

a. 0.56 g = _____ mg *b.* 256.3 mg = _____ g

---

a. 560 *b.* 0.2563

---

18 *a.* 0.7 kg = _____ g = _____ mg

b. 25.6 mg = _____ g = _____ kg

---

a. 700; 700,000 *b.* 0.0256; 0.0000256

---

19 *a.* 5.7 mg = _____ kg *b.* 23 mg = _____ g

c. 631 mg = _____ kg

---

a. 0.0000057 *b.* 0.023 *c.* 0.000631

---

20 *a.* 0.63 kg = _____ mg

b. 27.2 kg = _____ g

c. 6.35 kg = _____ mg

---

a. 630,000 *b.* 27,200 *c.* 6,350,000

You have finished the program for Sec. 7-3. Now take the posttest to see how much you have learned.

POSTTEST 7-3

Convert the following measurements to the units indicated:

1. 3.1 kilometers = _____ meter(s)
2. 112 centimeters = _____ kilometer(s)
3. 0.0671 meter = _____ millimeter(s)
4. 0.46 millimeter = _____ centimeter(s)
5. 0.0059 gram = _____ kilogram(s)
6. 22 milligrams = _____ gram(s)

When you have finished the posttest, correct it yourself using the answers on page 328. Then check all the questions you got wrong on the list below:

1. _____ 3. _____ 5. _____

2. _____ 4. _____ 6. _____

If you compare your performance with that on Pretest 7-3, you will see how much you have improved. If you made more than one mistake on this posttest, review the material and take the pretest again. Don't move on to the next section until you can get all the questions right.

Now proceed to the next section whose number you circled on page 300. If no other sections were circled, take Final Test 7 (page 326).

Section 7-4
Conversion of Fractions to Decimal Form and Introduction to Infinite Decimals

Do this section if you made a mistake in Questions 8 to 10 of Diagnostic Test 7. Begin by taking the pretest below, which will show you how much of Sec. 7-4 you already know and what you will be able to do by the time you have finished it.

PRETEST 7-4

Convert the following fractions to decimal forms (using infinite-decimal notation where necessary):

1. $\frac{218}{6}$ 2. $\frac{71}{4{,}000}$ 3. $\frac{12}{50}$

4. $\frac{88}{90{,}000}$ 5. $\frac{2{,}653}{8}$ 6. $\frac{523}{700}$

When you have finished the pretest, correct it yourself using the answers on page 328. Then check all the questions you got wrong on the list below:

1. ______ 3. ______ 5. ______

2. ______ 4. ______ 6. ______

If you made any mistakes (even if it was only *one* mistake), read the introduction and go through the program.

If you made no mistakes at all, skip Sec. 7-4, because you already know its material, and go on to take Final Test 7 (page 326).

Introduction

Section 7-4 shows how fractions can be converted to decimal form by division and how some rational numbers cannot be represented by decimals with a finite number of digits. You learn infinite-decimal notation as a method for representing some of these numbers.

FRAMES FOR SECTION 7-4

1 Many fractions with denominators which are not multiples of 10 can be converted into decimals by division. But this procedure needs some justification. If we divide 223 by 4, we get $55\frac{3}{4}$. Since

$$\frac{3}{4} = \frac{300}{400} = \frac{300}{4} \times \frac{1}{100} = \_\_\_\_ \times \frac{1}{100}$$

$$= \frac{?}{100} = \_\_\_\_ \text{ (in decimal form)}$$

So

$$\frac{223}{4} = 55 + \_\_\_\_ = \_\_\_\_ \text{ (in decimal form)}$$

---

75 75 0.75 0.75 55.75

---

2 Alternatively we could have written

$$\frac{223}{4} = \frac{22{,}300}{400} = \frac{22{,}300}{4} \times \frac{1}{100} = \_\_\_\_ \times \frac{1}{100} = \_\_\_\_ \text{ (in decimal form)}$$

5,575 55.75

---

3 This method involves moving the decimal point ____ places to the (left/right) ____ to convert 223 to 22,300 and then moving the decimal point back ____ places to the (left/right) ____ to convert 5,575 to 55.75.

---

two; right; two; left

---

4 It is quicker to keep the decimal point in the same position throughout the division. Notice that the division is essentially the same as for integers.

$$4\overline{)22300} \quad 5575 \qquad \text{and} \qquad 4\overline{)223.00} \quad 55.75$$

Use the same method to divide 549 and 618 by 4:

a. $4\overline{)549.00}$ *b.* $4\overline{)618.00}$

---

a. $4\overline{)549.00}$ 137.25 *b.* $4\overline{)618.0}$ 154.5

---

5 Use the same method to divide 2,313 by 2 and 4,391 by 8.

a. $2\overline{)2313.0}$ *b.* $8\overline{)4391.000}$

---

a. 2|2313.0
1156.5

b. 8|4391.000
548.875

---

6 Zero digits can be added on the end of the decimal for as long as the division remains incomplete. Thus:

a. In Frame 5*a*, one zero digit was added.

b. In Frame 4*a*, ______ zero digit(s) were added.

c. In Frame 5*b*, ______ zero digit(s) were added.

---

b. two *c.* three

---

7 Use the same method to divide 1,321 by 4 and 5,639 by 8.

4|1321.00
330.25

8|5635.000
704.375

---

8 Convert the following fractions to decimal forms:

a. $\frac{431}{8}$ *b.* $\frac{7{,}633}{8}$ *c.* $\frac{123}{5}$

---

a. 53.875 *b.* 954.125 *c.* 24.6

---

9 Convert the following fractions to decimal forms:

a. $\frac{2{,}374}{5}$ *b.* $\frac{670}{8}$ *c.* $\frac{3{,}135}{4}$

---

a. 474.8 *b.* 83.75 *c.* 783.75

---

10 Many fractions cannot be converted to decimal forms of finite length. Suppose, for example, we wish to divide 325 by 3:

$\frac{325}{3} =$ ______ (mixed-numeral form)

$\frac{3{,}250}{3} =$ ______ (mixed-numeral form)

$\frac{32{,}500}{3} =$ ______ (mixed-numeral form)

$\frac{325{,}000}{3} =$ ______ (mixed-numeral form)

$108\frac{1}{3}$ $1{,}083\frac{1}{3}$ $10{,}833\frac{1}{3}$ $108{,}333\frac{1}{3}$

---

11 Carry out the following division, but stop when you have eight digits in your answer:

$3\overline{)325.00000}$

What would happen if you continued?

$\begin{array}{r} 3\overline{)325.00000} \\ 108.33333 \end{array}$; you would go on adding 3s to your answer forever.

---

12 There are two ways of coping with this difficulty. In this chapter we use a special notation for *infinite decimals*. It is also possible to be content with an approximation. (We introduce the approximation method in Chap. 9.) We write the answer to Frame 11 in infinite-decimal notation as $108.\overline{3}$, using a line above the digit 3 to indicate that it is to be repeated an infinite number of times. Write the decimal form of $\frac{16}{3}$ in infinite-decimal form.

$5.\overline{3}$

---

13 Convert the following fractions to infinite-decimal form.

a. $\frac{17}{3}$ *b.* $\frac{11}{6}$ *c.* $\frac{112}{9}$

---

a. $5.\overline{6}$ *b.* $1.8\overline{3}$ *c.* $12.\overline{4}$

---

14 A line above two digits means that both digits should be repeated. Thus $5.3\overline{12}$ is short for 5.312121212. . . . . Convert the following fractions to infinite-decimal form:

a. $\frac{256}{11}$ *b.* $\frac{322}{6}$ *c.* $\frac{313}{11}$

---

a. $23.\overline{27}$ *b.* $53.\overline{6}$ *c.* $28.\overline{45}$

---

15 If the denominator contains a multiple of 10, it is easier to divide by 10 first. Thus

$\frac{119}{60} = \frac{11.9}{6} = \_\_\_\_$ and $\frac{4{,}138}{300} = \frac{41.38}{3} = \_\_\_\_$

$1.98\overline{6}$ $13.79\overline{3}$

---

16 Convert the following fractions to infinite-decimal form:

a. $\dfrac{125}{1{,}100}$ *b.* $\dfrac{627}{9{,}000}$

---

a. $\dfrac{1.25}{11} = 0.11\overline{36}$ *b.* $\dfrac{0.627}{9} = 0.069\overline{6}$

---

17 Convert the following fractions to infinite-decimal form:

a. $\dfrac{325}{11{,}000}$ *b.* $\dfrac{13}{6{,}000}$

---

a. $\dfrac{0.325}{11} = 0.029\overline{54}$ *b.* $\dfrac{0.013}{6} = 0.0021\overline{6}$

---

18 Convert the following fractions to infinite-decimal form:

a. $\dfrac{12}{110{,}000}$ *b.* $\dfrac{213}{9{,}000}$

---

a. $\dfrac{0.0012}{11} = 0.0001\overline{09}$ *b.* $\dfrac{0.213}{9} = 0.023\overline{6}$

---

19 Division by 7 is particularly lengthy because there are six repeating digits. Convert $\dfrac{26}{7}$ and $\dfrac{30}{7}$ to infinite-decimal form.

$3.\overline{714285}$ $4.\overline{285714}$

---

20 Convert the following fractions to infinite-decimal form:

a. $\dfrac{48}{7{,}000}$ *b.* $\dfrac{625}{30}$ *c.* $\dfrac{8}{900}$

---

a. $\dfrac{0.048}{7} = 0.006\overline{857142}$ *b.* $\dfrac{62.5}{3} = 20.8\overline{3}$ *c.* $\dfrac{0.08}{9} = 0.00\overline{8}$

---

21 Convert the following fractions to decimal form:

a. $\dfrac{22}{5{,}000}$ *b.* $\dfrac{639}{40}$ *c.* $\dfrac{5}{1{,}100}$

---

a. 0.0044 *b.* 15.675 *c.* $0.00\overline{45}$

---

22 Convert the following fractions to decimal form:

a. $\frac{21}{8{,}000}$ *b.* $\frac{321}{70}$ *c.* $\frac{5}{90{,}000}$

---

a. 0.002625 *b.* $4.5\overline{857142}$ *c.* $0.0000\overline{5}$

You have finished the program for Sec. 7-4. Now take the posttest to see how much you have learned.

POSTTEST 7-4

Convert the following fractions to decimal forms (using infinite-decimal notation where necessary):

1. $\frac{118}{3}$ 2. $\frac{69}{800}$ 3. $\frac{392}{500}$

4. $\frac{35}{60{,}000}$ 5. $\frac{1{,}931}{4}$ 6. $\frac{195}{7{,}000}$

When you have finished the posttest, correct it yourself using the answers on page 328. Then check all the questions you got wrong on the list below:

1. _____ 3. _____ 5. _____

2. _____ 4. _____ 6. _____

If you compare your performance with that on Pretest 7-4, you will see how much you have improved. If you made more than one mistake on this posttest, review the material and take the pretest again.

Now proceed to Final Test 7.

FINAL TEST 7

1. Convert 21.037 to mixed-numeral form.
2. Convert 0.00725 to fractional form.
3. Convert $\frac{113}{10{,}000}$ to decimal form.
4. $0.0694 \div 100{,}000 =$ _____
5. $0.00093 \times 1{,}000 =$ _____
6. 12.5 milligrams = _____ gram(s)
7. 0.052 kilometer = _____ centimeter(s)

Convert the following fractions to decimal forms (using infinite-decimal notation where necessary):

8. $\frac{419}{4}$ 9. $\frac{31}{600}$ 10. $\frac{129}{700}$

The answers to this test are on page 328. Check the questions you got wrong on the list below and compare your results with those on Diagnostic Test 7 to see how much you have improved.

1. ____	3. ____	6. ____	8. ____
2. ____	4. ____	7. ____	9. ____
	5. ____		10. ____

If you made more than one mistake, review the material and then take the diagnostic test again. Make sure that you can get all the questions right before you finish the chapter.

Now proceed to Chap. 8.

ANSWERS TO TESTS FOR CHAPTER 7

DIAGNOSTIC TEST 7

1. $39\frac{73}{1{,}000}$
2. $\frac{651}{1{,}000{,}000}$
3. 0.00235
4. 0.000516
5. 5.3
6. 0.00013
7. 630
8. 40.125
9. $0.002\overline{7}$
10. $3.3\overline{428571}$

PRETEST 7-1

1. $30\frac{53}{100}$
2. $212\frac{62}{1{,}000}$
3. $51\frac{39}{10{,}000}$
4. $\frac{18{,}007}{10}$
5. $\frac{96}{100{,}000}$
6. $\frac{20{,}012{,}795}{1{,}000{,}000}$

POSTTEST 7-1

1. $40\frac{273}{1{,}000}$
2. $329\frac{5}{100}$
3. $23\frac{78}{10{,}000}$
4. $\frac{7{,}906}{10}$
5. $\frac{57}{1{,}000{,}000}$
6. $\frac{3{,}002{,}398}{100{,}000}$

PRETEST 7-2

1. 32.019
2. 0.0091
3. 0.00428
4. 7.3
5. 0.0108
6. 970
7. 230
8. 0.0004239

POSTTEST 7-2

1. 562.07
2. 0.0317
3. 0.000327
4. 0.19
5. 0.0207
6. 790
7. 610
8. 0.0003127

PRETEST 7-3

1. 520,000
2. 0.00033
3. 520
4. 0.0035
5. 0.000065
6. 0.125

POSTTEST 7-3

1. 3,100
2. 0.00112
3. 67.1
4. 0.046
5. 0.0000059
6. 0.022

PRETEST 7-4

1. $36.\overline{3}$
2. 0.01775
3. 0.24
4. $0.0009\overline{7}$
5. 331.625
6. $0.74\overline{714285}$

POSTTEST 7-4

1. $39.\overline{3}$
2. 0.08625
3. 0.784
4. $0.00058\overline{3}$
5. 482.75
6. $0.027\overline{857142}$

FINAL TEST 7

1. $21\frac{37}{1,000}$
2. $\frac{725}{100,000}$
3. 0.0113
4. 0.000000694
5. 0.93
6. 0.0125
7. 5,200
8. 104.75
9. $0.051\overline{6}$
10. $0.18\overline{428571}$

Chapter 8
Operations with Decimals

This chapter introduces addition, subtraction, multiplication, and division with decimals. Section 8-1 covers the operations of addition and subtraction; Sec. 8-2 covers multiplication and division of decimals by integers and division of decimals by decimals; and Sec. 8-3 covers multiplication of decimals by decimals.

Chapter 8 requires completion of Chaps. 3, 5, 6, and 7.

DIAGNOSTIC TEST 8

1. $0.0019 - 0.04 - 0.0004 + 0.3030$
2. $1.0009 - 0.005 + 11.036 - 0.063$
3. 0.00526×730
4. $0.01325 \div 0.000053$
5. $0.1036 \div 3{,}700$
6. 0.00072×0.068
7. 21.67×0.0395

When you have finished the diagnostic test, correct it yourself using the answers on page 347. Then check all the questions you got wrong on the list below:

1. ______ 3. ______ 6. ______

2. ______ 4. ______ 7. ______

5. ______

If you made no mistakes at all, you need not do any work in this chapter and should go straight on to Chap. 9, Approximations, which begins on page 349. But if you made any mistakes (even if it was only *one* mistake), follow these instructions:

	section	page
If you made any mistakes in Question 1 or 2, circle Sec. 8-1.	8-1	330
If you made any mistakes in Questions 3 to 5, circle Sec. 8-2.	8-2	335
If you made any mistakes in Question 6 or 7, circle Sec. 8-3.	8-3	341

You should now work through the sections whose numbers you have circled, taking them in numerical order. The sections begin on the pages indicated.

Section 8-1
Addition and Subtraction of Decimals

Do this section if you made a mistake in Question 1 or 2 of Diagnostic Test 8. Begin by taking the pretest below, which will show you how much of Sec. 8-1 you already know and what you will be able to do by the time you have finished it.

PRETEST 8-1

1. $5.267 + 31.9063 + 0.00956$
2. $42.3 - 6.789$
3. $25.091 - 2.4689$
4. $107.38 - 6.24 + 0.032 - 0.0008$
5. $0.634 - 10.9782 + 12.39 - 1.235$

When you have finished the pretest, correct it yourself using the answers on page 347. Then check all the questions you got wrong on the list below:

1. ______ 3. ______ 5. ______

2. ______ 4. ______

If you made any mistakes (even if it was only *one* mistake), read the introduction and go through the program.

If you made no mistakes at all, skip Sec. 8-1, because you already know its material, and go on to the next section whose number you circled on page 329. If no further numbers were circled, take Final Test 8 (page 346).

Introduction

This section introduces the operations of addition and subtraction with decimals. You learn that they are essentially similar to operations with integers and how correct alignment of numbers avoids mistakes in positioning the decimal point.

FRAMES FOR SECTION 8-1

1 $10.367 + 5.2981 = \dfrac{?}{1{,}000} + \dfrac{?}{10{,}000}$

$$= \frac{? + ?}{10{,}000} = \frac{?}{10{,}000} = \text{\_\_\_\_\_\_ (in decimal form)}$$

$$\frac{10{,}367}{1{,}000} + \frac{52{,}981}{10{,}000} = \frac{103{,}670 + 52{,}981}{10{,}000} = \frac{156{,}651}{10{,}000} = 15.6651 \qquad \begin{bmatrix} 103{,}670 \\ 52{,}981 \\ \hline 156{,}651 \end{bmatrix}$$

---

2 $0.03321 + 16.157 = \dfrac{?}{100{,}000} + \dfrac{?}{1{,}000}$

$$= \frac{? + ?}{100{,}000} = \frac{?}{100{,}000} = \_\_\_\_\_ \text{ (in decimal form)}$$

$$\frac{3{,}321}{100{,}000} + \frac{16{,}157}{1{,}000} = \frac{3{,}321 + 1{,}615{,}700}{100{,}000} = \frac{1{,}619{,}021}{100{,}000} = 16.19021$$

$$\left[\begin{array}{r} 3{,}321 \\ 1{,}615{,}700 \\ \hline 1{,}619{,}021 \end{array}\right]$$

---

3 Notice that the first 3 in 0.03321 represents $3 \times \dfrac{1}{100}$.

a. What digit in 16.157 represents hundredths?

b. What digit in 16.19021 represents hundredths?

c. What do you notice about these three digits in the addition sum on the right of the answer to Frame 2?

---

a. 5 *b.* 9

c. They are in the same vertical line, i.e., above one another.

---

4 If we write these three decimals above one another with the 3, 5, and 9 in a vertical line

$$\begin{array}{l} \ \ 0.03321 \\ 16.157 \\ \hline 16.19021 \end{array}$$

we obtain an addition sum which is similar to that of Frame 2. What can you say about the decimal points in this addition?

They are in a vertical line.

---

5 If we rewrite the addition of Frame 1 with the same digits above one another, we get:

$$\begin{array}{r} 10\ 3670 \\ 5\ 2981 \\ \hline 15\ 6651 \end{array}$$

Insert decimal points appropriate to the addition carried out in Frame 1. What can you say about the positions of the decimal points?

10.3670
5.2981
15.6651

They are in a vertical line.

---

6 Provided that we maintain the decimal points in a vertical line, addition of decimals follows precisely the same pattern as addition of integers. Carry out the following additions:

a. 7.6230
0.0325

b. 21.3600
6.7423

---

a. 7.6555 *b.* 28.1023

---

7 The addition of zeros on the end of a decimal is optional, but it can help to position the digits correctly. Carry out the following additions (after arranging the numbers vertically and adding zeros where desired).

a. $2.635 + 0.03296$

b. $21.76 + 2.378$

---

a. 2.63500
0.03296
2.66796

b. 21.760
2.378
24.138

---

8 Carry out the following additions:

a. $0.07328 + 0.0005673$

b. $25.29 + 0.00763$

---

a. 0.0732800
0.0005673
0.0738473

b. 25.29000
0.00763
25.29763

---

9 The procedure is the same when more than two numbers are involved. Thus $0.0017 + 0.05 + 0.0003 + 0.2070$ is written as

0.0017
0.0500
0.0003
0.2070

Complete the addition.

0.2590

---

10 With each of the problems below, write down all the numbers one below the other with all the decimal points in a vertical line. Then carry out the addition and find the answer.

a. 0.1307 + 42.4 + 0.008 + 320.9216 = ______

```
 _._____
__._
 _.___
___._____
---------
___._____
```

b. 431.293 + 0.0007 + 6.981 = ______

c. 13.78 + 0.006 + 2.0134 = ______

---

a. 0.1307, 42.4000, 0.0080, 320.9216; sum 363.4603

b. 431.2930, 0.0007, 6.9810; sum 438.2747

c. 13.7800, 0.0060, 2.0134; sum 15.7994

---

11 *a.* 3,269.23 + 0.016 + 23.5432 = ______

b. 219.503 + 99.99 + 100.011 = ______

---

a. 3,269.2300, 0.0160, 23.5432; sum 3,292.7892

b. 219.503, 99.990, 100.011; sum 419.504

---

12 Subtraction of decimals follows the same pattern as subtraction of integers. Adding 0s to the end may be useful. Thus 332.963 − 229.3473 is written as

```
  ___.___0
− ___.____
  ---------
  ___.____
```

Complete the subtraction.

```
 332.9630
−229.3473
 103.6157
```

---

13 Use the same method to carry out the following subtractions:

a. 420.19 − 96.3528

b. 20.37 − 0.0293

---

a. 420.1900 − 96.3528 = 323.8372 *b.* 20.3700 − 0.0293 = 20.3407

---

14 Use the same method to carry out the following subtractions:

a. 2.769 − 0.3259 *b.* 0.763 − 0.002917

---

a. 2.7690 − 0.3259 = 2.4431 *b.* 0.763000 − 0.002917 = 0.760083

---

15 When there are both additions and subtractions, add the numbers to be added and the numbers to be subtracted separately and then subtract. Thus, to find 110.96 − 90.99 + 2.306 − 10.7,

$$\begin{array}{r} 110.96 \\ +\ \ 2.306 \\ \hline \_\_\_.\_\_\_ \end{array} \qquad \begin{array}{r} 90.99 \\ +\ 10.7 \\ \hline \_\_\_.\_\_ \end{array} \qquad \begin{array}{r} \_\_\_.\_\_\_ \\ -\ \_\_\_.\_\_\_ \\ \hline \_\_.\_\_\_ \end{array}$$

113.266 101.69 113.266 − 101.690 = 11.576

---

16 *a.* 112 − 13.68 + 0.34 − 6.392 = _____

b. 0.59 − 0.0083 + 7.806 − 1.2305 = _____

---

a. 112.00 + 0.34 = 112.340; 13.680 + 6.392 = 20.072; 112.340 − 20.072 = 92.268

b. 0.590 + 7.806 = 8.3960; 0.0083 + 1.2305 = 1.2388; 8.3960 − 1.2388 = 7.1572

---

17 *a.* 125 − 10.73 + 6.21 − 0.007

b. −0.62 + 113 − 0.191 + 6.7293

---

a. 125.00 + 6.21 = 131.210; 10.730 + 0.007 = 10.737; 131.210 − 10.737 = 120.473

b. 113.0000 + 6.7293 = 119.7293; 0.620 + 0.191 = 0.811; 119.7293 − 0.8110 = 118.9183

You have finished the program for Sec. 8-1. Now take the posttest to see how much you have learned.

POSTTEST 8-1

1. $2.915 + 25.6291 + 0.00964$
2. $25.7 - 6.218$
3. $29.052 - 4.2138$
4. $112.63 - 3.28 + 0.037 - 0.0009$
5. $0.762 - 11.4368 + 13.49 - 1.532$

When you have finished the posttest, correct it yourself using the answers on page 347. Then check all the questions you got wrong on the list below:

1. ______ 4. ______
2. ______ 5. ______
3. ______

If you compare your performance with that on Pretest 8-1, you will see how much you have improved. If you made more than one mistake on this posttest, review the material and take the pretest again. Don't move on to the next section until you can get all the questions right.

Now proceed to the next section whose number you circled on page 329. If no other sections were circled, take Final Test 8 (page 346).

Section 8-2 Multiplication by Integers and Division by Decimals

Do this section if you made a mistake in Questions 3 to 5 of Diagnostic Test 8. Begin by taking the pretest below, which will show you how much of Sec. 8-2 you already know and what you will be able to do by the time you have finished it.

PRETEST 8-2

1. 0.0389×23
2. $0.851 \times 1{,}700$
3. $0.0221 \div 1{,}700$
4. $62.1 \div 0.023$
5. $131.1 \div 0.00057$
6. $0.0133 \div 0.38$

When you have finished the pretest, correct it yourself using the answers on page 347. Then check all the questions you got wrong on the list below:

1. ______ 4. ______
2. ______ 5. ______
3. ______ 6. ______

If you made any mistakes (even if it was only *one* mistake), read the introduction and go through the program.

If you made no mistakes at all, skip Sec. 8-2, because you already know its material, and go on to the next section whose number you circled on page 329. If no further numbers were circled, take Final Test 8 (page 346).

Introduction

Section 8-2 introduces the multiplication and division of decimals by integers and shows it to be essentially similar to the multiplication and division of integers by integers. You then learn how division by a decimal can be transformed into an equivalent division by an integer by multiplying both numerator and denominator by an appropriate multiple of 10.

FRAMES FOR SECTION 8-2

1 0.07642 + 0.07642 + 0.07642 = ______

0.22926

---

2 Notice that the procedure you use is the same as for addition of integers and that you keep the decimal points in a vertical line.

$$\begin{array}{r} 0.07642 \\ 0.07642 \\ \underline{0.07642} \\ 0.22926 \end{array}$$

If we had multiplied by 3 instead of adding 3 times, the procedure would still have been the same as that used for the multiplication of integers:

a. $\begin{array}{r} 7{,}642 \\ \underline{\times 3} \\ 22{,}926 \end{array}$ *b.* $\begin{array}{r} 0.07642 \\ \underline{\times 3} \\ 022926 \end{array}$

Insert a decimal point in the answer to *b*. What can you say about the position of this decimal point?

0.22926 It is vertically below the decimal point in 0.07642.

---

3 Insert decimal points in the answers:

a.
```
 0.07642
     ×31
    7642
  229260
  236902
```

b.
```
  0.07642
     ×311
     7642
    76420
  2292600
  2376662
```

a. 2.36902 *b.* 23.76662

4 Insert decimal points in the answers:

a.
```
  23.721
     ×45
  118605
  948840
 1067445
```

b.
```
 0.00632
    ×35
   3160
  18960
  22120
```

a. 1,067.445 *b.* 0.22120

5 Carry out the following multiplications:

a. 0.167×23 *b.* 13.28×65

a.
```
 0.167
    23
   501
 3.340
 3.841
```

b.
```
  13.28
     65
  66.40
 796.80
 863.20
```

6 Carry out the following multiplications:

a. 0.00726×321 *b.* 0.0829×513

a.
```
 0.00726
     321
     726
   14520
 2.17800
 2.33046
```

b.
```
  0.0829
     513
    2487
    8290
 41.4500
 42.5277
```

7 Carry out the following multiplications:

a. 0.00000963 × 29 *b.* 0.000125 × 324

---

a. 0.00027927 *b.* 0.0405

---

8 Division of decimals by integers also follows the same procedure as division of integers by integers.

a. $4\overline{)836}$ = 209 *b.* $4\overline{)0.00836}$ = 000209

Insert the decimal point in the answer to division *b*. What can you say about the position of this decimal point?

0.00209 It's vertically above the decimal point in 0.00836.

---

9 Insert decimal points in the answers to the divisions below:

a. $3\overline{)0.022926}$ = 0007642

b. $35\overline{)2.2120}$ = 00632

```
   2 10
     112
     105
      70
      70
```

---

a. 0.007642 *b.* 0.0632

---

10 Carry out the following divisions:

a. 0.00063 ÷ 9 *b.* 0.03841 ÷ 23

---

a. $9\overline{)0.00063}$ = 0.00007 *b.* $23\overline{)0.03841}$ = 0.00167

```
     23
     154
     138
      161
      161
```

---

11 Carry out the following divisions:

a. 8.632 ÷ 65 *b.* 0.00236902 ÷ 31

---

a.
```
      0.1328
65 ) 8.6320
     6 5
     2 13
     1 95
       182
       130
        520
        520
```

b.
```
      0.00007642
31 ) 0.00236902
          217
           199
           186
            130
            124
              62
              62
```

---

12 Decimal denominators can be converted to integer denominators by multiplication and then divided by the same method. Thus:

a. $\frac{0.72}{0.04} = \frac{72}{4} =$ ______ *b.* $\frac{0.00072}{0.9} = \frac{0.0072}{9} =$ ______

---

a. 18 *b.* 0.0008

---

13 *a.* $\frac{0.000045}{0.03} = \frac{?}{3} =$ ______ *b.* $\frac{0.0000081}{0.009} = \frac{?}{9} =$ ______

---

a. $\frac{0.0045}{3} = 0.0015$ *b.* $\frac{0.0081}{9} = 0.0009$

---

14 *a.* $7.28 \div 56 =$ ______ *b.* $0.0728 \div 0.56 =$ ______

---

a. 0.13 *b.* 0.13

---

15 *a.* $0.002365 \div 0.055$ *b.* $0.0323 \div 0.19$

---

a. $\frac{2.365}{55} = 0.043$ *b.* $\frac{3.23}{19} = 0.17$

---

16 *a.* $62.1 \div 0.027$ *b.* $0.02067 \div 0.39$

---

a. $\frac{62,100}{27} = 2,300$ *b.* $\frac{2.067}{39} = 0.053$

---

17 In multiplying or dividing by integers ending in 0, it is usually easiest to multiply or divide by a multiple of 10 first. Thus

$\frac{62.1}{2700} = \frac{?}{27} =$ ______ and $\frac{0.02067}{390} = \frac{?}{39} =$ ______

$\frac{0.621}{27} = 0.023$ $\frac{0.002067}{39} = 0.000053$

18 *a.* 0.629 × 320 = _____ × 32 = _____ (*think carefully*)

b. 0.00836 × 800 = _____ × 8 = _____

a. 6.29; 201.28 *b.* 0.836; 6.688

19 *a.* 0.00469 × 410 = _____ *b.* 0.00546 ÷ 1,400 = _____

a. 0.0469 × 41 = 1.9229 *b.* 0.0000546 ÷ 14 = 0.0000039

20 *a.* 0.000613 × 5,700 = _____ *b.* 3.172 ÷ 61,000 = _____

a. 3.4941 *b.* 0.000052

You have finished the program for Sec. 8-2. Now take the posttest to see how much you have learned.

POSTTEST 8-2

1. 0.0065 × 219
2. 0.0923 × 580
3. 0.0221 ÷ 1,300
4. 60.9 ÷ 0.029
5. 131.1 ÷ 0.000023
6. 0.01411 ÷ 0.83

When you have finished the posttest, correct it yourself using the answers on page 347. Then check all the questions you got wrong on the list below:

1. _____ 4. _____

2. _____ 5. _____

3. _____ 6. _____

If you compare your performance with that on Pretest 8-2, you will see how much you have improved. If you made more than one mistake on this posttest, review the material and take the pretest again. Don't move on to the next section until you can get all the questions right.

Now proceed to the next section whose number you circled on page 329. If no other sections were circled, take Final Test 8 (page 346).

Section 8-3
Multiplication by Decimals

Do this section if you made a mistake in Question 6 or 7 of Diagnostic Test 8. Begin by taking the pretest below, which will show you how much of Sec. 8-3 you already know and what you will be able to do by the time you have finished it.

PRETEST 8-3

1. 902.39×41.3
2. 0.00063×0.74
3. 0.0791×113.8
4. $0.000632 \times 2{,}900$
5. 113.5×0.047
6. 0.0052×0.000061

When you have finished the pretest, correct it yourself using the answers on page 348. Then check all the questions you got wrong on the list below:

1. ______ 3. ______ 5. ______

2. ______ 4. ______ 6. ______

If you made any mistakes (even if it was only *one* mistake), read the introduction and go through the program.

If you made no mistakes at all, skip Sec. 8-3, because you already know its material, and go on to take Final Test 8 (page 346).

Introduction

This section covers the multiplication of decimals by decimals and shows how the traditional rule of counting the digits after the decimal point is derived.

FRAMES FOR SECTION 8-3

1 *a.* $100 \times 1{,}000 =$ ______

b. $1{,}000 \times 1{,}000 =$ ______

c. $10{,}000 \times 1{,}000 =$ ______

---

a. 100,000 *b.* 1,000,000 *c.* 10,000,000

---

2 *a.* How many 0s are there in 1,000? How many 0s are there in 100? How many 0s are there in 100 × 1,000, or 100,000?

$2 + 3 =$ ______

b. How many 0s are there in 1,000 × 1,000, or 1,000,000?

$3 + 3 =$ ______

c. How many 0s are there in 10,000? How many 0s are there in 10,000 × 1,000, or 10,000,000?

$4 + 3 =$ ______

---

a. three; two; five; 5 *b.* six; 6 *c.* four; seven; 7

---

3 *a.* Express 0.432 in fractional form.

b. How many 0s are there in the denominator?

c. How many digits are there to the right of the decimal point in 0.432?

---

a. $\frac{432}{1,000}$ *b.* three *c.* three

---

4 *a.* Express 0.0031 in fractional form.

b. How many 0s are there in the denominator?

c. How many digits are there to the right of the decimal point?

---

a. $\frac{31}{10,000}$ *b.* four *c.* four

---

5 $0.432 \times 0.0031 = \frac{432}{1000} \times \frac{31}{10,000} = \frac{?}{10,000,000} =$ ______ (in decimal form)

13,392 0.0013392

---

6 *a.* How many 0s are there in 1,000?

b. How many 0s are there in 10,000?

c. How many 0s are there in 10,000,000?

d. How many digits are there to the right of the decimal point in 0.432?

e. How many digits are there to the right of the decimal point in 0.0031?

f. How many digits are there to the right of the decimal point in 0.0001728?

---

a. three *b.* four *c.* seven *d.* three *e.* four *f.* seven

---

7 $0.0823 \times 0.52 = \frac{823}{?} \times \frac{52}{?} = \frac{?}{?} =$ ______ (in decimal form)

10,000 100 $\frac{42{,}796}{1{,}000{,}000} = 0.042796$

---

8 *a.* How many 0s are there in 10,000?

b. How many 0s are there in 100?

c. How many 0s are there in 1,000,000?

d. How many digits are there to the right of the decimal point in 0.0823?

e. How many digits are there to the right of the decimal point in 0.52?

f. How many digits are there to the right of the decimal point in 0.0013392?

---

a. four *b.* two *c.* six *d.* four *e.* two *f.* seven

---

9 $9.72 \times 0.4 = \frac{972}{?} \times \frac{4}{?} = \frac{?}{?} =$ ______ (in decimal form)

---

100 10 $\frac{3{,}888}{1{,}000}$ 3.888

---

10 *a.* How many 0s are there in 100?

b. How many 0s are there in 10?

c. How many 0s are there in 1,000?

d. How many digits are there to the right of the decimal point in 9.72?

e. How many digits are there to the right of the decimal point in 0.4?

f. How many digits are there to the right of the decimal point in 3.888?

---

a. two *b.* one *c.* three *d.* two *e.* one *f.* three

---

11 *a.* How many digits are there to the right of the decimal point in 53.4?

b. How many digits are there to the right of the decimal point in 0.0273?

c. Insert the decimal point in the answer:

$53.4 \times 0.0273 = 000145782$

---

a. one *b.* four *c.* 1.45782

---

12 From the examples of Frames 4 to 11 it can be seen that if the number of digits to the right of the decimal point in $A = a$, the number of digits to the right of the decimal point in $B = b$, the number of digits to the right of the decimal point in $C = c$, and $A \times B = C$, then a __ $b = c$ (insert correct sign).

+

---

13 Insert the decimal point in the answers of the following multiplications:

a. $5.34 \times 0.00273 = 000145782$ *b.* $0.972 \times 0.04 = 0003888$

c. $82.3 \times 0.052 = 00042796$

---

a. 0.0145782 *b.* 0.03888 *c.* 4.2796

---

14 Insert the decimal point in the answers of the following multiplications:

a. $4.32 \times 0.00031 = 0000013392$ *b.* $0.0972 \times 4 = 0003888$

c. $52 \times 0.00823 = 00042796$

---

a. 0.0013392 *b.* 0.3888 *c.* 0.42796

---

15 This gives us a simple two-stage method of multiplying decimals. Thus to multiply 0.067 by 0.21:

Stage 1: Multiply corresponding integers:

$67 \times 21 =$ ______

Stage 2: Insert decimal point in correct position:

$0.067 \times 0.21 =$ ______

1,407 0.01407

---

16 Use the same method to multiply 0.125 by 7.3:

a. $125 \times 73 =$ ______ *b.* $0.125 \times 7.3 =$ ______

---

a. 9,125 *b.* 0.9125

---

17 Use the same method to multiply 13.6 by 0.00062.

---

136 × 62 = 8,432 13.6 × 0.00062 = 0.008432

---

18 Multiply the following pairs of decimals:

a. 501.32 × 24.2 *b.* 50.132 × 0.242 *c.* 0.50132 × 0.0242

---

a. 12,131.944 *b.* 12.131944 *c.* 0.012131944

---

19 Particular care is needed when 0s are formed during the multiplication. Thus 375 × 84 = ______, and 0.375 × 0.84 is equal to:

a. 0.31500 *b.* 0.03150 *c.* 0.00315

31,500 *a*

---

20 Use this method to multiply 0.65 by 0.032.

---

65 × 32 = 2,080 0.65 × 0.032 = 0.02080

---

21 Multiply the following pairs of decimals:

a. 46.5 × 0.00112 *b.* 0.0465 × 0.0112

c. 465 × 11.2 *d.* 0.465 × 112

---

a. 0.052080 *b.* 0.00052080 *c.* 5,208.0 *d.* 52.080

---

22 Multiply the following pairs of decimals:

a. 0.00079 × 0.65 *b.* 0.0525 × 12.4 *c.* 193.2 × 0.068

---

a. 0.0005135 *b.* 0.65100 *c.* 13.1376

---

23 Care is also needed in multiplying by integers ending in 0, and it is perhaps easiest to remove the zero digit first. Thus

0.0726 × 240 = ______ × 24 = ______

0.6312 × 400 = ______ × 4 = ______

0.726 17.424 63.12 252.48

---

24 Multiply the following pairs of decimals:

a. 900 × 0.00005432 = 9 × ______ = ______

b. 0.03656 × 400 = ______

c. 0.0004396 × 50 = ______

---

a. 0.005432; 0.048888 *b.* 3.656 × 4 = 14.624
c. 0.004396 × 5 = 0.021980

---

25 Multiply the following pairs of decimals:

a. 0.0624 × 0.00375 *b.* 0.961 × 200

c. 32.6 × 0.037 *d.* 0.065 × 80

---

a. 0.000234000 *b.* 192.2 *c.* 1.2062 *d.* 5.20

You have finished the program for Sec. 8-3. Now take the posttest to see how much you have learned.

POSTTEST 8-3

1. 42.736 × 7.91
2. 0.00059 × 0.35
3. 0.00582 × 69.2
4. 0.000089 × 370
5. 260.7 × 0.031
6. 0.00076 × 0.059

When you have finished the posttest, correct it yourself using the answers on page 348. Then check all the questions you got wrong on the list below:

1. ______ 3. ______ 5. ______

2. ______ 4. ______ 6. ______

If you compare your performance with that on Pretest 8-3, you will see how much you have improved. If you made more than one mistake on this posttest, review the material and take the pretest again.

Now proceed to Final Test 8.

FINAL TEST 8

1. 0.020 − 9.403 + 1080 − 6.60
2. 4.0003 − 0.0381 + 12.16 − 0.00009
3. 0.00715 × 670
4. 0.2301 ÷ 0.00059
5. 0.02967 ÷ 6900
6. 0.00081 × 0.058
7. 32.94 × 0.00685

The answers to this test are on page 348. Check the questions you got wrong on the list below and compare your results with those on Diagnostic Test 8 to see how much you have improved.

1. ______	3. ______	6. ______
2. ______	4. ______	7. ______
	5. ______	

If you made more than one mistake, review the material and then take the diagnostic test again. Make sure that you can get all the questions right before you finish the chapter.

Now proceed to Chap. 9, Approximations.

ANSWERS TO TESTS FOR CHAPTER 8

DIAGNOSTIC TEST 8

1. 0.2645	2. 11.9689	3. 3.8398
4. 250	5. 0.000028	6. 0.00004896
7. 0.855965		

PRETEST 8-1

1. 37.18286
2. 35.511
3. 22.6221
4. 101.1712
5. 0.8108

POSTTEST 8-1

1. 28.55374
2. 19.482
3. 24.8382
4. 109.3861
5. 1.2832

PRETEST 8-2

1. 0.8947
2. 1,446.7
3. 0.000013
4. 2,700
5. 230,000
6. 0.035

POSTTEST 8-2

1. 1.4235
2. 53.534
3. 0.000017
4. 2,100
5. 5,700,000
6. 0.017

PRETEST 8-3

1. 37,268.707
2. 0.0004662
3. 9.00158
4. 1.8328
5. 5.3345
6. 0.0000003172

POSTTEST 8-3

1. 338.04176
2. 0.0002065
3. 0.402744
4. 0.03293
5. 8.0817
6. 0.00004484

FINAL TEST 8

1. 1064.017	2. 16.12211	3. 4.7905
4. 390	5. 0.0000043	6. 0.00004698
7. 0.225639		

Chapter 9
Approximations

This chapter deals with methods of making approximations and using them to simplify calculations. You will be shown how to estimate the magnitude of the resultant error. Section 9-1 introduces rounding off and Sec. 9-2 the concept of significant figures. Then Sec. 9-3 shows how approximate answers can be obtained by using one-figure approximations. Section 9-4 examines the error involved in making one- and two-figure approximations and introduces the concept of proportional error. In Sec. 9-5 you use approximations to simplify more complex expressions and learn how numbers can be slightly altered to permit division by common factors.

This chapter requires completion of Chaps. 7 and 8.

There is no diagnostic test for this chapter because most of it is compulsory. Sections 9-1 and 9-2 can be skipped if you can pass their pretests, but Secs. 9-3 to 9-5 are all compulsory. They have no pretests. You must proceed through each section in turn.

Section 9-1
Approximation by Rounding Off

Begin Sec. 9-1 by taking the pretest below, which will show you how much of Sec. 9-1 you already know and what you will be able to do by the time you have finished it.

PRETEST 9-1

Round off the following numbers to the nearest hundredth:

1. 17.8649 2. 0.055 3. 1.0091

Round off the following numbers to three decimal places:

4. 7.8165 5. 0.02350 6. 2.8997

Convert the following fractions to decimal forms correct to one decimal place:

7. $\frac{497}{6}$ 8. $\frac{41}{7}$

When you have finished the pretest, correct it yourself using the answers on page 382. Then check all the questions you got wrong on the list below:

1. _____ 4. _____ 7. _____

2. _____ 5. _____ 8. _____

3. _____ 6. _____

If you made any mistakes (even if it was only *one* mistake), read the introduction and go through the program.

If you made no mistakes at all, skip Sec. 9-1, because you already know its material, and go on to Sec. 9-2.

Introduction

This section introduces the process of rounding off a number to obtain an approximate but shorter form. Since we are often interested in answers of limited accuracy, shorter forms can be used to simplify much numerical work.

FRAMES FOR SECTION 9-1

1 Which of the following operations would you expect to be quicker?

a. 11.2 + 10.7 or *b.* 11.237 + 10.719

---

a

---

2 Often we are only interested in answers of limited accuracy; so we use approximate shortened forms of numbers in order to save time. Thus we may use 11.2 instead of 11.237 and 10.7 instead of 10.719. Reduce the numbers 13.732 and 15.941 to approximate three-digit forms.

13.7 15.9

---

3 This process of shortening a number to obtain an approximate form is called *rounding off*. In shortening 13.732 to 13.7, you were rounding off to the nearest (tenth/hundredth) ______. Rounding off 13.732 to the nearest tenth gives ______. Rounding off 13.732 to the nearest hundredth gives ______.

---

tenth 13.7 13.73

---

4 *a*. Is 13.732 closer to 13 or 14?

b. Rounding off 13.732 to the nearest whole number gives ______.

---

a. 14 *b*. 14

---

5 *a*. Is 10.719 closer to 10.71 or 10.72?

b. Rounding off 10.719 to the nearest hundredth gives ______.

---

a. 10.72 *b*. 10.72

---

6 *a*. Is 32.97 closer to 32.9 or 33.0?

b. Rounding off 32.97 to the nearest tenth gives ______.

---

a. 33.0 *b*. 33.0

---

7 Round off the following numbers to the nearest tenth:

a. 9.63 *b*. 11.542 *c*. 36.981 *d*. 0.74

---

a. 9.6 *b*. 11.5 *c*. 37.0 *d*. 0.7

---

8 Round off the following numbers to the nearest hundredth:

a. 11.542 *b.* 36.981 *c.* 8.638 *d.* 0.858 *e.* 0.033

---

a. 11.54 *b.* 36.98 *c.* 8.64 *d.* 0.86 *e.* 0.03

---

9 Round off the following numbers to the nearest whole number:

a. 36.981 *b.* 8.638 *c.* 11.842 *d.* 0.831

---

a. 37 *b.* 9 *c.* 12 *d.* 1

---

10 "Round off to one decimal place" means "round off until there is only one figure after the decimal point" or "round off to the nearest (tenth/hundredth) ______." Round off the following numbers to one decimal place:

a. 6.543 *b.* 27.692 *c.* 13.582 *d.* 0.543 *e.* 0.563

---

tenth *a.* 6.5 *b.* 27.7 *c.* 13.6 *d.* 0.5 *e.* 0.6

---

11 "Round off to two decimal places" means "round off until there are ______ figures after the decimal point" or "round off to the nearest (tenth/hundredth/thousandth) ______." Round off the following numbers to two decimal places:

a. 6.543 *b.* 27.692 *c.* 13.552 *d.* 0.563 *e.* 0.063

---

two; hundredth *a.* 6.54 *b.* 27.69 *c.* 13.55 *d.* 0.56 *e.* 0.06

---

12 Round off the following numbers to three decimal places:

a. 7.9642 *b.* 15.03944 *c.* 0.62872 *d.* 0.03271

---

a. 7.964 *b.* 15.039 *c.* 0.629 *d.* 0.033

---

13 There is one special case in rounding off which has to be settled by convention. If you have to round off a number like 6.85 to one decimal place, you don't know whether to make it ______ or ______ because 6.85 is exactly halfway between. A convention for cases like this is to select the alternative which gives the higher result. Thus 6.85 rounded off to one decimal place gives ______.

6.8 or 6.9 (either order) 6.9

---

14 This particular convention is used because it keeps errors to a minimum and requires the inspection of only one extra digit, because 6.85, 6.850, and 6.851 are all rounded off to 6.9. Use the convention to round off the following numbers to one decimal place:

a. 3.85 *b.* 3.95 *c.* 3.05

---

a. 3.9 *b.* 4.0 *c.* 3.1

---

15 Round off the following numbers to one decimal place:

a. 23.45 *b.* 4.15 *c.* 30.95 *d.* 0.35

---

a. 23.5 *b.* 4.2 *c.* 31.0 *d.* 0.4

---

16 Round off the following numbers to two decimal places:

a. 5.615 *b.* 31.365 *c.* 2.995 *d.* 0.645 *e.* 0.035

---

a. 5.62 *b.* 31.37 *c.* 3.00 *d.* 0.65 *e.* 0.04

---

17 Notice that this convention ensures that only one extra digit need be inspected during rounding off. In rounding off to two decimal places,

31.365 becomes ______

31.365 0001 becomes ______

31.364 9999 becomes ______

31.37 31.37 31.36

---

18 Round off the following numbers to three decimal places:

a. 32.3785 *b.* 19.2914999 *c.* 5.2345001

---

a. 32.379 *b.* 19.291 *c.* 5.235

---

19 Round off the following numbers to two decimal places:

a. 24.395001 *b.* 13.285 *c.* 0.50499 *d.* 1.9973

---

a. 24.40 *b.* 13.29 *c.* 0.50 *d.* 2.00

---

20 Round off the following numbers to one decimal place:

a. 5.983 *b.* 18.35 *c.* 0.976

---

a. 6.0 *b.* 18.4 *c.* 1.0

---

21 Round off the following numbers to two decimal places:

a. 0.0079 *b.* 1.0449 *c.* 0.097 *d.* 11.625

---

a. 0.01 *b.* 1.04 *c.* 0.10 *d.* 11.63

---

22 Round off the following numbers to three decimal places:

a. 0.0995 *b.* 0.0008 *c.* 3.7598 *d.* 17.1235

---

a. 0.100 *b.* 0.001 *c.* 3.760 *d.* 17.124

---

23 Rounding off is particularly useful in converting fractions to approximate decimal forms, as you have to continue the division for only one extra figure. For example, if we want $\frac{10}{7}$ to two decimal places, we have to proceed to only three decimal places:

$$7\underline{|10.000}$$
$$1.428 + \text{remainder}$$

and 1.428 rounded off to two decimal places is ______.

1.43

---

24 Convert $\frac{31}{11}$ to a decimal form which is correct to three decimal places.

$$11\underline{|31.0000}$$
$$2.8181 \qquad 2.818$$

---

25 Convert the following fractions to decimal forms correct to two decimal places:

a. $\frac{35}{9}$ *b.* $\frac{21}{8}$ *c.* $\frac{33}{7}$

---

a. 3.89 *b.* 2.63 *c.* 4.71

---

26 Convert the following fractions to decimal forms correct to one decimal place:

a. $\frac{159}{11}$ *b.* $\frac{231}{4}$ *c.* $\frac{451}{6}$

a. 14.5 *b.* 57.8 *c.* 75.2

You have finished the program for Sec. 9-1. Now take the posttest to see how much you have learned.

POSTTEST 9-1

Round off the following numbers to the nearest tenth:

1. 29.849
2. 0.75
3. 2.094

Round off the following numbers to three decimal places:

4. 15.2185
5. 0.01350
6. 7.3998

Convert the following fractions to decimal forms correct to two decimal places:

7. $\frac{235}{9}$
8. $\frac{69}{11}$

When you have finished the posttest, correct it yourself using the answers on page 382. Then check all the questions you got wrong on the list below:

1. _____	4. _____	7. _____
2. _____	5. _____	8. _____
3. _____	6. _____	

If you compare your performance with that on Pretest 9-1, you will see how much you have improved. If you made more than one mistake on this posttest, review the material and take the pretest again. Don't move on to the next section until you can get all the questions right.

Now proceed to Sec. 9-2.

Section 9-2
Significant Figures

Begin Sec. 9-2 by taking the pretest below, which will show you how much of Sec. 9-2 you already know and what you will be able to do by the time you have finished it.

PRETEST 9-2

How many significant figures have the following numbers?

1. 2.35
2. 0.007931
3. $1.58 \times 1{,}000$
4. 64.30

Convert the following numbers to two-significant-figure approximations:

5. 4.98
6. $1.53 \times 10{,}000$

Convert the following numbers to one-significant-figure approximations:

7. 75
8. 0.039

When you have finished the pretest, correct it yourself using the answers on page 382. Then check all the questions you got wrong on the list below:

1. ______	4. ______	7. ______
2. ______	5. ______	8. ______
3. ______	6. ______	

If you made any mistakes (even if it was only *one* mistake), read the introduction and go through the program.

If you made no mistakes at all, skip Sec. 9-2, because you already know its material, and go on to Sec. 9-3.

Introduction

This section introduces the term *significant figure* for a precise digit in a measurement. The accuracy of a measurement can then be defined by the number of significant figures and is independent of the units in which it happens to be expressed. You learn how rounding off reduces the number of significant figures and hence the accuracy.

FRAMES FOR SECTION 9-2

1 If we record a measurement as 4.53 meters, we mean that it is nearer to 4.53 than to 4.52 or 4.54 meters, i.e., that it is correct to ______ decimal places.

---

two

---

2 We could, however, record the same measurement as ______ centimeters; and in this case we would mean that it was closer to ______ than to ______ or ______ centimeters. Would it still be correct to two decimal places?

---

453 453 452 or 454 no

---

3 We can record the same measurement as 4.53 meters, which is correct to two decimal places; or as 453 centimeters, which is correct to the nearest whole number. But in both cases the number of precise digits in the measurement is ______.

---

three

---

4 We can record a measurement either as 47.95 centimeters, which is correct to ______ decimal place(s), or as 479.5 millimeters, which is correct to ______ decimal place(s). In both cases the number of precise digits in the measurement is ______.

---

two; one; four

---

5 The number of precise digits in a measurement is called the number of *significant figures.* Thus 4.53 has three significant figures, and 37.51 has four significant figures. How many significant figures have the following numbers?

a. 5.289 *b.* 45.3 *c.* 52.89

---

a. four *b.* three *c.* four

---

6 How many significant figures have the following numbers?

a. 3.29 *b.* 32.9 *c.* 3.298 *d.* 32.98

---

a. three *b.* three *c.* four *d.* four

---

7 How many significant figures have the following numbers?

a. 4.58 *b.* 3 *c.* 70.08 *d.* 21.6

---

a. three *b.* one *c.* four *d.* three

---

8 If you measure the length of an object, there (is/is not) ______ a limit to the precision of your measurements.

---

is

---

9 The way you record the results of a measurement should show the precision of the measurement. If you report a length of 12 cm, you mean that the length is nearer to 12 than it is to 11 or to 13 cm. If you report a length of 12.0 cm, you mean that the length is nearer to 12.0 than it is to 11.9 or to 12.1 cm. If you report a length of 12.00 cm, you mean that the length is nearer to 12.00 than it is to ______ or to ______ cm.

11.99 12.01 (either order)

---

10 If you report a length of 12.02 cm, you mean that the length is nearer to 12.02 than it is to ______ or to ______ cm. If you report a length of 12.020 cm, you mean that the length is nearer to 12.020 than it is to ______ or to ______ cm.

12.01 12.03 (either order) 12.019 12.021 (either order)

---

11 Reporting a length of 12.10 cm instead of 12.1 cm:

a. Makes no difference.

b. Indicates a different length.

c. Indicates that the measurement is more precise.

---

c

---

12 How many significant figures have the following numbers?

a. 12 *b.* 12.0 *c.* 12.00

---

a. two *b.* three *c.* four

---

13 How many significant figures have the following numbers?

a. 13.50 *b.* 6.0 *c.* 79.5200

---

a. four *b.* two *c.* six

---

14 With a number like 300 one cannot tell whether:

a. The zeros are only being used to indicate that the 3 represents 3×100 or

b. The zeros also represent precise digits.

In case *a* it has ______ significant figure(s), but in case *b* it has ______ significant figure(s).

---

one; three

---

15 If one wishes to avoid doubt, one can write 3×100 when there is one significant figure, 3.0×100 when there are ______ significant figures, and 3.00×100 when there are ______ significant figures. Write 400 and 220 in forms which indicate that there are two significant figures.

---

two three 4.0×100 2.2×100

---

16 *a.* Write 4,200 in a form which indicates that there are two significant figures.

b. Write 7,000 in a form which indicates that there are three significant figures.

$4.2 \times 1{,}000$ $7.00 \times 1{,}000$ This notation is not unlike powers-of-10 notation, which we introduce in the next chapter. The importance of indicating the precision of a measurement is one of the reasons why scientists like to use powers-of-10 notation.

---

17 How many significant figures have the following numbers?

a. 7×100 *b.* $6.70 \times 1{,}000$ *c.* 4,000.0

---

a. one *b.* three *c.* five

---

18 How many significant figures have the following numbers?

a. 6×10 *b.* 300.00 *c.* 4.0×100

---

a. one *b.* five *c.* two

---

19 A measurement of 3.10 millimeters can also be written as 0.310 centimeter or as 0.00310 meter. How many significant figures have the following numbers?

a. 3.10 *b.* 0.310 *c.* 0.00310

a. three *b.* three *c.* three Once a measurement has been made, its precision is fixed and cannot be altered by the choice of units for its description.

20 Notice in Frame 19 that a zero in front of a number (does/does not) ______ count as one of its significant figures but a zero after a number (does/does not) ______ count as one of its significant figures.

does not; does

21 How many significant figures have the following numbers?

a. 3.20 × 100 *b.* 32.0 *c.* 3.20 *d.* 0.320 *e.* 0.0320 *f.* 0.00320

three in every case

22 How many significant figures have the following numbers?

a. 8.50 × 1000 *b.* 9,000.0 *c.* 0.00620 *d.* 0.050 *e.* 0.0004 *f.* 423.10

a. three *b.* five *c.* three *d.* two *e.* one *f.* five

23 How many significant figures have the following numbers?

a. 0.0730 *b.* 0.08

c. 0.009100 *d.* 100.0

a. three *b.* one *c.* four *d.* four

24 How many significant figures have the following numbers?

a. 1.30 × 10,000 *b.* 0.0060

c. 0.531 *d.* 29.0

a. three *b.* two *c.* three *d.* three

25 Use rounding off to reduce the following numbers to two-significant-figure approximations:

a. 1.38 × 10,000 *b.* 0.00605

c. 0.531 *d.* 29.9

a. $1.4 \times 10{,}000$ *b.* 0.0061 *c.* 0.53 *d.* 30 or better 3.0×10

26 Use rounding off to reduce the following numbers to three-significant-figure approximations:

a. 2,351 *b.* 0.09155

c. 0.6399 *d.* 371.49

a. $2.35 \times 1{,}000$ or 23.5×100 *b.* 0.0916 *c.* 0.640 *d.* 371

27 Use rounding off to reduce the following numbers to one-significant-figure approximations:

a. 1,259 *b.* 0.973

c. 0.075 *d.* 34.9

a. $1 \times 1{,}000$ *b.* 1 *c.* 0.08 *d.* 3×10

28 Use rounding off to reduce the following numbers to one-significant-figure approximations:

a. 0.0099 *b.* 85000

c. 0.0379 *d.* 65

a. 0.01 *b.* $9 \times 10{,}000$ *c.* 0.04 *d.* 7×10

You have finished the program for Sec. 9-2. Now take the posttest to see how much you have learned.

POSTTEST 9-2

How many significant figures have the following numbers?

1. 2.478
2. 0.00032
3. $4.893 \times 10{,}000$
4. 2.90

Convert the following numbers to two-significant-figure approximations:

5. 0.0399
6. 2.61×1000

Convert the following numbers to one-significant-figure approximations:

7. 55
8. 6.97

When you have finished the posttest, correct it yourself using the answers on page 382. Then check all the questions you got wrong on the list below:

1. _____ 4. _____ 7. _____
2. _____ 5. _____ 8. _____
3. _____ 6. _____

If you compare your performance with that on Pretest 9-2, you will see how much you have improved. If you made more than one mistake on this posttest, review the material and take the pretest again. Don't move on to the next section until you can get all the questions right.

Now proceed to Sec. 9-3.

Section 9-3 The Use of One-figure Approximations

Introduction

This section is compulsory, so there is no pretest. It shows how approximate answers can be obtained for numerical calculations, a handy way of checking answers to long calculations. Start directly with the program.

FRAMES FOR SECTION 9-3

1 Which of the following decimals is closest to 4.1×3.05?

a. 10 *b.* 12 *c.* 15 *d.* 20

---

b

---

2 All you had to do to answer Frame 1 was to round off each number to one significant figure and multiply. Thus 4.1×3.05 is approximately equal to $4 \times 3 = 12$. Use the same method, i.e., rounding off to one significant figure, to derive approximate answers to the following:

a. 5.23×6.05 *b.* $7.14 \div 4.26$ *c.* 49.3×2.88

---

a. 30 *b.* 1.7 or 1.8 *c.* 150 (50×3)

---

3 Use the same method to derive approximate answers for

a. 5.241×698.3 *b.* $86.23 \div 2.81$

a. 3,500 or $3.5 \times 1{,}000$ *b.* 30

4 If you carried out a lengthy calculation like 5.241×698.3 and arrived at an answer that was substantially different from 3,500, would you suspect that:

a. Your lengthy calculation was wrong?

b. Your approximate answer of 3,500 was wrong?

This is a matter of opinion, but I would be more inclined to suspect a lengthy calculation than a simple approximation.

5 This is the advantage of very simple approximations. They may be inaccurate, but they are unlikely to be seriously wrong. Suppose three students attempt to calculate 112.78×5.23 and get answers of:

a. 823 *b.* 590 *c.* 82

Which of these answers would you expect to be most nearly correct?

b because $100 \times 5 = 500$ is an approximate answer

6 Approximations are very useful for checking calculations, particularly for checking the position of the decimal point. Thus $46 \times 36 = 1{,}656$, and $4.6 \times 3.6 \approx 5 \times 4 =$ ______ ($\approx$ is a symbol which means "is approximately equal to"), so 4.6×3.6 is equal to

a. 1.656 *b.* 16.56 *c.* 165.6

20 *b*

7 Notice that the approximate value of 20 obtained by rounding off to one significant figure was quite a way from 16.56 but still close enough to check the position of the decimal point. We can use the method of Frame 6 as the basis of a method for multiplying decimals. Thus to find 7.34×4.2

a. $734 \times 42 = 30{,}828$ *b.* $7.34 \times 4.2 \approx$ ______ *c.* $7.34 \times 4.2 =$ ______

28 30.828

8 Use the same method to find 8.16×6.4:

a. $816 \times 64 =$ ______ *b.* $8.16 \times 6.4 \approx$ ______ *c.* $8.16 \times 6.4 =$ ______

a. 52,224 *b.* 48 *c.* 52.224

9 Use the same method to find 34.5×1.5.

$345 \times 15 = 5{,}175$ $34.5 \times 1.5 \approx 30 \times 2 = 60$ $3.45 \times 1.5 = 51.75$

---

10 If one of the numbers is less than 1, the method is essentially similar. $0.08 \times 7 =$ ______, so $0.0832 \times 6.95 \approx$ ______.

0.56 0.56

---

11 Use one-significant-figure approximations to obtain approximate answers to the following:

a. $0.0093 \times 21.6 \approx$ ______ $\times 20 =$ ______ $\times 2 =$ ______

b. $371 \times 0.0073 \approx 400 \times$ ______ $= 4 \times$ ______ $=$ ______

---

a. 0.009; 0.09; 0.18 *b.* 0.007; 0.7; 2.8

---

12 Use one-significant-figure approximations to obtain approximate answers to the following:

a. 125×0.0078

b. 0.029×3.76

c. $5{,}291 \times 0.0000513$

---

a. $100 \times 0.008 = 0.8$ *b.* $0.03 \times 4 = 0.12$
c. $5{,}000 \times 0.00005 = 5 \times 0.05 = 0.25$

---

13 Use one-significant-figure approximations to obtain approximate answers to the following:

a. 29.5×0.00614

b. 0.0283×9817

c. 46.29×0.0000597

---

a. $30 \times 0.006 = 3 \times 0.06 = 0.18$ *b.* $0.03 \times 10{,}000$ (or 3×100) $= 300$
c. $50 \times 0.00006 = 5 \times 0.0006 = 0.0030$

---

14 $0.08 \times 0.006 = \frac{8}{?} \times \frac{6}{?} = \frac{48}{?} =$ ______ (in decimal form)

---

100 1,000 100,000 0.00048

---

15 *a.* Number of zeros in 100 = ______.

b. Number of zeros in 1,000 = ______.

c. Number of zeros in 100,000 = ______.

d. Number of digits after the decimal point in 0.08 = ______.

e. Number of digits after the decimal point in 0.006 = ______.

f. Number of digits after the decimal point in 0.00048 = ______.

g. $2 + 3 =$ ______.

---

a. two *b.* three *c.* five *d.* two *e.* three *f.* five *g.* 5

---

16 In Chap. 8, Operations with Decimals, we showed that the total number of digits after the decimal point was unchanged during the multiplications of decimals. Your work in Frames 14 and 15 should remind you of this and of its explanation. Use this result to answer $0.7 \times 0.0005 =$ ______.

0.00035 (0.7 has one digit after the decimal point; 0.0005 has four digits after the decimal point; 0.00035 has $4 + 1 = 5$ digits after the decimal point.)

---

17 *a.* $0.3 \times 0.000007 =$ ______

b. $0.02 \times 0.005 =$ ______

---

a. 0.0000021 *b.* 0.00010 (The 0 in $10 = 5 \times 2$ must be kept!)

---

18 Use one-significant-figure approximations to obtain approximate answers to the following:

a. 0.0763×0.814

b. 0.0231×0.00512

c. 0.0297×0.0183

---

a. $0.08 \times 0.8 = 0.064$ *b.* $0.02 \times 0.005 = 0.00010$
c. $0.03 \times 0.02 = 0.0006$

---

19 Use one-significant-figure approximations to obtain approximate answers to the following:

a. 0.523×0.000374

b. 0.000673×0.00712

c. 0.0097×0.00124

---

a. $0.5 \times 0.0004 = 0.00020$ *b.* $0.0007 \times 0.007 = 0.0000049$
c. $0.01 \times 0.001 = 0.00001$

---

20 In fact, $0.000673 \times 0.00712 = 0.00000479176$, which, when rounded off to two significant figures, becomes ______. Thus the second figure in 0.0000049 is not accurate. This is not surprising, since we were using one-figure approximations. It is in fact quite useful, and in most cases more accurate, to keep the second figure in an approximation. But we have to remember that 0.0000049 represents (choose *a* or *b*):

a. An approximation which is accurate to two significant figures.

b. An approximation which has two significant figures but only represents a rough estimate of an answer correct to two figures.

0.0000048 *b* (The problem of the accuracy of approximations is treated in Sec. 9-4.)

---

21 With divisions, the same method of approximation is used, and again we usually give two-figure answers even though the second figure is not quite accurate. Thus

$$\frac{4.1}{3.05} \approx \frac{4}{3} \approx 1.3$$

Similarly

$$\frac{7.23}{4.15} \approx \underline{\qquad} \qquad \text{and} \qquad \frac{8.32}{4.81} \approx \underline{\qquad}$$

$\frac{7}{4} = 1.75 \approx 1.8$ $\frac{8}{5} = 1.6$

---

22 With larger numbers, cancelation of common multiples of 10 can take place before division. Thus

$$\frac{5{,}023}{3{,}194} \approx \frac{5{,}\not{0}\not{0}\not{0}}{3{,}\not{0}\not{0}\not{0}} = \frac{5}{3} \approx \underline{\qquad}$$

and

$$\frac{4{,}963}{823} \approx \underline{\qquad}$$

1.7 $\frac{5{,}0\not{0}\not{0}}{8\not{0}\not{0}} = \frac{50}{8} \approx 6.3$

---

23 Use one-significant-figure approximations to obtain approximate answers to the following:

a. $\dfrac{693}{29}$ *b.* $\dfrac{7{,}014.5}{3.78}$ *c.* $\dfrac{231.8}{92.6}$

---

a. $\dfrac{70\not{0}}{3\not{0}} = \dfrac{70}{3} \approx 23$ *b.* $\dfrac{7{,}000}{4} \approx 1{,}800$ or $1.8 \times 1{,}000$ *c.* $\dfrac{20\not{0}}{9\not{0}} \approx 2.2$

---

24 Use one-significant-figure approximations to obtain approximate answers to the following:

a. $\dfrac{21.7}{39.8}$ *b.* $\dfrac{3.78}{996}$ *c.* $\dfrac{32.9}{5{,}123}$

---

a. $\dfrac{2\not{0}}{4\not{0}} \approx 0.5$ *b.* $\dfrac{4}{1{,}000} = 0.004$ *c.* $\dfrac{3\not{0}}{5{,}00\not{0}} = \dfrac{3}{500} = \dfrac{0.03}{5} = 0.006$

---

25 Use one-significant-figure approximations to obtain approximate answers to the following:

a. $\dfrac{59.7}{423}$ *b.* $\dfrac{0.054}{31.4}$ *c.* $\dfrac{0.00793}{929}$

---

a. $\dfrac{6\not{0}}{40\not{0}} = \dfrac{6}{40} = \dfrac{0.6}{4} = 0.15$ *b.* $\dfrac{0.05}{30} = \dfrac{0.005}{3} = 0.0017$

c. $\dfrac{0.008}{900} = \dfrac{0.00008}{9} = 0.0000089$

---

26 When the denominator is less than 1, the fraction can be converted to a form in which it is greater than 1. Thus

$$\frac{23.5}{0.0673} \approx \frac{20}{0.07} = \frac{2{,}000}{7} \approx \underline{\qquad}$$

$$\frac{7.3}{0.00529} \approx \underline{\qquad}$$

---

290 $\dfrac{7}{0.005} = \dfrac{7{,}000}{5} = 1.4 \times 1{,}000$

---

27 Use one-significant-figure approximations to obtain approximate answers to the following:

a. $\dfrac{273.3}{0.721}$ *b.* $\dfrac{0.0573}{0.099}$ *c.* $\dfrac{0.00813}{0.0749}$

$a.\ \frac{300}{0.7}=\frac{3{,}000}{7}\approx 4.3\times 100$ $\quad b.\ \frac{0.06}{0.1}=\frac{0.6}{1}=0.6$ $\quad c.\ \frac{0.008}{0.07}=\frac{0.8}{7}\approx 0.11$

28 Use one-significant-figure approximations to obtain approximate answers to the following:

$a.\ \frac{97.2}{0.0634}$ $\quad b.\ \frac{0.072}{0.0000916}$ $\quad c.\ \frac{0.00625}{0.793}$

$a.\ \frac{100}{0.06}=\frac{10{,}000}{6}\approx 1.7\times 1{,}000$ $\quad b.\ \frac{0.07}{0.00009}=\frac{7{,}000}{9}\approx 7.8\times 100$

$c.\ \frac{0.006}{0.8}=\frac{0.06}{8}=0.0075$

29 Use one-significant-figure approximations to obtain approximate answers to the following:

a. $0.09783 \div 0.000621$ *b.* $7.563 \div 0.00983$ *c.* $42.35 \div 0.0575$

$a.\ \frac{0.1}{0.0006}=\frac{1{,}000}{6}\approx 1.7\times 100$ $\quad b.\ \frac{8}{0.01}=8\times 100$

$c.\ \frac{40}{0.06}=\frac{4{,}000}{6}\approx 6.7\times 100$

You have finished the program for Sec. 9-3. Now take the posttest to see how much you have learned.

POSTTEST 9-3

Use one-significant-figure approximations to obtain approximate answers for the following calculations:

1. 123×65.9
2. 0.00748×0.0887
3. $36.7 \div 8971$
4. $0.063 \div 49.31$
5. $219.4 \div 0.0852$

When you have finished the posttest, correct it yourself using the answers on page 382. Then check all the questions you got wrong on the list below:

1. ______ 3. ______ 5. ______

2. ______ 4. ______

If you made any mistakes on this posttest, review the material. Don't move on to the next section until you can get all the questions right.

Now proceed to Sec. 9-4.

Section 9-4
Error and Proportional Error in Using Approximations

Introduction

This section is compulsory, so there is no pretest. It examines the error which results from one- and two-figure approximations and finds the maximum error that can arise during addition, subtraction, multiplication, and division. It expresses this in terms of proportional error, i.e., error/true value.

FRAMES FOR SECTION 9-4

1 3.7 is a two-digit number which approximates to 4 when reduced to one significant figure.

a. What is the lowest two-digit number which approximates to 4?

b. What is the highest two-digit number which approximates to 4?

---

a. 3.5 *b.* 4.4

---

2 *a.* What is the lowest two-digit number which approximates to 6 when reduced to one significant figure?

b. What is the highest two-digit number which approximates to 6 when reduced to one significant figure?

---

a. 5.5 *b.* 6.4

---

3 $4 \times 6 = 24$ could be used as an approximation for any of the four extreme cases:

a. $3.5 \times 5.5 =$ ______ *b.* $3.5 \times 6.4 =$ ______

c. $4.4 \times 5.5 =$ ______ *d.* $4.4 \times 6.4 =$ ______

---

a. 19.25 *b.* 22.40 *c.* 24.20 *d.* 28.16

---

4 In case *a* the error was $24 - 19.25 =$ ______, and in case *d* the error was $28.16 - 24 =$ ______. In case *b* the error was ______, and in case *c* the error was ______.

---

a. 4.75 *d.* 4.16 *b.* 1.60 *c.* 0.20

---

5 What is the error involved in using $2 \times 2 = 4$ as an approximation for:

a. 1.5×1.5 *b.* 2.4×2.4

---

a. Error $= 4 - 2.25 = 1.75$ *b.* Error $= 5.76 - 4 = 1.76$

---

6 What is the error involved in using $1 \times 1 = 1$ as an approximation for 1.4×1.4? **Note:** We do not use 1 as an approximation for 0.5, because 0.5 has only one significant figure.

Error $= 1.96 - 1 = 0.96$

---

7 A better indication of the inaccuracy of an approximation is the proportional error, or $\frac{\text{error}}{\text{true value}}$.

a. Using 4×6 as an approximation for 3.5×5.5 (Frames 3 and 4), the proportional error was

$$\frac{4.75}{19.25} \approx \frac{5}{20} = 0.25$$

b. Using 2×2 as an approximation for 1.5×1.5 (Frame 5), the proportional error was

$$\frac{1.75}{2.25} = \frac{7}{9} \approx 0.\_\_\_\_$$

c. Using 2×2 as an approximation for 2.4×2.4 (Frame 5), the proportional error was

$$\frac{1.76}{5.76} \approx \frac{2}{6} \approx 0.\_\_\_\_$$

d. Using 1×1 as an approximation for 1.4×1.4 (Frame 6), the proportional error was

$$\frac{?}{?} \approx 0.\_\_\_\_$$

---

b. 0.77 *c.* 0.33 *d.* $\frac{0.96}{1.96} \approx \frac{1}{2} = 0.50$

---

8 From these results we can infer that the greatest proportional error which can result from the use of one-significant-figure approximations in multiplying two numbers is _____.

0.77

---

9 Naturally the error will rarely be as large as this, but it is just as well to be aware of the possibilities! This error is still not large enough to prevent one-figure approximations from being useful in checking the position of a decimal point. But for other purposes a more accurate approximation is needed, so we use two significant figures. Give two-significant-figure approximations of the following numbers:

a. 0.973 *b.* 1.04 *c.* 1.05 *d.* 0.995

---

a. 0.97 *b.* 1.0 *c.* 1.1 *d.* 1.0

---

10 If we use two-significant-figure approximations, which of the following situations is likely to give rise to the greatest error?

a. 1.0×1.0 for 0.995×0.995

b. 1.0×1.0 for 1.04×1.04

c. 1.1×1.1 for 1.05×1.05

---

c because the difference is the greatest:
$1.0 - 0.995 = 0.005$ $1.04 - 1.0 = 0.04$ $1.10 - 1.05 = 0.05$

---

11 Which of the following situations is likely to give rise to the greatest proportional error?

a. 1.1×1.1 for 1.05×1.05

b. 1.2×1.2 for 1.15×1.15

c. 9.9×9.9 for 9.85×9.85

---

a because the difference is proportionally the greatest:

$\frac{0.05}{1.05} \approx 0.05$ $\frac{0.05}{1.15} \approx 0.04$ $\frac{0.05}{9.85} \approx 0.005$

---

12 *a.* $1.1 \times 1.1 =$ ______.

b. $1.05 \times 1.05 =$ ______.

c. Error in using 1.1×1.1 as an approximation for $1.05 \times 1.05 =$ ______.

---

a. 1.21 *b.* 1.1025 *c.* 0.1075

---

13 Proportional error in using 1.1×1.1 as an approximation for $1.05 \times 1.05 =$

$\frac{\text{error}}{\text{true value}} = \frac{?}{?} \approx 0.$______.

$$\frac{0.1075}{1.1025} = \frac{1{,}075}{11{,}025} \approx \frac{1{,}100}{11{,}000} = 0.10$$

14 For which two of the following, will the approximation $\frac{1.1}{1.5}$ be most accurate?

a. $\frac{1.14}{1.54}$ *b.* $\frac{1.14}{1.45}$ *c.* $\frac{1.05}{1.54}$ *d.* $\frac{1.05}{1.45}$

a and *d* If numerator and denominator are both increased or both decreased by the same amount, there will be much less change in the fraction than if one increases and the other decreases.

15 The greatest proportional error in using two-figure approximations for divisions will occur in the substitution of:

a. $\frac{1.1}{1}$ for $\frac{1.05}{1.04}$ or *b.* $\frac{1}{1.1}$ for $\frac{1.04}{1.05}$

Since $\frac{1.05}{1.04} \approx 1$, the error in case *a* will be $1.1 - 1 =$ ______, and in case *b* will be $1 - 0.91 =$ ______. The proportional error in case *a* will be $\frac{?}{1} =$ ______, and in case *b* will be $\frac{?}{1} =$ ______.

0.1 0.09 $\frac{0.1}{1} = 0.1$ $\frac{0.09}{1} = 0.09$

16 So for both multiplication and division the greatest proportional error that can arise from the use of two-figure approximations is 0.1. Now let's look at addition. In which of the following situations will the greatest proportional error arise when two-figure approximations are used?

a. $1.05 + 1.05$ *b.* $1.15 + 1.15$

c. $1.04 + 1.04$ *d.* $1.14 + 1.14$

a because the error is largest in comparison with the true value.

17 What is the proportional error in this case?

Error $= 2.2 - 2.1 = 0.1$ Proportional error $= \frac{0.1}{2.1} \approx 0.05$

18 So 0.05 is the maximum possible error that can arise from the use of two-figure approximations during addition. But let us consider subtraction. If $A = 3.24$ and $B = 3.25$, the two-figure approximation for A is ______, and the two-figure approximation for B is ______.

3.2 3.3

---

19 If we use these approximations, we get

$B - A = 3.3 - 3.2 = 0.1$

But the true value of $B - A$ is $3.25 - 3.24 =$ ______.

Error $= 0.1 -$ ______ $=$ ______

$$\text{Proportional error} = \frac{\text{error}}{\text{true value}} = \frac{?}{?} = \_\_\_\_\_\_$$

0.01 $0.1 - 0.01 = 0.09$ $\frac{0.09}{0.01} = 9$

---

20 Now let us consider the situation when $A = 3.249$ and $B = 3.250$ and we use two-figure approximations. True value of $A - B$ is ______. The error is $0.1 -$ ______ $=$ ______ $\approx$ ______, and the proportional error is ______.

0.001 $0.001 = 0.099 \approx 0.1$ $\frac{0.1}{0.001} = 100$

---

21 What will be the proportional error when $A = 3.2499$ and $B = 3.250$ and we use two-figure approximations?

1,000 The error remains at 0.1, and the true value becomes 0.0001.

---

22 If we add 9s to A until it becomes the infinite decimal $3.24\overline{9}$ and keep $B = 3.25$, the difference approaches zero. In this case, what will happen to the proportional error if we continue to use two-figure approximations?

It will become infinitely large.

---

23 If A and B are two numbers and we use two-figure approximations, what are the maximum possible errors for

a. $A + B$

b. $A - B$

c. $A \times B$

d. $A \div B$

You may refer back to previous frames if you wish.

a. 0.05 *b.* infinitely large *c.* 0.1 *d.* 0.1

24 Let us consider just a few more examples of proportional error. If we use two-figure approximations to find 1.45 × 1.95, what is the proportional error? Give your answer correct to one decimal place.

$$\text{Approximate value} = 1.5 \times 2.0 = 3.0$$

$$\text{True value} = 1.45 \times 1.95 = 2.8275$$

$$\text{Error} = 0.01725$$

$$\text{Proportional error} = \frac{0.01725}{2.8275} \approx \frac{0.017}{2.8} = \frac{0.17}{28} \approx 0.006$$

25 If we use one-figure approximations to find 1.45 × 1.95, what is the proportional error? Give your answer correct to one decimal place.

$$\text{Approximate value} = 1 \times 2 = 2$$

$$\text{Error} = 0.8275$$

$$\text{Proportional error} = \frac{0.8275}{2.8275} \approx \frac{0.83}{2.8} = \frac{8.3}{28} \approx 0.3$$

26 If we use two-figure approximations to find 3.45 + 4.55, what is the proportional error? Give your answer correct to one decimal place.

$$\text{Approximate value} = 3.5 + 4.6 = 8.1$$

$$\text{True value} = 8.0$$

$$\text{Proportional error} = \frac{0.1}{8.0} \approx 0.01$$

27 If we use one-figure approximations to find 6.4 ÷ 1.6, what is the proportional error? Give your answer correct to two decimal places.

$$\text{Approximate value} = 6 \div 2 = 3$$

$$\text{True value} = 4$$

$$\text{Proportional error} = \tfrac{1}{4} = 0.25$$

28 If we use two-figure approximations to find 2.55 − 2.44, what is the proportional error? Give your answer correct to one decimal place.

$$\text{Approximate value} = 2.6 - 2.4 = 0.2$$

$$\text{True value} = 0.11$$

$$\text{Proportional error} = \frac{0.09}{0.11} = \frac{9}{11} \approx 0.8$$

You have finished the program for Sec. 9-4. Now take the posttest to see how much you have learned.

POSTTEST 9-4

What is the proportional error when one-figure approximations are used in the following calculations?

1. $8.4 \div 5.6$
2. $2.51 + 3.50$

What is the proportional error when two-figure approximations are used in the following calculations?

3. $1.653 - 1.645$
4. 11.8×23.11

5. What operation is most likely to give rise to a very large proportional error when approximations are used, addition, subtraction, multiplication, or division?

When you have finished the posttest, correct it yourself using the answers on page 382. Then check all the questions you got wrong on the list below:

1. ______ 4. ______
2. ______ 5. ______
3. ______

If you made any mistakes on this posttest, review the material. Don't move on to the next section until you can get all the questions right.

Now proceed to Sec. 9-5.

Section 9-5
The Use of Two-figure Approximations to Evaluate Large Expressions

Introduction

This section is compulsory, so there is no pretest. You learn how to use approximations in evaluating expressions containing more than two digits and how numbers can be slightly altered to permit division by common factors.

FRAMES FOR SECTION 9-5

1 Section 9-4 showed that the proportional error involved in making one-significant-figure approximations in the multiplication of two numbers can be as high as 0.77 though it is usually much lower. When three numbers are involved, the proportional error can be twice as large. Let us take the case involving the maximum possible error, $2 \times 2 \times 2$ as an approximation for $1.5 \times 1.5 \times 1.5$:

$2 \times 2 \times 2 = 8$ and $1.5 \times 1.5 \times 1.5 = 3.375$

Error $= 8 - 3.375 =$ ______

$$\text{Proportional error} = \frac{\text{error}}{\text{true value}} = \frac{?}{3.375} \approx \_\_\_\_\_\_$$

4.625 $\quad \dfrac{4.625}{3.375} \approx 1.4$

---

2 When more than three numbers are being multiplied, the proportional error can become greater still; so it is usual to use two-significant-figure approximations when more than two numbers are involved. But you should never use approximate values for numbers being (added/subtracted/multiplied/divided) ______ if the numbers are of similar magnitude. Give reasons for your choice.

---

subtracted The proportional error can become infinitely large if approximations are used for A and B when $A - B$ is almost equal to 0.

---

3 So to obtain an approximate value for an expression like $(1.123 - 1.057) \times 1.216$, *first* perform the subtraction and *then* introduce two-figure approximations.

$(1.123 - 1.057) \times 1.216 =$ ______ $\times 1.216 \approx$ ______ $\times$ ______

$=$ ______ $\approx$ ______ (two figures)

$0.066 \times 1.216 \approx 0.066 \times 1.2 = 0.0792 \approx 0.079$

4 Find approximate (two-figure) values for the following expressions:

a. $2.369 \times (2.792 - 2.783)$

b. $\dfrac{5.621}{2.691 - 2.573}$

a. $2.369 \times 0.009 \approx 2.4 \times 0.009 = 0.0216 \approx 0.022$

b. $\dfrac{5.621}{0.118} \approx \dfrac{5.6}{0.12} = \dfrac{560}{12} \approx 47$

5 Where division is involved, approximated expressions can frequently be simplified by canceling common factors. Thus

$$\frac{63.4 \times 76.8}{53.5 \times 33.1} \approx \frac{\overset{7}{\cancel{63}} \times \overset{7}{\cancel{77}}}{\underset{6}{\cancel{54}} \times \underset{3}{\cancel{33}}} \qquad \text{Dividing by 9 and 11}$$

$$= \frac{?}{?} \approx \_\_\_\_\_ \text{ (two figures)}$$

$\dfrac{49}{18} \approx 2.7$

6 Using the same method, we see that

$$\frac{44.5 \times 95.75}{63.93 \times 35.21} \approx \frac{45 \times 96}{64 \times 35} = \frac{?}{?} \approx \_\_\_\_\_ \text{ (two figures)}$$

$\dfrac{27}{14} \approx 1.9$

7 Simplify the following expression by the same method:

$$\frac{83.51 \times 48.4}{71.8 \times 18.4}$$

$\dfrac{84 \times 48}{72 \times 18} = \dfrac{28}{9} \approx 3.1$

8 If the two-figure approximations are all converted to integers, there will be no danger of misplacing decimal points during canceling. Thus

$$\frac{0.0273 \times 63.82}{0.00242 \times 2081} \approx \frac{0.027 \times 64}{0.0024 \times 2100} = \frac{? \times 64}{24 \times 2100} = \frac{?}{?} \approx \_\_\_\_\_ \text{ (two figures)}$$

---

270 $\quad \frac{24}{70}$ or $\frac{12}{35} \approx 0.34$

---

9 Notice that in converting an expression to a form containing integers only, the number of 10s by which the numerator is multiplied must equal the number of 10s by which the denominator is multiplied (and multiplying by 10 is equivalent to moving the decimal point one place to the right). Thus:

a. $\dfrac{0.41 \times 0.0039}{4.6 \times 0.59} = \dfrac{41 \times 39}{46 \times ?}$

b. $\dfrac{610 \times 0.079}{0.37 \times 0.058} = \dfrac{610 \times ?}{37 \times 58}$

---

a. 59,000 *b.* 7,900

---

10 Find an approximate value for the expression below by the method we have been using:

$$\frac{0.809 \times 0.07655}{3.578 \times 0.4176} \approx \frac{0.81 \times 0.077}{3.6 \times 0.42} = \frac{81 \times 77}{36 \times ?} = \frac{?}{?} \approx \text{\_\_\_\_\_\_} \text{ (two figures)}$$

4,200 $\quad \frac{33}{800} \approx 0.041$

---

11 Find an approximate value for the expression $\dfrac{0.746 \times 56.3}{0.01555 \times 0.0448}$.

$$\frac{0.75 \times 56}{0.016 \times 0.045} = \frac{750{,}000 \times 56}{16 \times 45} = \frac{350{,}000}{6} \approx 5.8 \times 10{,}000$$

---

12 When there are no common factors, a simple approximation can be obtained by making slight alterations in the numbers concerned. This replaces them by numbers which do have common factors that can be canceled. Thus:

$$\frac{37 \times 43}{83 \times 46} \approx \frac{36 \times 42}{84 \times 45} = \frac{?}{?} \approx \text{\_\_\_\_\_\_} \text{ (one figure)}$$

$\frac{2}{5} = 0.4$

---

13 Find an approximate value (one significant figure only) for the expression $\dfrac{79 \times 73}{65 \times 41}$.

$0.066 \times 1.216 \approx 0.066 \times 1.2 = 0.0792 \approx 0.079$

4 Find approximate (two-figure) values for the following expressions:

a. $2.369 \times (2.792 - 2.783)$

b. $\dfrac{5.621}{2.691 - 2.573}$

a. $2.369 \times 0.009 \approx 2.4 \times 0.009 = 0.0216 \approx 0.022$

b. $\dfrac{5.621}{0.118} \approx \dfrac{5.6}{0.12} = \dfrac{560}{12} \approx 47$

5 Where division is involved, approximated expressions can frequently be simplified by canceling common factors. Thus

$$\frac{63.4 \times 76.8}{53.5 \times 33.1} \approx \frac{\overset{7}{\cancel{63}} \times \overset{7}{\cancel{77}}}{\underset{6}{\cancel{54}} \times \underset{3}{\cancel{33}}} \qquad \text{Dividing by 9 and 11}$$

$$= \frac{?}{?} \approx \underline{\qquad} \quad \text{(two figures)}$$

$\dfrac{49}{18} \approx 2.7$

6 Using the same method, we see that

$$\frac{44.5 \times 95.75}{63.93 \times 35.21} \approx \frac{45 \times 96}{64 \times 35} = \frac{?}{?} \approx \underline{\qquad} \quad \text{(two figures)}$$

$\dfrac{27}{14} \approx 1.9$

7 Simplify the following expression by the same method:

$$\frac{83.51 \times 48.4}{71.8 \times 18.4}$$

$\dfrac{84 \times 48}{72 \times 18} = \dfrac{28}{9} \approx 3.1$

8 If the two-figure approximations are all converted to integers, there will be no danger of misplacing decimal points during canceling. Thus

$$\frac{0.0273 \times 63.82}{0.00242 \times 2081} \approx \frac{0.027 \times 64}{0.0024 \times 2100} = \frac{? \times 64}{24 \times 2100} = \frac{?}{?} \approx \underline{\qquad} \quad \text{(two figures)}$$

270 $\quad \frac{24}{70}$ or $\frac{12}{35} \approx 0.34$

---

9 Notice that in converting an expression to a form containing integers only, the number of 10s by which the numerator is multiplied must equal the number of 10s by which the denominator is multiplied (and multiplying by 10 is equivalent to moving the decimal point one place to the right). Thus:

a. $\dfrac{0.41 \times 0.0039}{4.6 \times 0.59} = \dfrac{41 \times 39}{46 \times ?}$

b. $\dfrac{610 \times 0.079}{0.37 \times 0.058} = \dfrac{610 \times ?}{37 \times 58}$

---

a. 59,000 $\quad$ *b.* 7,900

---

10 Find an approximate value for the expression below by the method we have been using:

$$\frac{0.809 \times 0.07655}{3.578 \times 0.4176} \approx \frac{0.81 \times 0.077}{3.6 \times 0.42} = \frac{81 \times 77}{36 \times ?} = \frac{?}{?} \approx \underline{\qquad} \text{ (two figures)}$$

4,200 $\quad \frac{33}{800} \approx 0.041$

---

11 Find an approximate value for the expression $\dfrac{0.746 \times 56.3}{0.01555 \times 0.0448}$.

$$\frac{0.75 \times 56}{0.016 \times 0.045} = \frac{750,000 \times 56}{16 \times 45} = \frac{350,000}{6} \approx 5.8 \times 10,000$$

---

12 When there are no common factors, a simple approximation can be obtained by making slight alterations in the numbers concerned. This replaces them by numbers which do have common factors that can be canceled. Thus:

$$\frac{37 \times 43}{83 \times 46} \approx \frac{36 \times 42}{84 \times 45} = \frac{?}{?} \approx \underline{\qquad} \text{ (one figure)}$$

$\frac{2}{5} = 0.4$

---

13 Find an approximate value (one significant figure only) for the expression $\dfrac{79 \times 73}{65 \times 41}$.

≈ 2; my method was $\dfrac{80 \times 72}{64 \times 40} = \dfrac{9}{4} \approx 2$. Many other methods are possible, but those which are any good should give the answer correct to one significant figure.

---

14 Find an approximate value (one significant figure only) for the expression $\dfrac{830 \times 46}{1{,}700 \times 310}$.

≈ 0.07; my method was $\dfrac{840 \times 45}{1{,}800 \times 300} = \dfrac{7}{100} = 0.07$. Many other methods are possible and, if reasonable, should give the answer correct to one significant figure.

---

15 Find an approximate value (to one significant figure only) for the expression $\dfrac{0.089 \times 0.97}{0.0047 \times 0.053}$.

$\approx 3 \times 100$; my method was $\dfrac{90 \times 9{,}600}{48 \times 54} = \dfrac{1{,}000}{3} \approx 3 \times 100$. Your answer should be correct to one significant figure though your method may have been different.

---

16 Find an approximate value (to one significant figure only) for the expression $\dfrac{0.0391 \times 0.728}{0.573 \times 6.94}$.

$\dfrac{0.039 \times 0.73}{0.57 \times 6.9} = \dfrac{39 \times 73}{57 \times 6900} \approx 0.007$ This time I haven't disclosed my method. I hope your method gave you the right answer.

---

17 Find an approximate value (to one significant figure only) for the expression $\dfrac{0.712 \times 343}{294 \times 0.00671}$.

$\dfrac{0.71 \times 340}{290 \times 0.0067} = \dfrac{7{,}100 \times 340}{290 \times 67} \approx 100$ Your method probably gave you a figure greater than 100 but still closer to 100 than to 200.

You have finished the program for Sec. 9-5. Now take the posttest to see how much you have learned.

POSTTEST 9-5

Find approximate values correct to two significant figures for the following expressions:

1. $4.651 \times (5.884 - 5.867)$

2. $\dfrac{18.34 \times 44.61}{9.88 \times 14.79}$

3. $\dfrac{0.0549 \times 0.837}{14.48 \times 0.006565}$

Find approximate values correct to one significant figure for the following expressions:

4. $\dfrac{520 \times 78}{1{,}700 \times 370}$

5. $\dfrac{0.393 \times 4.14}{0.0891 \times 0.937}$

When you have finished the posttest, correct it yourself using the answers on page 382. Then check all the questions you got wrong on the list below:

1. _____ 4. _____

2. _____ 5. _____

3. _____

If you made any mistakes on this posttest, review the material. Don't move on until you can get all the questions right. Now proceed to Final Test 9.

FINAL TEST 9

Round off the following numbers to three decimal places:

1. 8.91250

2. 3.1998

3. Convert $\dfrac{535}{11}$ to a decimal form correct to one decimal place.

How many significant figures have the following numbers?

4. $4.760 \times 10{,}000$

5. 0.004932

Convert the following numbers to two-significant-figure approximations:

6. 0.09235
7. 6.459

Use one-significant-figure approximations to obtain approximate answers for the following calculations:

8. 0.0087×0.935
9. $11.5 \div 0.0629$

What is the proportional error when approximations are used in the following calculations?

10. $3.6 \div 5.4$ (one-figure approximations)
11. $2.757 - 2.746$ (two-figure approximations)

Find approximate values for the following expressions:

12. $\dfrac{0.637 \times 5.571}{0.0963 \times 0.0317}$ (correct to two significant figures)

13. $\dfrac{42.78 \times 0.6143}{0.0934 \times 2.91}$ (correct to one significant figure)

The answers to this test are on page 382. Check the questions you got wrong on the list below:

1.	______	5.	______	8.	______	11.	______
2.	______	6.	______	9.	______	12.	______
3.	______	7.	______	10.	______	13.	______
4.	______						

If you made more than one mistake, review the material. Make sure that you can get all the questions right before you finish the chapter.

Now proceed to Chap. 10.

ANSWERS TO TESTS FOR CHAPTER 9

PRETEST 9-1

1. 17.86
2. 0.06
3. 1.01
4. 7.817
5. 0.024
6. 2.900
7. 82.8
8. 5.9

POSTTEST 9-1

1. 29.8
2. 0.8
3. 2.1
4. 15.219
5. 0.014
6. 7.400
7. 26.11
8. 6.27

PRETEST 9-2

1. 3
2. 4
3. 3
4. 4
5. 5.0
6. $1.5 \times 10{,}000$
7. 8×10 (or 80)
8. 0.04

POSTTEST 9-2

1. 4
2. 2
3. 4
4. 3
5. 0.040
6. $2.6 \times 1{,}000$
7. 6×10 (or 60)
8. 7

POSTTEST 9-3

1. 7,000 or $7 \times 1{,}000$
2. 0.00063
3. 0.0044
4. 0.0012
5. 2,200 or $2.2 \times 1{,}000$

POSTTEST 9-4

1. 0.11
2. 0.16 or 0.17
3. 11.5
4. 0.012
5. subtraction

POSTTEST 9-5

1. 0.080
2. 5.5
3. 0.50
4. 0.06 or 0.07
5. 20

FINAL TEST 9

1. 8.913
2. 3.200
3. 48.6
4. four
5. four
6. 0.092
7. 6.5
8. 0.0081
9. 170
10. 0.20
11. 8.1
12. 1,200 or $1.2 \times 1{,}000$
13. 100

Chapter 10
Powers of 10

In this chapter you will learn that decimals can be written in a standard exponent form $A \times 10^b$, where A is between 1 and 10 and b is an integer. This convenient form for carrying out multiplications and divisions of decimals leads easily to the use of the slide rule and logarithms.

Section 10-1 introduces positive integer exponents and shows how numbers in exponent form can be multiplied and divided. Section 10-2 defines negative integer exponents and the zero exponent, and extends the laws of multiplication and division which were developed for positive integer exponents in Sec. 10-1 to cover all integral exponents.

In Sec. 10-3 you learn how to convert numbers in integer, decimal, or nonstandard exponent form into standard exponent form. Section 10-4 shows how to use standard exponent forms in the multiplication and division of decimals, and Sec. 10-5 shows how numbers in standard exponent form can be added and subtracted.

This chapter requires knowledge of Chaps. 7, 8, and 9.

DIAGNOSTIC TEST 10

1. Give the integer or decimal forms of the following numbers:

   10^1 $\quad$ 10^0 $\quad$ 10^{-3}

Give the answers to the following in exponential notation:

2. $10^8 \times 10^{-11}$
3. $10^{-9} \div 10^{-12}$
4. $10^6 \div 10^9$

5. Write 3.1×10^{-2} in decimal form.

Write the following in standard exponent form, i.e., in the form $A \times 10^b$, where A is between 1 and 10 and b is an integer.

6. 1,792
7. 193×10^6
8. 0.24×10^{-9}

Carry out the following operations, putting your answers into standard exponent form:

9. $(3 \times 10^{-5}) \times (4 \times 10^{-4})$ 10. $(4 \times 10^{-4}) \div (8 \times 10^{2})$

Use powers of 10 to solve the following problems. Give the answers in standard exponent form.

11. $\dfrac{4{,}800 \times 0.0000144}{25{,}600}$ 12. $\dfrac{0.00000162 \times 0.11}{0.00099}$

Carry out the following operations, putting your answers into standard exponent form:

13. $(5.43 \times 10^{1}) - (6.98 \times 10^{-1})$ 14. $(7.2 \times 10^{-5}) + (9.43 \times 10^{-4})$

15. $(8.43 \times 10^{7}) - (2.71 \times 10^{5})$

When you have finished the diagnostic test, correct it yourself using the answers on page 419. Then check all the questions you got wrong on the list below:

1.	_____	5.	_____	9.	_____	13.	_____
2.	_____	6.	_____	10.	_____	14.	_____
3.	_____	7.	_____	11.	_____	15.	_____
4.	_____	8.	_____	12.	_____		

If you made no mistakes at all, you need not do any work in this chapter and should go straight on to Chap. 11. But if you made any mistakes (even if it was only *one* mistake), follow these instructions:

	section	page
If you made any mistakes in Questions 1 to 4, circle Secs. 10-1 and 10-2.	10-1 10-2	385 391
If you made any mistakes in Questions 5 to 8, circle Sec. 10-3.	10-3	396
If you made any mistakes in Questions 9 to 12, circle Sec. 10-4.	10-4	404
If you made any mistakes in Questions 13 to 15, circle Sec. 10-5.	10-5	413

You should now work through the sections whose numbers you have circled, taking them in numerical order. The sections begin on the pages indicated.

Section 10-1
Multiplication and Division of Positive Integer Powers of 10

Do this section if you made a mistake in Questions 1 to 4 of Diagnostic Test 10. Begin by taking the pretest below, which will show you how much of Sec. 10-1 you already know and what you will be able to do by the time you have finished it.

PRETEST 10-1

1. Write 16^5 in multiplication notation, i.e., ______ × ______ × ______, etc.
2. Write 23 × 23 × 23 × 23 in exponential notation.
3. Name the base and the exponent of 7^{11}.
4. Give the integer forms of 10^6 and 10^1.

Give the answers to the following in exponential notation:

5. $10^{17} \times 10^{32}$
6. $10^{16} \times 10^{25}$
7. $10^9 \div 10^8$
8. $10^{13} \div 10^7$

When you have finished the pretest, correct it yourself using the answers on page 419. Then check all the questions you got wrong on the list below:

1. ______	4. ______	7. ______
2. ______	5. ______	8. ______
3. ______	6. ______	

If you made any mistakes (even if it was only *one* mistake), read the introduction and go through the program.

If you made no mistakes at all, skip Sec. 10-1, because you already know its material, and go on to Sec. 10-2.

Introduction

After introducing exponential notation and the terms *base* and *exponent*, this section concentrates on *powers of* 10, that is, exponents with a base of 10. It derives laws for the multiplication and division of powers of 10 with positive integer exponents and shows you how to apply them.

FRAMES FOR SECTION 10-1

1 In mathematics special notation is often a great help in expressing something briefly and concisely. Multiplication notation, for example, enables us to summarize a series of similar additions. Thus, instead of $14 + 14 + 14 + 14 + 14$, we write ____ $\times$ 14 or 14 $\times$ ____. How would you write $21 + 21 + 21 + 21$ in multiplication notation?

---

5 5 4×21 or 21×4

---

2 To summarize a series of similar additions, we use (multiplication/division) ____ notation, and to summarize a series of similar multiplications, we use *exponential* notation. In exponential notation $7 \times 7 \times 7 \times 7$ is written as 7^4. Note that the 4 is written half a line (above/below) ____ the 7 and comes (before/after) ____ the 7. How would you write $3 \times 3 \times 3 \times 3 \times 3 \times 3$ in exponential notation?

---

multiplication; above; after; 3^6

---

3 *a.* How would you write $8 \times 8 \times 8$ in exponential notation?

b. How would you write 11^5 in multiplication notation?

---

a. 8^3 *b.* $11 \times 11 \times 11 \times 11 \times 11$

---

4 *a.* Write 5^2 in multiplication notation.

b. Write $2 \times 2 \times 2 \times 2 \times 2 \times 2 \times 2$ in exponential notation.

---

a. 5×5 *b.* 2^7

---

5 Write the following in multiplication notation:

a. 3^4 *b.* 4^3

What is the value of:

c. 3^4 *d.* 4^3

e. Are 3^4 and 4^3 the same?

---

a. $3 \times 3 \times 3 \times 3$ *b.* $4 \times 4 \times 4$ *c.* $3^4 = 81$ *d.* $4^3 = 64$
e. No, their values are different.

---

6 Write the following in multiplication form and give their numerical values:

a. 3^2 *b.* 2^3 *c.* 5^2 *d.* 2^5

---

a. $3 \times 3 = 9$ *b.* $2 \times 2 \times 2 = 8$ *c.* $5 \times 5 = 25$
d. $2 \times 2 \times 2 \times 2 \times 2 = 32$

---

7 Write the following in exponential form:

a. $5 \times 5 \times 5$ *b.* $3 \times 3 \times 3 \times 3 \times 3$

c. $7 \times 7 \times 7 \times 7$ *d.* $4 \times 4 \times 4 \times 4 \times 4 \times 4 \times 4$

---

a. 5^3 *b.* 3^5 *c.* 7^4 *d.* 4^7

---

8 In 6^8, 6 is called the *base,* and 8 is called the *exponent.* Give the base and exponent of the following numbers:

a. 7^4 *b.* 4^7 *c.* 2^3 *d.* 3^4

---

a. base 7, exponent 4 *b.* base 4, exponent 7
c. base 2, exponent 3 *d.* base 3, exponent 4

---

9 Give the base and exponent of the following numbers:

a. 31^6 *b.* 4^{15} *c.* 10^8

---

a. base 31, exponent 6 *b.* base 4, exponent 15
c. base 10, exponent 8

---

10 This chapter on powers of 10 is concerned only with exponents whose base is 10. Many of the properties of these exponents also apply to other exponents, but this will not be discussed. Give the values of the following:

a. 10^2 *b.* 10^3 *c.* 10^4 *d.* 10^5

---

a. 100 *b.* 1,000 *c.* 10,000 *d.* 100,000

---

11 Just as $1{,}000 = 10 \times 10 \times 10 = 10^{—}$ and $100 = 10 \times 10 = 10^{—}$ we might expect to write 10 in exponential notation as $10^{—}$.

10^3 10^2 10^1

---

12 Give the values of the following:

a. 10^7 *b.* 10^{10} *c.* 10^1

d. Write 1,000,000,000 in exponent form.

---

a. 10,000,000 *b.* 10,000,000,000 *c.* 10 *d.* 10^9

---

13 $10^2 \times 10^5 =$ ________ (integer) $\times$ ________ (integer) $=$ ________ (integer) $=$ ________ (exponent form)

$100 \times 100{,}000 = 10{,}000{,}000 = 10^7$

---

14 $10^3 \times 10^4 =$ ________ (integer) $\times$ ________ (integer) $=$ ________ (integer) $=$ ________ (exponent form)

$3 + 4 =$ ______

$1{,}000 \times 10{,}000 = 10{,}000{,}000 = 10^7$ 7

---

15 $10^2 \times 10^3 =$ ________ (integer) $\times$ ________ (integer) $=$ ________ (exponent form)

$2 + 3 =$ ______

$100 \times 1{,}000 = 10^5$ 5

---

16 *a.* $10^1 \times 10^5 =$ ________ $\times$ ________ (integers) $=$ ________ (exponent form)

$1 + 5 =$ ______

b. $10^2 \times 10^6 =$ ________ $\times$ ________ (integers) $=$ ________ (exponent form)

$2 + 6 =$ ______

c. $10^4 \times 10^4 =$ ________ $\times$ ________ (integers) $=$ ________ (exponent form)

$4 + 4 =$ ______

---

a. $10 \times 100{,}000 = 10^6$; 6 *b.* $100 \times 1{,}000{,}000 = 10^8$; 8

c. $10{,}000 \times 10{,}000 = 10^8$; 8

---

17 *a.* $10^4 \times 10^5 = 10^{-}$ *b.* $10^3 \times 10^8 = 10^{-}$

$4 + 5 =$ ______ $8 + 3 =$ ______

c. $10^9 \times 10^{12} = 10^{-}$

$9 + 12 =$ ______

---

a. 10^9; 9 *b.* 10^{11}; 11 *c.* 10^{21}; 21

---

18 *a.* $10^{31} \times 10^{42} = 10^{—}$

b. $10^9 \times 10^{27} = 10^{—}$

c. $10^{11} \times 10^{35} = 10^{—}$

---

a. 10^{73} *b.* 10^{36} *c.* 10^{46}

---

19. If a, b, c, d are positive integers, then

a. $10^4 \times 10^a = 10^{—}$

b. $10^b \times 10^7 = 10^{—}$

c. $10^c \times 10^d = 10^{—}$

---

a. 10^{4+a} *b.* 10^{b+7} *c.* 10^{c+d}

---

20 This gives us a rule for the multiplication of powers of 10: $10^m \times 10^n = 10^{m+n}$, where m and n are positive integers. Can you find a similar rule for the division of powers of 10? If m is greater than n and m and n are positive integers, $10^m \div 10^n =$ ______. Test your rule on a number of examples.

10^{m-n} Did you work this out, or just guess it?

---

21 Let us check this rule on a few examples:

a. $10^7 \div 10^4 = \frac{?}{?}$ (integers) $= 10^{—}$

$7 - 4 =$ ______

b. $10^8 \div 10^3 = \frac{?}{?}$ (integers) $= 10^{—}$

$8 - 3 =$ ______

c. $10^6 \div 10^5 = \frac{?}{?}$ (integers) $= 10^{—}$

$6 - 5 =$ ______

---

a. $\frac{10{,}000{,}000}{10{,}000} = 10^3$; 3 *b.* $\frac{100{,}000{,}000}{1{,}000} = 10^5$; 5

c. $\frac{1{,}000{,}000}{100{,}000} = 10^1$; 1

---

22 *a.* $10^{18} \div 10^{6} = 10^{-}$

b. $10^{11} \div 10^{a} = 10^{-}$

c. $10^{b} \div 10^{c} = 10^{-}$

where a, b, and c are positive integers, a is less than 11, and c is less than b.

---

a. 10^{12} *b.* 10^{11-a} *c.* 10^{b-c}

---

23 So the rules for the multiplication and division of powers of 10 with positive integer exponents are: if m and n are positive integers,

a. $10^{m} \times 10^{n} =$ ______.

b. $10^{m} \div 10^{n} =$ ______ provided that m is greater than n.

---

a. 10^{m+n} *b.* 10^{m-n}

---

24 *a.* $10^{5} \times 10^{37} = 10^{-}$

b. $10^{42} \div 10^{17} = 10^{-}$

c. $10^{16} \div 10^{15} = 10^{-}$

---

a. 10^{42} *b.* 10^{25} *c.* 10^{1}

You have finished the program for Sec. 10-1. Now take the posttest to see how much you have learned.

POSTTEST 10-1

1. Write 13^{6} in multiplication notation, i.e., ______ × ______ × ______ etc.
2. Write 35 × 35 × 35 × 35 × 35 in exponential notation.
3. Name the base and the exponent of 9^{12}.
4. Give the values of 10^{7} and 10^{1}.

Give the answers to the following in exponential notation:

5. $10^{15} \times 10^{44} =$ ______
6. $10^{13} \times 10^{28} =$ ______
7. $10^{18} \div 10^{17} =$ ______
8. $10^{14} \div 10^{8} =$ ______

When you have finished the posttest, correct it yourself using the answers on page 419. Then check all the questions you got wrong on the list below:

1. ______	4. ______	7. ______
2. ______	5. ______	8. ______
3. ______	6. ______	

If you compare your performance with that on Pretest 10-1, you will see how much you have improved. If you made more than one mistake on this posttest, review the material and take the pretest again. Don't move on to the next section until you can get all the questions right.

Now proceed to Sec. 10-2.

Section 10-2 Multiplication and Division of All Integral Powers of 10

Do this section if you made a mistake in Questions 1 to 4 of Diagnostic Test 10. Begin by taking the pretest below, which will show you how much of Sec. 10-2 you already know and what you will be able to do by the time you have finished it.

PRETEST 10-2

Give the integer or decimal forms of the following numbers:

1. 10^{-4}
2. 10^{0}

Give the answers to the following in exponential notation:

3. $10^{7} \times 10^{-8}$
4. $10^{-5} \times 10^{-6}$
5. $10^{-3} \div 10^{-3}$
6. $10^{4} \div 10^{7}$
7. $10^{-7} \div 10^{-9}$
8. $10^{8} \div 10^{-5}$
9. $10^{-11} \div 10^{7}$
10. $10^{5} \times 10^{-5}$

When you have finished the pretest, correct it yourself using the answers on page 419. Then check all the questions you got wrong on the list below:

1.	______	5.	______	8.	______
2.	______	6.	______	9.	______
3.	______	7.	______	10.	______
4.	______				

If you made any mistakes (even if it was only *one* mistake), read the introduction and go through the program.

If you made no mistakes at all, skip Sec. 10-2, because you already know its material, and go on to the next section whose number you circled on page 384. If no further numbers were circled, take Final Test 10 (page 418).

Introduction

This section uses multiplication notation to find the result of dividing 10^m by 10^n when n is greater than m and shows that the use of 10^{-n} for $1/10^n$ is consistent with the rule that $10^m \div 10^n = 10^{m-n}$. Negative exponents are defined on this basis, and 10^0 is defined as 1. The laws $10^m \times 10^n = 10^{m+n}$ and $10^m \div 10^n = 10^{m-n}$ can then be shown to hold for all integral values of m and n.

FRAMES FOR SECTION 10-2

1 Section 10-1 dealt only with positive integer exponents, which it introduced as a notation for multiple multiplication. Thus $10^3 = 10 \times 10 \times 10 =$ _____, and $10^6 = 10 \times 10 \times$ _______________ = _________.

1,000 $\quad 10 \times 10 \times 10 \times 10 = 1{,}000{,}000$

---

2 This doesn't seem to leave much scope for a definition of negative exponents, but it may have led you to wonder what happens to the rule $10^m \div 10^n = 10^{m-n}$ when n is greater than m. Let's investigate.

a. $10^2 \div 10^3 = \dfrac{10 \times 10}{10 \times 10 \times 10} =$ _____ $\qquad 2 - 3 =$ _____

b. $10^2 \div 10^4 = \dfrac{?}{?} =$ _____ $\qquad 2 - 4 =$ _____

c. $10^2 \div 10^5 = \dfrac{?}{?} =$ _____ $\qquad 2 - 5 =$ _____

---

a. $\dfrac{1}{10}$; -1 $\qquad$ *b.* $\dfrac{10 \times 10}{10 \times 10 \times 10 \times 10} = \dfrac{1}{100}$ or $\dfrac{1}{10^2}$; -2

c. $\dfrac{10 \times 10}{10 \times 10 \times 10 \times 10 \times 10} = \dfrac{1}{1{,}000}$ or $\dfrac{1}{10^3}$; -3

---

3 These examples suggest that we can define negative exponents as follows:

$10^{-1} = \dfrac{1}{10} \qquad 10^{-2} = \dfrac{1}{10^2} \qquad 10^{-3} = \dfrac{1}{?} \qquad 10^{-4} =$ _____

$10^3 \qquad \dfrac{1}{10^4}$

---

4 The general definition of a negative exponent would then be given by $10^{-n} =$ _____, where n is a positive integer.

$\frac{1}{10^n}$

---

5 We choose this definition of negative exponents in order to be able to apply the rule $10^m \div 10^n = 10^{m-n}$ to all positive integer values of m and n. But can we also apply it to negative integer values of m and n? Let's see.

$$10^5 \div 10^{-2} = 10^5 \div \frac{1}{?} \left(\text{because } 10^{-2} \text{ is defined as } \frac{1}{?}\right)$$

$$= 10^5 \times \_\_\_\_\_\_ = 10^{-}$$

$5 - (-2) =$ ______

10^2 10^2 10^2 10^7 7

---

6 Or to take another example,

$$10^{-3} \div 10^4 = \frac{1}{?} \div 10^4 = \frac{1}{?} \times \frac{1}{?} = \frac{1}{?} = 10^{-}$$

$-3 - 4 =$ ______

$\frac{1}{10^3} \div 10^4 = \frac{1}{10^3} \times \frac{1}{10^4} = \frac{1}{10^7} = 10^{-7}$ -7

---

7 Finally, let's look at an example with both exponents negative.

a. $10^{-6} \div 10^{-2} = \frac{1}{?} \div \frac{1}{?} = \frac{1}{?} \times \_\_\_\_\_\_ = \frac{1}{?} = 10^{-}$

b. 6 (2) = ______

---

a. $\frac{1}{10^6} \div \frac{1}{10^2} = \frac{1}{10^6} \times 10^2 = \frac{1}{10^4} = 10^{-4}$ *b.* -4

---

8 It looks as if the rule $10^m \div 10^n = 10^{m-n}$ applies to both positive and negative integer values of m and n; and this can be proved to be true. Now let's look at the other rule, $10^m \times 10^n = 10^{m+n}$.

a. $10^5 \times 10^{-3} = 10^5 \times \frac{1}{?} = 10^5 \div 10^{-} = 10^{-}$

$5 + (-3) =$ ______

b. $10^{-2} \times 10^{-4} = \frac{1}{?} \times \frac{1}{?} = \frac{1}{?} = 10^{-}$

$-2 + (-4) =$ ______

---

a. $10^5 \times \frac{1}{10^3} = 10^5 \div 10^3 = 10^2$ 2

b. $\frac{1}{10^2} \times \frac{1}{10^4} = \frac{1}{10^6} = 10^{-6}$ -6

---

9 Both rules can in fact be proved to be true. So we know that if m and n are integers other than 0,

$10^m \times 10^n =$ ______ and $10^m \div 10^n =$ ______

10^{m+n} 10^{m-n}

---

10 We have now defined exponents for all integers except 0. To be consistent 10^0 should equal $10^4 \div 10^4$ because $4 - 4 =$ ______, and 10^0 should also equal $10^3 \div 10^3$ because $3 - 3 =$ ______. Now $10{,}000 \div 10{,}000 =$ ______, and $1{,}000 \div 1{,}000 =$ ______, so we define $10^0 =$ ______.

---

0 0 1 1 1

---

11 This definition also fits if we look at the decimal equivalents of powers of 10.

$10^3 = 1{,}000$

$10^2 =$ ______

$10^1 =$ ______

$10^0 =$ ______

$10^{-1} = 0.1$

$10^{-2} =$ ______ (decimal form)

$10^{-3} =$ ______ (decimal form)

---

100 10 1 0.01 0.001

---

12 *a.* $10^5 \div 10^5 = 10^{—}$

b. $10^6 \times 10^{-6} = 10^{—}$

c. $10^0 \div 10^3 = 10^{—}$

---

a. 10^0 *b.* 10^0 *c.* 10^{-3}

---

13 *a.* $10^0 \times 10^{17} = 10^{-}$

b. $10^{11} \div 10^{10} = 10^{-}$

c. $10^{-77} \times 10^{77} = 10^{-}$

---

a. 10^{17} *b.* 10^1 *c.* 10^0

---

14 *a.* $10^{-5} \div 10^{-7} = 10^{-}$

b. $10^{-14} \times 10^3 = 10^{-}$

c. $10^6 \div 10^6 = 10^{-}$

---

a. 10^2 *b.* 10^{-11} *c.* 10^0

---

15 *a.* $10^0 =$ ______

b. $10^1 =$ ______

c. $10^{-3} =$ ______ (decimal form)

---

a. 1 *b.* 10 *c.* 0.001

---

16 *a.* $10^{-3} \div 10^{19} =$ ______

b. $10^{-7} \times 10^{-24} =$ ______

c. $10^{47} \div 10^{-11} =$ ______

---

a. 10^{-22} *b.* 10^{-31} *c.* 10^{58}

You have finished the program for Sec. 10-2. Now take the posttest to see how much you have learned.

POSTTEST 10-2

Give the integer decimal forms of the following numbers:

1. 10^{-5}
2. 10^0

Give the answers to the following in exponential notation:

3. $10^9 \times 10^{-6}$
4. $10^{-4} \times 10^{-7}$
5. $10^{-5} \div 10^{-5}$
6. $10^6 \div 10^8$
7. $10^{-3} \div 10^{-8}$
8. $10^{11} \div 10^{-6}$
9. $10^{-7} \div 10^6$
10. $10^4 \div 10^{-4}$

When you have finished the posttest, correct it yourself using the answers on page 419. Then check all the questions you got wrong on the list below:

1. ______	5. ______	8. ______
2. ______	6. ______	9. ______
3. ______	7. ______	10. ______
4. ______		

If you compare your performance with that on Pretest 10-2, you will see how much you have improved. If you made more than one mistake on this posttest, review the material and take the pretest again. Don't move on to the next section until you can get all the questions right.

Now proceed to the next section whose number you circled on page 384. If no other sections were circled, take Final Test 10 (page 418).

Section 10-3 Conversion of Decimals into Standard Exponent Form

Do this section if you made a mistake in Questions 5 to 8 of Diagnostic Test 10. Begin by taking the pretest below, which will show you how much of Sec. 10-3 you already know and what you will be able to do by the time you have finished it.

PRETEST 10-3

The following numbers are in standard exponent form. Write them as ordinary numbers:

1. 4.2×10^{-5}
2. 1.29×10^{4}
3. 6.7×10^{0}
4. 1.23×10^{-7}

Write the following in standard exponent form, using powers of 10:

5. 1,463
6. 0.48×10^{-8}
7. 289×10^{4}
8. 0.071×10^{5}
9. $3,260 \times 10^{-7}$
10. 0.0052

When you have finished the pretest, correct it yourself using the answers on page 420. Then check all the questions you got wrong on the list below:

1. ______	5. ______	8. ______
2. ______	6. ______	9. ______
3. ______	7. ______	10. ______
4. ______		

If you made any mistakes (even if it was only *one* mistake), read the introduction and go through the program.

If you made no mistakes at all, skip Sec. 10-3, because you already know its material, and go on to the next section whose number you circled on page 384. If no further numbers were circled, take Final Test 10 (page 418).

Introduction

This section shows that all integers or decimals can be converted to the form $A \times 10^b$, where A is between 1 and 10 and b is an integer, and calls this the *standard exponent form*. You learn how to convert numbers from integer or decimal form into standard exponent form and vice versa and then how to convert numbers in nonstandard exponent forms into standard exponent forms.

FRAMES FOR SECTION 10-3

1 Convert the following powers of 10 to integer or decimal forms:

a. 10^4 *b.* 10^1 *c.* 10^0 *d.* 10^{-4}

---

a. 10,000 *b.* 10 *c.* 1 *d.* 0.0001

---

2 Convert the following numbers to exponent forms:

a. 100,000 *b.* 100 *c.* 0.01 *d.* 0.0000001

---

a. 10^5 *b.* 10^2 *c.* 10^{-2} *d.* 10^{-7}

---

3 Any number can be expressed in the form $A \times 10^b$, where A is between 1 and 10 and b is an integer. We call this its *standard exponent form*. Thus

$300 = 3 \times 10^2$ $(A = 3, b = 2)$

$5{,}000 = 5 \times 10^3$ $(A = \_\_, b = \_\_)$

$A = 5, b = 3$

---

4 $60{,}000 = 6 \times 10^4$ $(A = \_\_, b = \_\_)$

$400 = 4 \times 10^b$ $(A = \_\_, b = \_\_)$

$A = 6, b = 4$ $A = 4, b = 2$

---

5 $431 = 4.31 \times 10^b$ $(A = \_\_, b = \_\_)$

$431.32 = 4.3132 \times 10^b$ $(A = \_\_, b = \_\_)$

$4{,}312.6 = 4.3126 \times 10^b$ $(A = \_\_, b = \_\_)$

How is b related to the number of places you move the decimal point?

$A = 4.31, b = 2$ $A = 4.3132, b = 2$ $A = 4.3126, b = 3$ It is the same number.

---

6 $32.5 = A \times 10^b$ $(A = \_\_, b = \_\_)$

$532.2 = A \times 10^b$ $(A = \_\_, b = \_\_)$

$67.98 = A \times 10^b$ $(A = \_\_, b = \_\_)$

Note: A is between 1 and 10, and b is an integer.

$A = 3.25, b = 1$ $A = 5.322, b = 2$ $A = 6.798, b = 1$

---

7 In Frames 3 to 6 you were converting numbers (greater/less) ______ than 10 to their standard exponent form, and in each case you moved the decimal point b places to the (left/right) ______.

---

greater; left

---

8 Now let us convert some numbers less than 1 to their standard exponent forms, that is, $A \times 10^b$, where A is between 1 and 10 and b is an integer.

a. $0.03 = 3 \times 10^{-2}$ $(A = 3, b = -2)$

b. $0.0023 = 2.3 \times 10^{-3}$ $(A = \_\_, b = \_\_)$

c. $0.45 = 4.5 \times 10^{-1}$ $(A = \_\_, b = \_\_)$

---

b. 2.3; −3 *c.* 4.5; −1

---

9 In 8*a* you moved the decimal point ______ (how many?) places to the (left/right) ______, and b was ______. In 8*b* you moved the decimal point ______ places to the (left/right) ______, and b was ______. In 8*c* you moved the decimal point ______ places to the (left/right) ______, and b was ______.

two; right; -2 three; right; -3 one; right; -1

10 In general when converting numbers less than 1 to their standard exponent forms, $A \times 10^b$, you move the decimal point ______ (how many?) places to the (left/right) ______.

b; right

11 *a.* $0.062 = 6.2 \times 10^b$ ($A =$ __, $b =$ __)

b. $0.00346 = 3.46 \times 10^b$ ($A =$ __, $b =$ __)

c. $0.00005 = 5 \times 10^b$ ($A =$ __, $b =$ __)

a. 6.2; -2 *b.* 3.46; -3 *c.* 5; -5

12 *a.* $0.367 = A \times 10^b$ ($A =$ __, $b =$ __)

b. $0.000279 = A \times 10^b$ ($A =$ __, $b =$ __)

c. $0.0059 = A \times 10^b$ ($A =$ __, $b =$ __)

a. 3.67; -1 *b.* 2.79; -4 *c.* 5.9; -3

13 Express the following numbers in standard exponent form:

a. 0.00567 *b.* 0.000006 *c.* 31 *d.* 3,125

a. 5.67×10^{-3} *b.* 6×10^{-6} *c.* 3.1×10^1 *d.* 3.125×10^3

14 Express the following numbers in standard exponent form:

a. 620.53 *b.* 0.00739 *c.* 7,000 *d.* 0.095

a. 6.2053×10^2 *b.* 7.39×10^{-3} *c.* 7×10^3 *d.* 9.5×10^{-2}

15 The following numbers are in standard exponent form. Express them in decimal or integer form.

	Decimal or integer form
3.65×10^4	36,500
a. 4.617×10^{-2}	______
b. 6.201×10^{-1}	______
c. 6.08×10^2	______

a. 0.04617 *b.* 0.6201 *c.* 608

16 The following numbers are in standard exponent form. Express them in decimal or integer form.

	Decimal or integer form
a. 3.1×10^5	______
b. 4.006×10^{-3}	______
c. 5.14×10^1	______
d. 4.14×10^{-4}	______

a. 310,000 *b.* 0.004006 *c.* 51.4 *d.* 0.000414

17 In 16*a* the exponent was ______, and the decimal point was moved ______ place(s) to the (left/right) ______. In 16*b* the exponent was ______, and the decimal point was moved ______ place(s) to the (left/right) ______.

5; five; right —3; three; left

18 Suppose $c \times 10^d$ is a number in standard exponent form and you wish to convert it to a decimal or integer form. If d is positive, you convert it by moving the decimal point ______ place(s) to the (left/right) ______, and if d is negative, you convert it by moving the decimal point ______ place(s) to the (left/right) ______.

d; right; $-d$; left

19 The same method applies to the conversion of numbers in nonstandard exponent form. Convert the following numbers to decimal or integer form:

a. 124×10^2 *b.* 0.63×10^3

c. 0.089×10^{-2} *d.* $3{,}146 \times 10^{-4}$

a. 12,400 *b.* 630 *c.* 0.00089 *d.* 0.3146

20 The following numbers are in nonstandard exponent form. First convert them to decimal or integer form, then to standard exponent form.

	Decimal or integer form	*Standard exponent form*
a. 32.5×10^3	32,500	3.25×10^4
b. 11×10^{-2}		
c. 0.76×10^4		
d. 25×10^1		
e. 36.7×10^0		
f. 0.03×10^{-2}		

b. 0.11; 1.1×10^{-1} *c.* 7,600; 7.6×10^3 *d.* 250; 2.5×10^2
e. 36.7; 3.67×10^1 *f.* 0.0003; 3×10^{-4}

21 Convert the following numbers to standard exponent form:

a. 64.3×10^2 *b.* 135.6×10^{-4}

c. 0.072×10^{-2} *d.* 0.634×10^3

e. 0.0085×10^0

a. 6.43×10^3 *b.* 1.356×10^{-2} *c.* 7.2×10^{-4} *d.* 6.34×10^2
e. 8.5×10^{-3}

22 The easiest way to convert a number in nonstandard exponent form, $H \times 10^k$, to a number in standard exponent form, $A \times 10^b$, is shown below for $H = 84.8$, $k = 2$.

a. Change H to a standard exponent form to get the value of A.

$H = 8.48 = 8.48 \times 10^1$ (A = ______)

b. Collect powers of 10 to get the value of the exponent b.

$H \times 10^k = (8.48 \times 10^1) \times 10^2 = 8.48 \times 10^3$ (b = ______)

a. 8.48 *b.* 3

23 Use this method to convert the following numbers to standard exponent form:

a. $785.6 \times 10^{-4} = 7.856 \times 10^{m} \times 10^{-4}$ ($m =$ ______)

$= 7.856 \times 10^{b}$ ($b =$ ______)

b. $0.097 \times 10^{3} = 9.7 \times 10^{m} \times 10^{3}$ ($m =$ ______)

$= 9.7 \times 10^{b}$ ($b =$ ______)

c. $0.00563 \times 10^{-2} = 5.63 \times 10^{m} \times 10^{-2}$ ($m =$ ______)

$= 5.63 \times 10^{b}$ ($b =$ ______)

---

a. 2; −2 *b.* −2; 1 *c.* −3; −5

---

24 Convert the following to standard exponent form by the method just shown:

a. $0.088 \times 10^{-2} =$ ______ × ______ × ______ = ______ × ______

b. $0.0073 \times 10^{2} =$ ______ × ______ × ______ = ______ × ______

c. $0.000035 \times 10^{-7} =$ ______ × ______ × ______ = ______ × ______

---

a. $8.8 \times 10^{-2} \times 10^{-2} = 8.8 \times 10^{-4}$

b. $7.3 \times 10^{-3} \times 10^{2} = 7.3 \times 10^{-1}$

c. $3.5 \times 10^{-5} \times 10^{-7} = 3.5 \times 10^{-12}$

---

25 Convert the following numbers to standard exponent form:

a. $4{,}961 \times 10^{-8}$ *b.* 32.95×10^{-3}

c. 0.0731×10^{6} *d.* 0.00495×10^{-4}

---

a. 4.961×10^{-5} *b.* 3.295×10^{-2} *c.* 7.31×10^{4} *d.* 4.95×10^{-7}

---

26 Convert the following numbers to standard exponent form:

a. $75{,}203 \times 10^{-6}$ *b.* 0.00076×10^{3}

c. 137.9×10^{5} *d.* 0.895×10^{-5}

---

a. 7.5203×10^{-2} *b.* 7.6×10^{-1} *c.* 1.379×10^{7} *d.* 8.95×10^{-6}

You have finished the program for Sec. 10-3. Now take the posttest to see how much you have learned.

POSTTEST 10-3

The following numbers are in standard exponent form. Write them as ordinary numbers:

1. 2.95×10^{-3}
2. 6.9×10^{0}
3. 4.6×10^{4}
4. 3.26×10^{-5}

Write the following in standard exponent form, using powers of 10:

5. 1,329
6. 0.32×10^{-7}
7. 326×10^{3}
8. 0.068×10^{4}
9. $2,930 \times 10^{-5}$
10. 0.00985

When you have finished the posttest, correct it yourself using the answers on page 420. Then check all the questions you got wrong on the list below:

1. ______ 5. ______ 8. ______

2. ______ 6. ______ 9. ______

3. ______ 7. ______ 10. ______

4. ______

If you compare your performance with that on Pretest 10-3, you will see how much you have improved. If you made more than one mistake on this posttest, review the material and take the pretest again. Don't move on to the next section until you can get all the questions right.

Now proceed to the next section whose number you circled on page 384. If no other sections were circled, take Final Test 10 (page 418).

Section 10-4
The Use of Powers of 10 in the Multiplication and Division of Decimals

Do this section if you made a mistake in Questions 9 to 12 of Diagnostic Test 10. Begin by taking the pretest below, which will show you how much of Sec. 10-4 you already know and what you will be able to do by the time you have finished it.

PRETEST 10-4

Carry out the following operations, putting answers into standard exponent form:

1. $(4 \times 10^{5}) \times (2 \times 10^{-4})$
2. $(3 \times 10^{-5}) \times (5 \times 10^{-1})$
3. $(8 \times 10^{5}) \div (4 \times 10^{-3})$
4. $(2 \times 10^{-2}) \div (8 \times 10^{-5})$

Use powers-of-10 notation to do the following problems, putting answers into standard exponent form:

5. $\dfrac{221 \times 0.004}{0.13}$
6. $\dfrac{0.221 \times 400}{0.00013}$
7. $\dfrac{15600 \times 0.000112}{0.028}$
8. $\dfrac{0.0156 \times 11200}{280}$

When you have finished the pretest, correct it yourself using the answers on page 420. Then check all the questions you got wrong on the list below:

1. ______ 4. ______ 7. ______

2. ______ 5. ______ 8. ______

3. ______ 6. ______

If you made any mistakes (even if it was only *one* mistake), read the introduction and go through the program.

If you made no mistakes at all, skip Sec. 10-4, because you already know its material, and go on to the next section whose number you circled on page 384. If no further numbers were circled, take Final Test 10 (page 418).

Introduction

After you are shown how numbers in standard exponent form can be multiplied or divided by treating the powers-of-10 components separately, you learn how to multiply and divide through conversion to standard exponent form.

FRAMES FOR SECTION 10-4

1 Multiplication of powers of 10 is defined by the law $10^m \times 10^n =$ ______, where m and n are integers.

10^{m+n}

---

2 $10^3 \times 10^4 = 10^a$ $(a = \_)$ $10^a =$ ______ (what decimal?)

$10^{-3} \times 10^4 = 10^b$ $(b = \_)$ $10^b =$ ______ (what decimal?)

$10^{-3} \times 10^{-4} = 10^c$ $(c = \_)$ $10^c =$ ______ (what decimal?)

$10^3 \times 10^{-4} = 10^d$ $(d = \_)$ $10^d =$ ______ (what decimal?)

---

7 10,000,000 1 10 −7 0.0000001 −1 0.1

---

3 Powers-of-10 notation is extremely useful in simplifying the multiplication and division of decimals. Thus

$$\begin{aligned} 0.002 \times 0.0004 &= (2 \times 10^{-3}) \times (4 \times 10^{-4}) \\ &= (2 \times 4) \times (10^{-3} \times 10^{-4}) \\ &= 8 \times 10^m \qquad (m = \_) \\ &= \_\_\_\_\_\_ \text{ (what decimal?)} \end{aligned}$$

−7 0.0000008

4
$$\begin{aligned} 0.03 \times 0.0003 &= (3 \times 10^a) \times (3 \times 10^b) \qquad (a = \_,\ b = \_) \\ &= (3 \times 3) \times (10^a \times 10^b) \\ &= 9 \times 10^c \qquad (c = \_) \\ &= \_\_\_\_\_\_ \text{ (what decimal?)} \end{aligned}$$

−2 −4 −6 0.000009

5
$$\begin{aligned} 0.632 \times 0.002 &= (6.32 \times 10^a) \times (2 \times 10^b) \qquad (a = \_,\ b = \_) \\ &= (6.32 \times 2) \times (10^a \times 10^b) \\ &= \_\_\_\_\_\_ \times 10^c \qquad (c = \_) \\ &= \_\_\_\_\_\_ \text{ (what decimal?)} \end{aligned}$$

−1 −3 12.64 −4 0.001264

---

6 In using this method, the decimals are first converted to standard exponent form. Thus

$0.024 \times 0.0012 = (\_\_\_\_ \times 10^a) \times (\_\_\_\_ \times 10^b) \qquad (a = \_, b = \_)$

and then the decimals and powers of 10 are multiplied separately:

$\_\_\_\_ \times 10^c \qquad (c = \_) = \_\_\_\_$ (what decimal?)

$(2.4 \times 10^{-2}) \times (1.2 \times 10^{-3})$ $\quad -2 \quad -3 \quad 2.88 \times 10^{-5} \quad -5$
0.0000288

---

7 Use powers-of-10 notation to carry out the multiplication 0.64×0.0011. Make sure that you first convert the decimals to their standard exponent form. Give your answer as a decimal.

$$\begin{aligned} 0.64 \times 0.0011 &= (6.4 \times 10^{-1}) \times (1.1 \times 10^{-3}) \\ &= (6.4 \times 1.1) \times (10^{-1} \times 10^{-3}) \\ &= 7.04 \times 10^{-4} = 0.000704 \end{aligned}$$

---

8 Use powers-of-10 notation to carry out the multiplication 0.0035×0.076. Make sure that you first convert the decimals to their standard exponent forms. Give your answer as a decimal.

$$\begin{aligned} 0.0035 \times 0.076 &= (3.5 \times 10^{-3}) \times (7.6 \times 10^{-2}) \\ &= (3.5 \times 7.6) \times (10^{-3} \times 10^{-2}) \\ &= 26.6 \times 10^{-5} \\ &= 0.000266 \end{aligned}$$

---

9 Use powers-of-10 notation to carry out the following multiplications:

a. $93 \times 0.05 \times 0.06 = (\_\_\_\_ \times 10^a) \times (\_\_\_\_ \times 10^b) \times (\_\_\_\_ \times 10^c)$

$(a = \_\_\_\_, b = \_\_\_\_, c = \_\_\_\_)$

$= \_\_\_\_ \times 10^d \qquad (d = \_\_\_\_)$

$= \_\_\_\_$ (decimal form)

b. $0.093 \times 0.5 \times 0.006 = \_\_\_\_$

c. $9.3 \times 500 \times 0.06 = \_\_\_\_$

---

a. 9.3; 5; 6; 1; −2; −2; 279; −3; 0.279
b. $(9.3 \times 10^{-2}) \times (5 \times 10^{-1}) \times (6 \times 10^{-3}) = 279 \times 10^{-6} = 0.000279$
c. $9.3 \times (5 \times 10^{2}) \times (6 \times 10^{-2}) = 279$

---

10 Use powers-of-10 notation to carry out the following multiplications:

a. $0.07 \times 3.56 \times 0.008$

b. $0.7 \times 0.0356 \times 0.00008$

c. $0.007 \times 35.6 \times 0.0008$

---

a. $0.07 \times 3.56 \times 0.008 = (7 \times 10^{-2}) \times 3.56 \times (8 \times 10^{-3})$
$= (7 \times 3.56 \times 8) \times (10^{-2} \times 10^{-3})$
$= 199.36 \times 10^{-5} = 0.0019936$

b. $0.7 \times 0.0356 \times 0.00008 = (7 \times 10^{-1}) \times (3.56 \times 10^{-2}) \times (8 \times 10^{-5})$
$= (7 \times 3.56 \times 8) \times (10^{-1} \times 10^{-2} \times 10^{-5})$
$= 199.36 \times 10^{-8} = 0.0000019936$

c. $0.007 \times 35.6 \times 0.0008 = (7 \times 10^{-3}) \times (3.56 \times 10) \times (8 \times 10^{-4})$
$= (7 \times 3.56 \times 8) \times (10^{-3} \times 10 \times 10^{-4})$
$= 199.36 \times 10^{-6} = 0.00019936$

---

11 Division of powers of 10 is defined by the law $10^m \div 10^n =$ ______, where m and n are integers.

10^{m-n}

---

12 *a.* $10^3 \div 10^4 = 10^a$ ($a =$ _) *b.* $10^{-3} \div 10^4 = 10^b$ ($b =$ _)

c. $10^{-3} \div 10^{-4} = 10^c$ ($c =$ _) *d.* $10^3 \div 10^{-4} = 10^d$ ($d =$ _)

---

a. 10^{-1}; -1 *b.* 10^{-7}; -7 *c.* 10^1; 1 *d.* 10^7; 7

---

13 $0.002 \div 0.0004 = \dfrac{2 \times 10^{-3}}{4 \times 10^{-4}} = \dfrac{2}{4} \times \dfrac{10^{-3}}{10^{-4}} = 0.5 \times 10^m$ ($m =$ ______)

$=$ ______

---

1 5

---

14 $0.03 \div 0.0003 = \dfrac{3 \times 10^a}{3 \times 10^b}$ ($a =$ ______, $b =$ ______)

$= \dfrac{3}{3} \times \dfrac{10^a}{10^b} = 1 \times 10^c$ ($c =$ ______)

$=$ ______

-2 -4 2 100

15 $0.632 \div 0.002 = \dfrac{6.32 \times 10^a}{2 \times 10^b}$ $(a = \_\_\_\_,\ b = \_\_\_\_)$

$= \dfrac{6.32}{2} \times \dfrac{10^a}{10^b} = \_\_\_\_ \times 10^c$ $(c = \_\_\_\_)$

$= \_\_\_\_$

−1 −3 3.16 2 316

---

16 As with multiplication, the decimals are first converted to standard exponent form, and then the powers of 10 are divided separately.

$0.005 \div 0.25 = \dfrac{\_ \times 10^a}{\_ \times 10^b}$ $(a = \_\_\_\_,\ b = \_\_\_\_)$

$= \dfrac{?}{?} \times \dfrac{10^a}{10^b}$

$= \_\_\_\_ \times 10^c$ $(c = \_\_\_\_)$

$= \_\_\_\_$ (decimal form)

$\dfrac{5 \times 10^a}{2.5 \times 10^b}$ −3 −1 $\dfrac{5}{2.5}$ 2 −2 0.02

---

17 $0.0008 \div 0.02 = \dfrac{\_ \times 10^a}{\_ \times 10^b}$ $(a = \_\_\_\_,\ b = \_\_\_\_)$

$= \dfrac{?}{?} \times \dfrac{10^a}{10^b}$

$= \_\_\_\_ \times 10^c$ $(c = \_\_\_\_)$

$= \_\_\_\_$ (decimal form)

$\dfrac{8 \times 10^a}{2 \times 10^b}$ −4 −2 $\dfrac{8}{2}$ 4 −2 0.04

---

18 Use powers-of-10 notation to carry out the division $0.0084 \div 0.0012$. Make sure that you first convert the decimals to their standard exponent forms. Give your answer as a decimal or an integer.

$\dfrac{8.4 \times 10^{-3}}{1.2 \times 10^{-3}} = \dfrac{8.4}{1.2} \times \dfrac{10^{-3}}{10^{-3}} = 7 \times 10^0 = 7$

---

19 Use powers-of-10 notation to carry out the divisions:

a. $0.195 \div 0.013$ *b.* $1.95 \div 0.0013$ *c.* $0.0195 \div 130$

---

a. $\dfrac{1.95 \times 10^{-1}}{1.3 \times 10^{-2}} = \dfrac{1.95}{1.3} \times \dfrac{10^{-1}}{10^{-2}} = 1.5 \times 10^{1} = 15$

b. $\dfrac{1.95 \times 10^{0}}{1.3 \times 10^{-3}} = \dfrac{1.95}{1.3} \times \dfrac{10^{0}}{10^{-3}} = 1.5 \times 10^{3} = 1{,}500$

c. $\dfrac{1.95 \times 10^{-2}}{1.3 \times 10^{2}} = \dfrac{1.95}{1.3} \times \dfrac{10^{-2}}{10^{2}} = 1.5 \times 10^{-4} = 0.00015$

---

20 In the case of a combined multiplication and division the same procedure can be used. Thus

$$\frac{170 \times 0.015}{0.51}$$

$\updownarrow$ Change to standard exponent form

$$\frac{(1.7 \times 10^{a}) \times (\_\_\_ \times 10^{b})}{\_\_\_ \times 10^{c}} \qquad (a = \_,\ b = \_,\ c = \_)$$

$\updownarrow$ Separate decimals and powers of 10

$$\frac{1.7 \times \_\_\_}{\_\_\_} \times \frac{10^{a} \times 10^{b}}{10^{c}} = \_\_\_ \times 10^{d} \qquad (d = \_)$$

$$= \_\_\_ \text{ (decimal form)}$$

$$\frac{(1.7 \times 10^{2}) \times (1.5 \times 10^{-2})}{5.1 \times 10^{-1}} \qquad (a = 2,\ b = -2,\ c = -1)$$

$\updownarrow$

$$\frac{1.7 \times 1.5}{5.1} \times \frac{10^{2} \times 10^{-2}}{10^{-1}} = 0.5 \times 10^{1} \qquad (d = 1)$$

$$= 5$$

21

$$\frac{3{,}400 \times 0.0019}{0.038}$$

$\updownarrow$

$$\frac{(\_\_\_ \times 10^{a}) \times (\_\_\_ \times 10^{b})}{\_\_\_ \times 10^{c}} \qquad (a = \_,\ b = \_,\ c = \_)$$

$\updownarrow$

$$\frac{\_\_\_ \times \_\_\_}{\_\_\_} \times \frac{10^{a} \times 10^{b}}{10^{c}} = \_\_\_ \times 10^{d} \qquad (d = \_)$$

$$= \_\_\_ \text{ (decimal form)}$$

$$\frac{(3.4 \times 10^3) \times (1.9 \times 10^{-3})}{3.8 \times 10^{-2}} \qquad (a = 3,\ b = -3,\ c = -2)$$

$$\updownarrow$$

$$\frac{3.4 \times 1.9}{3.8} \times \frac{10^3 \times 10^{-3}}{10^{-2}} = 1.7 \times 10^2 \qquad (d = 2)$$

$$= 170$$

22 Use powers-of-10 notation to calculate

$$\frac{320 \times 0.000081}{0.036}$$

$$\frac{320 \times 0.000081}{0.036} = \frac{(3.2 \times 10^2) \times (8.1 \times 10^{-5})}{3.6 \times 10^{-2}} = \frac{3.2 \times 8.1}{3.6} \times \frac{10^2 \times 10^{-5}}{10^{-2}}$$

$$= 7.2 \times 10^{-1} = 0.72$$

23 Use powers-of-10 notation to calculate the following:

a. $\dfrac{14.7 \times 0.00321}{0.00025}$ *b.* $\dfrac{14{,}700 \times 0.321}{0.025}$ *c.* $\dfrac{147 \times 32.1}{2{,}500}$

a. $$\frac{14.7 \times 0.00321}{0.00025} = \frac{(1.47 \times 10^1) \times (3.21 \times 10^{-3})}{2.5 \times 10^{-4}} = \frac{1.47 \times 3.21}{2.5} \times \frac{10^1 \times 10^{-3}}{10^{-4}}$$

$$= 1.88748 \times 10^2 = 188.748$$

b. $$\frac{14{,}700 \times 0.321}{0.025} = \frac{(1.47 \times 10^4) \times (3.21 \times 10^{-1})}{2.5 \times 10^{-2}} = \frac{1.47 \times 3.21}{2.5} \times \frac{10^4 \times 10^{-1}}{10^{-2}}$$

$$= 1.88748 \times 10^5 = 188{,}748$$

c. $$\frac{147 \times 32.1}{2500} = \frac{(1.47 \times 10^2) \times (3.21 \times 10^1)}{2.5 \times 10^3} = \frac{1.47 \times 3.21}{2.5} \times \frac{10^2 \times 10^1}{10^3}$$

$$= 1.88748 \times 10^0 = 1.88748$$

24 Use powers-of-10 notation to calculate the following. Give your answers in decimal form.

a. $\dfrac{8{,}100 \times 11.9 \times 0.0132}{0.136 \times 29.7}$ *b.* $\dfrac{810 \times 0.000119 \times 0.132}{13.6 \times 0.0297}$

c. $\dfrac{8.1 \times 0.00119 \times 0.0000132}{0.0136 \times 0.297}$

---

a. $$\frac{(8.1 \times 10^3) \times (1.19 \times 10^1) \times (1.32 \times 10^{-2})}{(1.36 \times 10^{-1}) \times (2.97 \times 10^1)}$$

$$= \frac{8.1 \times 1.19 \times 1.32}{1.36 \times 2.97} \times \frac{10^3 \times 10^1 \times 10^{-2}}{10^{-1} \times 10^1} = 3.15 \times 10^2 = 315$$

b. $$\frac{(8.1 \times 10^2) \times (1.19 \times 10^{-4}) \times (1.32 \times 10^{-1})}{(1.36 \times 10^1) \times (2.97 \times 10^{-2})}$$

$$= \frac{8.1 \times 1.19 \times 1.32}{1.36 \times 2.97} \times \frac{10^2 \times 10^{-4} \times 10^{-1}}{10^1 \times 10^{-2}} = 3.15 \times 10^{-2} = 0.0315$$

c. 0.0000315

---

25 Use powers-of-10 notation to calculate the following. Give your answers in *standard exponent form.*

a. $\dfrac{7{,}200 \times 0.084 \times 0.00063}{0.0108 \times 12{,}600}$ *b.* $\dfrac{0.014 \times 1600 \times 10500}{0.98 \times 0.75}$

---

a. $$\frac{(7.2 \times 10^3) \times (8.4 \times 10^{-2}) \times (6.3 \times 10^{-4})}{(1.08 \times 10^{-2}) \times (1.26 \times 10^4)}$$

$$= \frac{7.2 \times 8.4 \times 6.3}{1.08 \times 1.26} \times \frac{10^3 \times 10^{-2} \times 10^{-4}}{10^{-2} \times 10^4}$$

$$= 280 \times 10^{-5} = 2.8 \times 10^{-3}$$

b. $$\frac{(1.4 \times 10^{-2}) \times (1.6 \times 10^3) \times (1.05 \times 10^4)}{(9.8 \times 10^{-1}) \times (7.5 \times 10^{-1})}$$

$$= \frac{1.4 \times 1.6 \times 1.05}{9.8 \times 7.5} \times \frac{10^{-2} \times 10^3 \times 10^4}{10^{-1} \times 10^{-1}}$$

$$= 0.032 \times 10^7 = 3.2 \times 10^5$$

---

26 Use powers-of-10 notation to calculate the following. Give your answers in standard exponent form correct to three significant figures.

a. $\dfrac{9{,}000 \times 800 \times 0.0081}{120 \times 0.11}$ *b.* $\dfrac{0.0013 \times 1{,}200 \times 0.15}{850 \times 0.09}$

---

a. $$\frac{(9 \times 10^3) \times (8 \times 10^2) \times (8.1 \times 10^{-3})}{(1.2 \times 10^2) \times (1.1 \times 10^{-1})} = \frac{9 \times 8 \times 8.1}{1.2 \times 1.1} \times \frac{10^3 \times 10^2 \times 10^{-3}}{10^2 \times 10^{-1}}$$

$$\approx 442 \times 10^1 = 4.42 \times 10^3$$

b. $$\frac{(1.3 \times 10^{-3}) \times (1.2 \times 10^3) \times (1.5 \times 10^{-1})}{(8.5 \times 10^2) \times (9 \times 10^{-2})}$$

$$= \frac{1.3 \times 1.2 \times 1.5}{8.5 \times 9} \times \frac{10^{-3} \times 10^3 \times 10^{-1}}{10^2 \times 10^{-2}}$$

$$\approx 0.0306 \times 10^{-1} = 3.06 \times 10^{-3}$$

You have finished the program for Sec. 10-4. Now take the posttest to see how much you have learned.

POSTTEST 10-4

Carry out the following operations, putting answers into standard exponent form:

1. $(3 \times 10^4) \times (3 \times 10^{-5})$
2. $(5 \times 10^{-6}) \times (9 \times 10^{-2})$
3. $(8 \times 10^4) \div (2 \times 10^{-5})$
4. $(2 \times 10^{-3}) \div (4 \times 10^{-4})$

Use powers-of-10 notation to do the following problems, putting answers into standard exponent form:

5. $\dfrac{221 \times 0.003}{0.17}$
6. $\dfrac{0.221 \times 300}{0.00017}$
7. $\dfrac{14{,}400 \times 0.000325}{0.045}$
8. $\dfrac{0.0144 \times 32{,}500}{450}$

When you have finished the posttest, correct it yourself using the answers on page 420. Then check all the questions you got wrong on the list below:

1. ______	4. ______	7. ______
2. ______	5. ______	8. ______
3. ______	6. ______	

If you compare your performance with that on Pretest 10-4, you will see how much you have improved. If you made more than one mistake on this posttest, review the material and take the pretest again. Don't move on to the next section until you can get all the questions right.

Now proceed to the next section whose number you circled on page 384. If no other sections were circled, take Final Test 10 (page 418).

Section 10-5
Addition and Subtraction of Numbers in Standard Exponent Form

Do this section if you made a mistake in Questions 13 to 15 of Diagnostic Test 10. Begin by taking the pretest below, which will show you how much of Sec. 10-5 you already know and what you will be able to do by the time you have finished it.

PRETEST 10-5

Carry out the following operations, putting your answers into standard exponent form:

1. $(1.3 \times 10^1) - (5 \times 10^{-3})$
2. $(6.42 \times 10^7) - (3.51 \times 10^4)$
3. $(2.31 \times 10^{-7}) + (7.9 \times 10^{-6})$
4. $(4.6 \times 10^4) + (9.9 \times 10^5)$
5. $(5.6 \times 10^{-1}) + (3.4 \times 10^3)$
6. $(6.4 \times 10^{-3}) - (6.78 \times 10^{-4})$

When you have finished the pretest, correct it yourself using the answers on page 420. Then check all the questions you got wrong on the list below:

1. ______ 3. ______ 5. ______

2. ______ 4. ______ 6. ______

If you made any mistakes (even if it was only *one* mistake), read the introduction and go through the program.

If you made no mistakes at all, skip Sec. 10-5, because you already know its material, and go on to take Final Test 10 (page 418).

Introduction

This section shows that numbers in standard exponent form usually have to be converted to decimal or integer form before they can be added and subtracted. When exponents are of similar magnitude, however, additions and subtractions can be simplified by taking out a common factor.

FRAMES FOR SECTION 10-5

1 In Sec. 10-4 we showed that numbers in standard exponent form could be multiplied by separating off the powers-of-10 components. Thus:

$(5 \times 10^4) \times (6 \times 10^{-3}) = (5 \times 6) \times (\_\_\_\_ \times \_\_\_\_) = 30 \times \_\_\_\_ = 3 \times \_\_\_\_$

$10^4 \times 10^{-3}$ 10^1 10^2

---

2 This method depends on multiplication being associative, i.e., on the fact that $(a \times b) \times c = (a \times c) \times b$. But if we want to add $(5 \times 10^4) + (6 \times 10^{-3})$, we have a mixture of operations whose order cannot be changed. We have to (add/multiply) _____ before we can (add/multiply) _____.

multiply; add

3 This means converting the numbers to their integer or decimal forms before adding them.

$(5 \times 10^4) + (6 \times 10^{-3}) =$ _____ + _____ = _____

50,000 + 0.006 = 50,000.006

4 Use the same method to add the following numbers:

a. $(3 \times 10^2) + (7 \times 10^{-1}) =$ _____ + _____ = _____

b. $(4 \times 10^0) + (5 \times 10^{-2}) =$ _____

c. $(7 \times 10^1) + (6 \times 10^3) =$ _____

a. 300; 0.7; 300.7 *b.* 4.05 *c.* 6,070

5 Add the following numbers:

a. $(4.23 \times 10^{-2}) + (5.61 \times 10^3)$

b. $(5.49 \times 10^{-3}) + (7.534 \times 10^0)$

c. $(7.256 \times 10^1) + (6.293 \times 10^{-1})$

a. 5,610.0423 *b.* 7.53949 *c.* 73.1893

6 Add the following numbers:

a. $(6.34 \times 10^{-3}) + (7.28 \times 10^{-4})$

b. $(7.183 \times 10^{-2}) + (6.53 \times 10^0)$

c. $(7.98 \times 10^1) + (3.491 \times 10^2)$

a. 0.007068 *b.* 6.60183 *c.* 428.9

7 Subtraction of numbers in exponent form is essentially similar.

a. $(4.27 \times 10^2) - (5.16 \times 10^1) =$ _____ − _____ = _____

b. $(5.98 \times 10^{-1}) - (7.43 \times 10^{-3}) =$ _____

c. $(2.75 \times 10^0) - (6.318 \times 10^{-2}) =$ _____

a. 427; 51.6; 375.4 *b.* 0.59057 *c.* 2.68682

8 Subtract the following numbers:

a. $5.06 - (7.32 \times 10^{-2})$ *b.* $(5.91 \times 10^{1}) - 6.735$

c. $(6.72 \times 10^{-2}) - (5.317 \times 10^{-3})$

a. 4.9868 *b.* 52.365 *c.* 0.061883

9 Give your answers to the following in standard exponent form:

a. $(5.32 \times 10^{9}) - (6.21 \times 10^{8}) =$ ______

b. $(3.67 \times 10^{-8}) + (4.56 \times 10^{-7}) =$ ______

c. $(7.139 \times 10^{-9}) - (4.29 \times 10^{-10}) =$ ______

a. 4.699×10^{9} *b.* 4.927×10^{-7} *c.* 6.71×10^{-9}

10 Problems like those in Frame 9 can be shortened by taking out a common factor. Thus

$(5.32 \times 10^{9}) - (6.21 \times 10^{8}) = 10^{8}[(5.32 \times 10^{1}) - 6.21]$

$= 10^{8}$ (______ − ______) = ______ $\times 10^{8}$

= ______ (standard exponent form)

53.2 6.21 46.99 4.699×10^{9}

11 Similarly

$(3.67 \times 10^{-8}) + (4.56 \times 10^{-7}) = 10^{-7}[(3.67 \times 10^{-1}) + 4.56]$

$= 10^{-7}$(______ + ______) = ______ $\times 10^{-7}$

$(7.139 \times 10^{-9}) - (4.29 \times 10^{-10}) = 10^{-9}$ ____________

$= 10^{-9}$(______ − ______) = ______ $\times 10^{-9}$

$0.367 + 4.56 = 4.927$ $[7.139 - (4.29 \times 10^{-1})]$ $7.139 - 0.429 = 6.71$

12 Use the method of factoring to calculate the following and give your answers in *standard exponent form.*

a. $(4.26 \times 10^{6}) + (7.53 \times 10^{7}) = 10^{6}$(______________________________) = ?

b. $(5.31 \times 10^{-8}) - (7.94 \times 10^{-9}) = 10^{-8}$(______________________________) = ?

c. $(6.219 \times 10^{-12}) + (4.53 \times 10^{-10}) = 10^{-10}$(______________________________) = ?

a. $10^6[4.26 + (7.53 \times 10)] = 10^6(4.26 + 75.3) = 79.56 \times 10^6 = 7.956 \times 10^7$
b. $10^{-8}[5.31 - (7.94 \times 10^{-1})] = 10^{-8}(5.31 - 0.794) = 4.516 \times 10^{-8}$
c. $10^{-10}[(6.219 \times 10^{-2}) + 4.53] = 10^{-10}(0.06219 + 4.53) = 4.59219 \times 10^{-10}$

13 Use the method of factoring to calculate the following and give your answers in *standard exponent form.*

a. $(4.916 \times 10^{-7}) - (3.43 \times 10^{-10}) = 10^{-7}($________________$) =$ ______
b. $(5.632 \times 10^{-8}) + (4.915 \times 10^{-6}) =$ ________________
c. $(5.437 \times 10^{7}) - (6.39 \times 10^{4}) =$ ________________

a. $10^{-7}[4.916 - (3.43 \times 10^{-3})] = 4.91257 \times 10^{-7}$
b. $10^{-6}[(5.632 \times 10^{-2}) + 4.915] = 4.97132 \times 10^{-6}$
c. $10^{4}[(5.437 \times 10^{3}) - 6.39] = 5{,}430.61 \times 10^{4} = 5.43061 \times 10^{7}$

14 Use the method of factoring to calculate the following and give your answers in standard exponent form.

a. $(7.621 \times 10^{3}) + (5.319 \times 10^{6}) =$ ________________
b. $(5.719 \times 10^{-6}) - (3.417 \times 10^{-9}) =$ ________________
c. $(4.325 \times 10^{13}) - (6.15 \times 10^{11}) =$ ________________

a. $10^{3}[7.621 + (5.319 \times 10^{3})] = 5{,}326.621 \times 10^{3} = 5.326621 \times 10^{6}$
b. $10^{-6}[5.719 - (3.417 \times 10^{3})] = 5.715583 \times 10^{-6}$
c. $10^{11}[(4.325 \times 10^{2}) - 6.15] = 426.35 \times 10^{11} = 4.2635 \times 10^{13}$

15 Use the method of factoring to calculate the following and give your answers in standard exponent form.

a. $(1.619 \times 10^{-5}) - (7.328 \times 10^{-6}) =$ ______
b. $(2.759 \times 10^{15}) - (6.91 \times 10^{13}) =$ ______
c. $(7.593 \times 10^{-9}) + (6.21 \times 10^{-6}) =$ ______

a. $10^{-5}[1.619 - (7.328 \times 10^{-1})] = 0.8862 \times 10^{-5} = 8.862 \times 10^{-6}$
b. $10^{13}[(2.759 \times 10^{2}) - 6.91] = 268.99 \times 10^{13} = 2.6899 \times 10^{15}$
c. $10^{-6}[(7.593 \times 10^{-3}) + 6.21] = 6.217593 \times 10^{-6}$

16 When the exponents are very different in magnitude or of opposite signs, it is usually easiest to proceed by conversion to the decimal form without bothering to extract factors. Calculate the following without factoring and give your answers in standard exponent form.

a. $(4.63 \times 10^{-2}) + (3.51 \times 10^{2}) =$ ______

b. $1.79 - (8.27 \times 10^{-1}) =$ ______

---

a. $0.0463 + 351 = 351.0463 = 3.510463 \times 10^{2}$
b. $1.79 - 0.827 = 0.963 = 9.63 \times 10^{-1}$

---

17 Calculate the following without factoring and give your answers in standard exponent form.

a. $9.32 + (7.94 \times 10^{-1})$

b. $(7.69 \times 10^{3}) - (4.51 \times 10^{-4})$

---

a. $9.32 + 0.794 = 10.114 = 1.0114 \times 10^{1}$
b. $7{,}690 - 0.0451 = 7{,}689.9549 = 7.6899549 \times 10^{3}$

You have finished the program for Sec. 10-5. Now take the posttest to see how much you have learned.

POSTTEST 10-5

Carry out the following operations, putting your answers into standard exponent form:

1. $(1.4 \times 10^{1}) - (8 \times 10^{-3})$
2. $(3.92 \times 10^{8}) - (4.67 \times 10^{5})$
3. $(3.5 \times 10^{-6}) + (5.94 \times 10^{-5})$
4. $(5.7 \times 10^{3}) + (9.9 \times 10^{4})$
5. $(4.3 \times 10^{-1}) + (6.7 \times 10^{4})$
6. $(3.6 \times 10^{-4}) - (7.69 \times 10^{-5})$

When you have finished the posttest, correct it yourself using the answers on page 420. Then check all the questions you got wrong on the list below:

1. ______ 3. ______ 5. ______

2. ______ 4. ______ 6. ______

If you compare your performance with that on Pretest 10-5, you will see how much you have improved. If you made more than one mistake on this posttest, review the material and take the pretest again.

Now proceed to Final Test 10.

FINAL TEST 10

1. Give the integer or decimal forms of the following numbers:

10^0 10^1 10^{-4}

Give the answers to the following in exponential notation:

2. $10^7 \times 10^{-12}$
3. $10^{-8} \div 10^{-11}$
4. $10^5 \div 10^8$

5. Write 4.3×10^{-3} in decimal form.

Write the following in standard exponent form, i.e., in the form $A \times 10^b$, where A is between 1 and 10 and b is an integer.

6. 47,823
7. 279×10^5
8. 0.065×10^{-7}

Carry out the following operations, putting your answers into standard exponent form:

9. $(8 \times 10^{-6}) \times (4 \times 10^{-2})$
10. $(3 \times 10^{-5}) \div (6 \times 10^3)$

Use powers of 10 to solve the following problems. Give the answers in standard exponent form:

11. $\dfrac{4500 \times 0.0000375}{22{,}500}$
12. $\dfrac{0.0000115 \times 0.0108}{0.000414}$

Carry out the following operations, putting your answers into standard exponent form:

13. $(4.753 \times 10^1) - (3.5 \times 10^{-2})$
14. $(2.4 \times 10^{-9}) + (6.36 \times 10^{-8})$
15. $(6.12 \times 10^8) - (8.79 \times 10^6)$

The answers to this test are on page 420. Check the questions you got wrong on the list below and compare your results with those on Diagnostic Test 10 to see how much you have improved.

1. ____ 5. ____ 9. ____ 13. ____
2. ____ 6. ____ 10. ____ 14. ____
3. ____ 7. ____ 11. ____ 15. ____
4. ____ 8. ____ 12. ____

If you made more than one mistake, review the material and then take the diagnostic test again. Make sure that you can get all the questions right before you finish the chapter.

Now proceed to Chap. 11.

ANSWERS TO TESTS FOR CHAPTER 10

DIAGNOSTIC TEST 10

1. 10; 1; 0.001
2. 10^{-3}
3. 10^{3}
4. 10^{-3}
5. 0.031
6. 1.792×10^{3}
7. 1.93×10^{8}
8. 2.4×10^{-10}
9. 1.2×10^{-8}
10. 5×10^{-7}
11. 2.7×10^{-6}
12. 1.8×10^{-4}
13. 5.3602×10^{1}
14. 1.015×10^{-3}
15. 8.4029×10^{7}

PRETEST 10-1

1. $16 \times 16 \times 16 \times 16 \times 16$
2. 23^{4}
3. base 7, exponent 11
4. 1,000,000; 10
5. 10^{49}
6. 10^{41}
7. 10^{1}
8. 10^{6}

POSTTEST 10-1

1. $13 \times 13 \times 13 \times 13 \times 13 \times 13$
2. 35^{5}
3. base 9, exponent 12
4. 10,000,000; 10
5. 10^{59}
6. 10^{41}
7. 10^{1}
8. 10^{6}

PRETEST 10-2

1. 0.0001
2. 1
3. 10^{-1}
4. 10^{-11}
5. 10^{0}
6. 10^{-3}
7. 10^{2}
8. 10^{13}
9. 10^{-18}
10. 10^{0}

POSTTEST 10-2

1. 0.00001
2. 1
3. 10^{3}
4. 10^{-11}
5. 10^{0}
6. 10^{-2}
7. 10^{5}
8. 10^{17}
9. 10^{-13}
10. 10^{8}

PRETEST 10-3

1. 0.000042
2. 12,900
3. 6.7
4. 0.000000123
5. 1.463×10^{3}
6. 4.8×10^{-9}
7. 2.89×10^{6}
8. 7.1×10^{3}
9. 3.260×10^{-4}
10. 5.2×10^{-3}

POSTTEST 10-3

1. 0.00295
2. 6.9
3. 46,000
4. 0.0000326
5. 1.329×10^{3}
6. 3.2×10^{-8}
7. 3.26×10^{5}
8. 6.8×10^{2}
9. 2.93×10^{-2}
10. 9.85×10^{-3}

PRETEST 10-4

1. 8×10^{1}
2. 1.5×10^{-5}
3. 2×10^{8}
4. 2.5×10^{2}
5. 6.8×10^{0}
6. 6.8×10^{5}
7. 6.24×10^{1}
8. 6.24×10^{-1}

POSTTEST 10-4

1. 9×10^{-1}
2. 4.5×10^{-7}
3. 4×10^{9}
4. 5×10^{0}
5. 3.9×10^{0}
6. 3.9×10^{5}
7. 1.04×10^{2}
8. 1.04×10^{0}

PRETEST 10-5

1. 1.2995×10^{1}
2. 6.41649×10^{7}
3. 8.131×10^{-6}
4. 1.036×10^{6}
5. 3.40056×10^{3}
6. 5.722×10^{-3}

POSTTEST 10-5

1. 1.3992×10^{1}
2. 3.91533×10^{8}
3. 6.29×10^{-5}
4. 1.047×10^{5}
5. 6.700043×10^{4}
6. 2.831×10^{-4}

FINAL TEST 10

1. 1; 10; 0.0001
2. 10^{-5}
3. 10^{3}
4. 10^{-3}
5. 0.0043
6. 4.7823×10^{4}
7. 2.79×10^{7}
8. 6.5×10^{-9}
9. 3.2×10^{-7}
10. 5×10^{-9}
11. 7.5×10^{-6}
12. 3.0×10^{-4}
13. 4.7495×10^{1}
14. 6.6×10^{-8}
15. 6.0321×10^{8}

Chapter 11
Simple Equations

In this chapter you will learn methods for the solution of simple first-degree equations in one variable. Section 11-1 introduces the concept of a solution and the concept of equivalent equations as equations which have the same solution. You will see that addition of the same number to both sides of an equation gives an equivalent equation. Section 11-2 shows that multiplication of both sides of an equation by the same number gives an equivalent equation, and Sec. 11-3 that addition of the same multiple of the variable to both sides of an equation gives an equivalent equation. Section 11-4 extends these methods to the solution of equations with several terms.

This chapter requires knowledge of Chaps. 5 and 6.

DIAGNOSTIC TEST 11

Solve the following equations and check your answers:

1. $a - 11 = 3$
2. $y + 4 = -8$
3. $z - 3 = -5$
4. $-16k = -4$
5. $12m = \frac{3}{4}$
6. $\frac{4}{5}b = -\frac{2}{3}$
7. $-\frac{3}{7}n = \frac{5}{8}$
8. $6y - 8 - 3y = 8 - 2y - 5$
9. $5d + 4 + d = 6 - 4d - 15 + 7d$
10. $11 - 3d + 6 = 5d + 4 - 2d + 1$

When you have finished the diagnostic test, correct it yourself using the answers on page 450. Then check all the questions you got wrong on the list below.

1. ______	4. ______	8. ______
2. ______	5. ______	9. ______
3. ______	6. ______	10. ______
	7. ______	

If you made no mistakes at all, you need not do any work in this chapter and should go on straight to Chap. 12. But if you made any mistakes (even if it was only *one* mistake), follow these instructions:

	section	page
If you made any mistakes in Questions 1 to 3, circle Sec. 11-1.	11-1	423
If you made any mistakes in Questions 4 to 7, circle Sec. 11-2.	11-2	428
If you made any mistakes in Questions 8 to 10, circle Secs. 11-3 and 11-4.	11-3 11-4	435 443

You should now work through the sections whose numbers you have circled, taking them in numerical order. The sections begin on the pages indicated.

Section 11-1
The Addition Transformation with Numerals

Do this section if you made a mistake in Questions 1 to 3 of Diagnostic Test 11. Begin by taking the pretest below, which will show you how much of Sec. 11-1 you already know and what you will be able to do by the time you have finished it.

PRETEST 11-1

Convert the following equations to equivalent equations of the form $z =$?:

1. $z + 5 = -8 \leftrightarrow z =$ 2. $z - 3 = 6 \leftrightarrow z =$ 3. $z + 4 = 2 \leftrightarrow z =$

Give the solutions of the following equations:

4. $k - 5 = -9$ 5. $y - 2 = 7$ 6. $q + 3 = -4$

When you have finished the pretest, correct it yourself using the answers on page 450. Then check all the questions you got wrong on the list below:

1. ______ 3. ______ 5. ______

2. ______ 4. ______ 6. ______

If you made any mistakes (even if it was only *one* mistake), read the introduction and go through the program.

If you made no mistakes at all, skip Sec. 11-1, because you already know its material, and go on to the next section whose number you circled on page 422. If no further numbers were circled, take Final Test 11 (page 449).

Introduction

This section defines a *solution* of an equation as any value of the variable which makes the equation into a true statement. It then defines equations which have the same solution as *equivalent* and shows that adding the same number to each side of an equation gives an equivalent equation. You then learn to use this property to solve equations of the general type $x + a = b$.

FRAMES FOR SECTION 11-1

1 If $x + 4 = 8$ is a true statement about x, then $x =$ ______.

4

---

2 *a.* 4 is called a *solution* of $x + 4 = 8$ because if $x = 4$, we get $4 + 4 = 8$, which is a (true/false) ______ statement.

b. 5 is not a solution of $x + 4 = 8$ because if $x = 5$, we get $5 + 4 = 8$, which is a (true/false) ______ statement.

c. Is $3 + 4 = 8$ a true statement?

d. Is 3 a solution of $x + 4 = 8$?

---

a. true *b.* false *c.* no *d.* no

---

3 Give one solution of $x + 8 = 10$. Can you think of another solution?

---

2 I hope not, because 2 is the only solution!

---

4 The simplest equation which has 2 as its only solution is $x = 2$ because $2 = 2$ is a (true/false) ______ statement and $1 = 2$, $3 = 2$, $4 = 2$, etc., are (true/false) ______ statements.

---

true; false

---

5 Because $x + 8 = 10$ and $x = 2$ have the same solution, they are said to be *equivalent*, and we write

$x + 8 = 10 \leftrightarrow x = 2$

a. Is $x + 5 = 7$ equivalent with $x = 2$?

b. Is $x + 4 = 7$ equivalent with $x = 2$?

c. Is $x + 4 = 7$ equivalent with $x = 3$?

d. Is $x + 5 = 7$ equivalent with $x = 3$?

---

a. yes *b.* no *c.* yes *d.* no

---

6 Which of the following statements are true?

a. $x + 3 = 8 \leftrightarrow x = 5$

b. $x + 3 = 8 \leftrightarrow x = 4$

c. $x + 1 = 9 \leftrightarrow x = 7$

d. $x + 2 = 7 \leftrightarrow x = 5$

---

a and *d*

---

7 Which of the following statements are true?

a. $x + 1 = 7 \leftrightarrow x = 6$ *b.* $x + 2 = 8 \leftrightarrow x = 6$

c. $x + 3 = 9 \leftrightarrow x = 6$ *d.* $x + 4 = 10 \leftrightarrow x = 6$

e. $x + 5 = 11 \leftrightarrow x = 6$

---

all of them

---

8 Notice that if you add the same number to each side of an equation, you obtain an equivalent equation. If we start with $x + 0 = 6$ and add 1 to each side, we get $x + 1 = 7$. If we add 2, we get $x + 2 = 8$.

a. If we add 3, we get $x + 3 =$ ______.

b. If we add 4, we get ______ + ______ = ______.

c. If we add 5, we get ______.

---

a. 9 *b.* $x + 4 = 10$ *c.* $x + 5 = 11$

---

9 Suppose we start with $x + 1 = 7$ and add 10 to both sides. This gives

$x + 1 + 10 = 7 + 10$ or $x +$ ______ = ______

---

11 17

---

10 *a.* Is $6 + 11 = 17$ a true statement?

b. Is 6 a solution of $x + 11 = 17$?

c. Is $x + 11 = 17$ equivalent to $x + 1 = 7$?

---

a. yes *b.* yes *c.* yes

---

11 Suppose we start with $x + 1 = 7$ and add -10 to both sides. This gives $x + 1 + (-10) = 7 + (-10)$ or $x +$ ______ = ______.

(-9) -3

---

12 *a.* Is $6 + (-9) = -3$ a true statement?

b. Is 6 a solution of $x + (-9) = -3$?

c. Is $x + (-9) = -3$ equivalent to $x + 1 = 7$?

---

a. yes *b.* yes *c.* yes

---

13 In general, whenever you add the same number to each side of an equation, you obtain a(n) ______ equation.

equivalent

14 Adding a number to each side can be a good method of solving an equation, i.e., finding its solution.

a. $x-2=7 \leftrightarrow x-2+2=7+2 \leftrightarrow x=$ ______ adding 2 to each side. Solution is ______.

b. $y+5=-2 \leftrightarrow y+5+$ ______ $=-2+$ ______ $\leftrightarrow y=$ ______ adding −5 to each side. Solution is ______.

c. $d-4=-4 \leftrightarrow d-4+$ ______ $=-4+$ ______ $\leftrightarrow d=$ ______ adding 4 to each side. Solution is ______.

a. 9; 9 *b.* (−5); (−5); −7; −7 *c.* 4; 4; 0; 0

15 *a.* $k+3=-2 \leftrightarrow k=$ ______ adding −3 to each side. Solution is ______.

b. $x-5=-3 \leftrightarrow x=$ ______ adding ______ to each side. Solution is ______.

c. $b+7=-11 \leftrightarrow b=$ ______ adding ______ to each side. Solution is ______.

a. −5; −5 *b.* 2; 5; 2 *c.* −18; −7; −18

16 Use the method of adding a number to each side to solve the following equations:

a. $q+6=-1 \leftrightarrow q=$ ______ adding ______ to each side. Solution is ______.

b. $m-7=6 \leftrightarrow m=$ ______. Solution is ______.

c. $x+12=-5 \leftrightarrow x=$ ______. Solution is ______.

a. −7; −6; −7 *b.* 13; 13 *c.* −17; −17

17 Solve the following equations:

a. $x+13=-4 \leftrightarrow x=$ ______. Solution is ______.

b. $y-11=-8 \leftrightarrow y=$ ______. Solution is ______.

a. −17; −17 *b.* 3; 3

18 Solve the following equations:

a. $m - 10 = -20$ *b.* $p + 17 = -4$

---

a. -10 *b.* -21

You have finished the program for Sec. 11-1. Now take the posttest to see how much you have learned.

POSTTEST 11-1

Convert the following equations to equivalent equations of the form $z = ?$:

1. $z + 7 = -9$
2. $z - 4 = 11$
3. $z + 5 = 7$

Give the solutions of the following equations:

4. $k - 6 = -4$
5. $y - 3 = 10$
6. $q + 8 = -5$

When you have finished the posttest, correct it yourself using the answers on page 450. Then check all the questions you got wrong on the list below:

1. ______ 3. ______ 5. ______

2. ______ 4. ______ 6. ______

If you compare your performance with that on Pretest 11-1, you will see how much you have improved. If you made more than one mistake on this posttest, review the material and take the pretest again. Don't move on to the next section until you can get all the questions right.

Now proceed to the next section whose number you circled on page 422. If no other sections were circled, take Final Test 11 (page 449).

Section 11-2
The Multiplication Transformation with Numerals

Do this section if you made a mistake in Questions 4 to 7 of Diagnostic Test 11. Begin by taking the pretest below, which will show you how much of Sec. 11-2 you already know and what you will be able to do by the time you have finished it.

PRETEST 11-2

Convert the following equations to equivalent equations of the form $k = ?$:

1. $8k = -4$ 2. $-\frac{3}{8}k = \frac{3}{5}$ 3. $-20k = -5$

Give the solutions of the following equations:

4. $6y = \frac{17}{2}$ 5. $-\frac{4}{5}z = -7$ 6. $\frac{3}{4}q = 11$

When you have finished the pretest, correct it yourself using the answers on page 451. Then check all the questions you got wrong on the list below:

1. ______ 3. ______ 5. ______

2. ______ 4. ______ 6. ______

If you made any mistakes (even if it was only *one* mistake), read the introduction and go through the program.

If you made no mistakes at all, skip Sec. 11-2, because you already know its material, and go on to the next section whose number you circled on page 422. If no further numbers were circled, take Final Test 11 (page 449).

Introduction

Section 11-2 introduces the notation $3z$ for $3 \times z$, $-5f$ for $-5 \times f$, $-\frac{2}{3}y$ for $-\frac{2}{3} \times y$, etc. You then learn that multiplying both sides of an equation by the same number yields an equivalent equation. You use this property to solve equations of the general type

$$\frac{a}{b}y = \frac{c}{d}$$

FRAMES FOR SECTION 11-2

1 $7 \times k$ is written $7k$, $\frac{5}{6} \times q$ as $\frac{5}{6}q$, etc. How would you write the following?

a. $3 \times q$ *b.* $\frac{1}{6} \times y$ *c.* $\frac{7}{5} \times z$

a. $3q$ *b.* $\frac{1}{8}y$ *c.* $\frac{7}{9}z$

2 Since the operations "negative of" and "multiplying by -1" are equivalent, $-3q$, the negative of $3q$, is the same as $3q \times (-1) = -1 \times 3q$. Hence

a. $-3q = -1 \times 3 \times q =$ ______ $\times q$

How would you write the following in their shortest form?

b. $-5 \times k$ *c.* $-\frac{1}{8} \times z$ *d.* $-\frac{7}{9} \times w$

a. -3 *b.* $-5k$ *c.* $-\frac{1}{8}z$ *d.* $-\frac{7}{9}w$

3 Since $3 \times p = p \times 3$, they are both written as $3p$. How would you write the following?

a. $q \times 5$ *b.* $k \times (-8)$

c. $-\frac{1}{3} \times m$ *d.* $z \times (-2)$

a. $5q$ *b.* $-8k$ *c.* $-\frac{1}{3}m$ *d.* $-2z$

4 *a.* Is $3 \times 4 = 12$ a true statement?

b. Is 4 a solution of $3y = 12$?

c. Is $3 \times 3 = 12$ a true statement?

d. Is 3 a solution of $3y = 12$?

a. yes *b.* yes *c.* no *d.* no

5 Since $3y = 12$ and $y = 4$ have the same solution, they are said to be *equivalent* and we write $3y = 12 \leftrightarrow y = 4$.

a. Is $5y = 20$ equivalent with $y = 4$?

b. Is $4y = 20$ equivalent with $y = 4$?

c. Is $4y = 20$ equivalent with $y = 5$?

d. Is $5y = 20$ equivalent with $y = 5$?

a. yes *b.* no *c.* yes *d.* no

6 Which of the following statements are true?

a. $6y = 24 \leftrightarrow y = 4$ *b.* $6y = 24 \leftrightarrow y = 6$

c. $5y = 35 \leftrightarrow y = 5$ *d.* $5y = 35 \leftrightarrow y = 7$

---

a and *d*

---

7 Which of the following statements are true?

a. $2y = 6 \leftrightarrow y = 3$ *b.* $3y = 9 \leftrightarrow y = 3$

c. $4y = 12 \leftrightarrow y = 3$ *d.* $5y = 15 \leftrightarrow y = 3$

e. $6y = 18 \leftrightarrow y = 3$

---

all of them

---

8 Notice that if you multiply each side of an equation by the same number, you obtain an equivalent equation. If we start with $y = 3$ and multiply each side by 2, we get $2y = 6$.

a. If we multiply by 3, we get $3y = 9$.

b. If we multiply by 4, we get $4y =$ ______.

c. If we multiply by 5, we get __$y =$ ______.

d. If we multiply by 6, we get ______.

---

b. 12 *c.* 5; 15 *d.* $6y = 18$

---

9 Suppose we start with $y = 3$ and multiply each side by 11. This gives $11y =$ ______.

a. Is 3 a solution of this new equation?

b. Is the new equation equivalent to $y = 3$?

---

33 *a.* yes *b.* yes

---

10 Suppose we start with $y = 3$ and multiply each side by -4.

a. This gives $-4y =$ ______.

b. Is 3 a solution of this new equation?

c. Is the new equation equivalent to $y = 3$?

---

a. -12 *b.* yes *c.* yes

---

11 Suppose we start with $2y = 6$ and multiply each side by $\frac{1}{6}$.

a. This gives $\frac{1}{3}y =$ ______.

b. Is 3 a solution of this new equation?

c. Is the new equation equivalent to $2y = 6$?

a. 1 *b.* yes *c.* yes

12 In general, whenever you multiply each side of an equation by the same number, you obtain a(n) ______ equation (provided that you don't multiply by 0).

equivalent

13 Let's see what happens when you multiply by 0. $y = 3$ becomes $y \times 0 = 3 \times 0 = 0$. Now $y \times 0 = 0$ is a rather special type of equation.

a. Is $1 \times 0 = 0$ a true statement?

b. Is $2 \times 0 = 0$ a true statement?

c. Is $3 \times 0 = 0$ a true statement?

d. Is $4 \times 0 = 0$ a true statement?

e. Is there any value of y for which $y \times 0 = 0$ is false?

a. yes *b.* yes *c.* yes *d.* yes *e.* no

14 *a.* Is 1 a solution of $y \times 0 = 0$?

b. Is 2 a solution of $y \times 0 = 0$?

c. Is 3 a solution of $y \times 0 = 0$?

d. Is 4 a solution of $y \times 0 = 0$?

e. Is there any value of y which is not a solution of $y \times 0 = 0$?

a. yes *b.* yes *c.* yes *d.* yes *e.* no

15 *a.* $y = 3$ has only one solution, ______. But $y \times 0 = 3 \times 0$ has (two/three/four/many) ______ solutions.

b. Are $y = 3$ and $y \times 0 = 3 \times 0$ equivalent?

a. 3; many *b.* no

16 In general, whenever you multiply each side of an equation by the same number, you obtain a(n) ______ equation, provided that the number is not ______.

equivalent; 0

17 Solve the following equations by using multiplication:

a. $4z = 12 \leftrightarrow 4z \times \frac{1}{4} = 12 \times \frac{1}{4} \leftrightarrow z =$ ______. Multiplying each side by the reciprocal of 4, which is $\frac{1}{4}$, the solution is ______.

b. $\frac{1}{3}y = 7 \leftrightarrow \frac{1}{3}y \times$ ______ $= 7 \times$ ______ $\leftrightarrow y =$ ______. Multiplying each side by the reciprocal of $\frac{1}{3}$, which is 3, the solution is ______.

c. $10d = 5 \leftrightarrow 10d \times$ ______ $= 5 \times$ ______ $\leftrightarrow d =$ ______. Multiplying each side by the reciprocal of 10, which is ______, the solution is ______.

---

a. 3; 3 *b.* $\frac{1}{3}y \times 3 = 7 \times 3 \leftrightarrow y = 21$; 21
c. $10d \times \frac{1}{10} = 5 \times \frac{1}{10} \leftrightarrow d = \frac{1}{2}$; $\frac{1}{10}$; $\frac{1}{2}$

---

18 Solve the following equations by using multiplication:

a. $-5y = 20 \leftrightarrow -5y \times$ ______ $= 20 \times$ ______ $\leftrightarrow y =$ ______. Multiplying each side by the reciprocal of -5, which is $-\frac{1}{5}$, the solution is ______.

b. $-\frac{1}{4}g = -11 \leftrightarrow -\frac{1}{4}g \times$ ______ $= -11 \times$ ______ $\leftrightarrow g =$ ______. Multiplying each side by the reciprocal of $-\frac{1}{4}$, which is -4, the solution is ______.

c. $-\frac{28}{3}q = 7 \leftrightarrow -\frac{28}{3}q \times$ ______ $= 7 \times$ ______ $\leftrightarrow q =$ ______. Multiplying each side by the reciprocal of $-\frac{28}{3}$, which is ______, the solution is ______.

---

a. $-5y \times (-\frac{1}{5}) = 20 \times (-\frac{1}{5}) \leftrightarrow y = -4$; -4
b. $-\frac{1}{4}g \times (-4) = -11 \times (-4) \leftrightarrow g = 44$; 44
c. $-\frac{28}{3}q \times (-\frac{3}{28}) = 7 \times (-\frac{3}{28}) \leftrightarrow q = -\frac{3}{4}$; $-\frac{3}{28}$; $-\frac{3}{4}$

---

19 Solve the following equations by using multiplication:

a. $\frac{5}{6}y = 12 \leftrightarrow \frac{5}{6}y \times$ ______ $= 12 \times$ ______ $\leftrightarrow y =$ ______. Multiplying each side by $\frac{6}{5}$, the solution is ______.

b. $-\frac{8}{3}y = 2 \leftrightarrow -\frac{8}{3}y \times$ ______ $= 2 \times$ ______ $\leftrightarrow y =$ ______. Multiplying each side by ______, the solution is ______.

c. $7y = -11 \leftrightarrow 7y \times$ ______ $= -11 \times$ ______ $\leftrightarrow y =$ ______. Multiplying each side by ______, the solution is ______.

---

a. $\frac{5}{6}y \times \frac{6}{5} = 12 \times \frac{6}{5} \leftrightarrow y = \frac{72}{5}$; $\frac{72}{5}$
b. $-\frac{8}{3}y \times (-\frac{3}{8}) = 2 \times (-\frac{3}{8}) \leftrightarrow y = -\frac{3}{4}$; $-\frac{3}{4}$
c. $7y \times \frac{1}{7} = -11 \times \frac{1}{7} \leftrightarrow y = -\frac{11}{7}$; $\frac{1}{7}$; $-\frac{11}{7}$

---

20 Solve the following equations by using multiplication:

a. $-3k = 16 \leftrightarrow -3k \times$ ______ $= 16 \times$ ______ $\leftrightarrow k =$ ______. Multiplying each side by ______, the solution is ______.

b. $\frac{2}{3}z = 5 \leftrightarrow \frac{2}{3}z \times$ ______ $= 5 \times$ ______ $\leftrightarrow z =$ ______. Solution is ______.

c. $-\frac{3}{7}p = -11 \leftrightarrow -\frac{3}{7}p \times$ ______ $= -11 \times$ ______ $\leftrightarrow p =$ ______. Solution is ______.

---

a. $-3k \times (-\frac{1}{3}) = 16 \times (-\frac{1}{3}) \leftrightarrow k = -\frac{16}{3}; -\frac{1}{3}; -\frac{16}{3}$
b. $\frac{2}{3}z \times \frac{3}{2} = 5 \times \frac{3}{2} \leftrightarrow z = \frac{15}{2}; \frac{15}{2}$
c. $-\frac{3}{7}p \times (-\frac{7}{3}) = -11 \times (-\frac{7}{3}) \leftrightarrow p = \frac{77}{3}; \frac{77}{3}$

---

21 Solve the following equations:

a. $4z = -22 \leftrightarrow 4z \times$ ______ $= -22 \times$ ______ $\leftrightarrow z =$ ______. Solution is ______.

b. $-\frac{5}{4}z = 30 \leftrightarrow -\frac{5}{4}z \times$ ______ $= 30 \times$ ______ $\leftrightarrow z =$ ______. Solution is ______.

---

a. $4z \times \frac{1}{4} = -22 \times \frac{1}{4} \leftrightarrow z = -\frac{11}{2}; -\frac{11}{2}$
b. $-\frac{5}{4}z \times (-\frac{4}{5}) = 30 \times (-\frac{4}{5}) \leftrightarrow z = -24; -24$

---

22 Solve the following equations:

a. $5z = -23 \leftrightarrow 5z \times$ ______ $= -23 \times$ ______ $\leftrightarrow z =$ ______. Solution is ______.

b. $7z = -\frac{5}{2} \leftrightarrow 7z \times$ ______ $= -\frac{5}{2} \times$ ______ $\leftrightarrow z =$ ______. Solution is ______.

---

a. $5z \times \frac{1}{5} = -23 \times \frac{1}{5} \leftrightarrow z = -\frac{23}{5}; -\frac{23}{5}$
b. $7z \times \frac{1}{7} = -\frac{5}{2} \times \frac{1}{7} \leftrightarrow z = -\frac{5}{14}; -\frac{5}{14}$

---

23 Solve the following equations:

a. $-4z = \frac{7}{3} \leftrightarrow -4z \times$ ______ $= \frac{7}{3} \times$ ______ $\leftrightarrow z =$ ______. Solution is ______.

b. $-\frac{3}{5}z = \frac{4}{7} \leftrightarrow -\frac{3}{5}z \times$ ______ $= \frac{4}{7} \times$ ______ $\leftrightarrow z =$ ______. Solution is ______.

---

a. $-4z \times (-\frac{1}{4}) = \frac{7}{3} \times (-\frac{1}{4}) \leftrightarrow z = -\frac{7}{12}; -\frac{7}{12}$
b. $-\frac{3}{5}z \times (-\frac{5}{3}) = \frac{4}{7} \times (-\frac{5}{3}) \leftrightarrow z = -\frac{20}{21}; -\frac{20}{21}$

---

24 Solve the following equations:

a. $-6d = 21$ *b.* $-12d = \frac{7}{5}$

---

a. $-\frac{7}{2}$ *b.* $-\frac{7}{60}$

---

25 Solve the following equations:

a. $\frac{3}{11}q = -9$ *b.* $\frac{4}{5}b = -\frac{3}{11}$

---

a. -33 *b.* $-\frac{15}{44}$

---

26 Solve the following equations:

a. $\frac{5}{4}y = -\frac{3}{8}$ *b.* $-\frac{6}{7}z = -\frac{9}{10}$

---

a. $-\frac{3}{10}$ *b.* $\frac{21}{20}$

You have finished the program for Sec. 11-2. Now take the posttest to see how much you have learned.

POSTTEST 11-2

Convert the following equations to equivalent equations of the form $p = ?$:

1. $3p = -12$ 2. $-\frac{4}{9}p = \frac{5}{11}$ 3. $-8p = -2$

Give the solutions of the following equations:

4. $-5k = \frac{13}{3}$ 5. $-\frac{5}{6}y = -4$ 6. $\frac{6}{7}z = 11$

When you have finished the posttest, correct it yourself using the answers on page 451. Then check all the questions you got wrong on the list below:

1. _____ 3. _____ 5. _____

2. _____ 4. _____ 6. _____

If you compare your performance with that on Pretest 11-2, you will see how much you have improved. If you made more than one mistake on this posttest, review the material and take the pretest again. Don't move on to the next section until you can get all the questions right.

Now proceed to the next section whose number you circled on page 422. If no other sections were circled, take Final Test 11 (page 449).

Section 11-3
The Addition Transformation with Variables

Do this section if you made a mistake in Questions 8 to 10 of Diagnostic Test 11. Begin by taking the pretest below, which will show you how much of Sec. 11-3 you already know and what you will be able to do by the time you have finished it.

PRETEST 11-3

Solve the following equations and check your answers by substitution:

1. $-4y + 6 = 3$	2. $2y - 4 = -8$
3. $-6 = 4 - 3y$	4. $k = -5k + 6$
5. $3z + 8 = 5z$	6. $2z - 9 = -4z$

When you have finished the pretest, correct it yourself using the answers on page 451. Then check all the questions you got wrong on the list below:

1. ______	3. ______	5. ______
2. ______	4. ______	6. ______

If you made any mistakes (even if it was only *one* mistake), read the introduction and go through the program.

If you made no mistakes at all, skip Sec. 11-3, because you already know its material, and go on to Sec. 11-4.

Introduction

Adding the same number to both sides of an equation is called the *addition transformation* (AT), and multiplying both sides of an equation by the same number is called the *multiplication transformation* (MT). You use combinations of both transformations to solve equations of the general type $ax + b = c$. The AT is shown to be valid for multiples of x as well as for numbers, and you use this property to solve equations of the general type $ax = bx + c$. The section also introduces the method of checking a solution by substitution in the original equation.

FRAMES FOR SECTION 11-3

1 Convert the following equations to equivalent equations of the form $y = ?$:

a. $y + 2 = 4 \leftrightarrow y =$ ____ *b.* $y - 3 = 8 \leftrightarrow y =$ ____

c. $3y = 12 \leftrightarrow y =$ ____ *d.* $\frac{y}{8} = 2 \leftrightarrow y =$ ____

---

a. 2 *b.* 11 *c.* 4 *d.* 16

---

2 *a.* In Frame 1*a* you had to add ____ to both sides.

b. In 1*b* you added ____.

c. In 1*c* you multiplied both sides by ____.

d. In 1*d* you multiplied both sides by ____.

---

a. −2 *b.* 3 *c.* $\frac{1}{3}$ *d.* 8

---

3 Adding the same number to both sides of an equation is called an *addition transformation* (AT), and multiplying both sides of an equation by the same number is called a *multiplication transformation* (MT).

a. To convert $3y = 5$ to $y =$ ____, you use the ____ transformation with $\frac{1}{3}$.

b. To convert $y + 5 = 7$ to $y =$ ____, you use the ____ transformation with −5.

---

a. $\frac{5}{3}$; multiplication *b.* 2; addition

---

4 An equation of the form $3y + 7 = 5$ can be solved by using both transformations:

a. Using the ____ transformation with −7,

$3y + 7 = 5 \leftrightarrow 3y = -2$

b. Using the ____ transformation with $\frac{1}{3}$,

$3y = -2 \leftrightarrow y =$ ____

---

a. addition *b.* multiplication; $-\frac{2}{3}$

---

5 Convert the following equations to the form $k = ?$ by using AT followed by the MT:

a.
$$\begin{array}{c} 3k + 2 = 4 \\ \updownarrow \text{AT} \\ 3k = \_\_ \\ \updownarrow \text{MT} \\ k = \_\_ \end{array}$$

b.
$$\begin{array}{c} 6 = 5k - 1 \\ \updownarrow \text{AT} \\ \_\_ = 5k \\ \updownarrow \text{MT} \\ \_\_ = k \end{array}$$

c.
$$\begin{array}{c} 5k - 1 = 7 \\ \updownarrow \text{AT} \\ \_\_ = \_\_ \\ \updownarrow \text{MT} \\ k = \_\_ \end{array}$$

d.
$$\begin{array}{c} 2 = 4k + 6 \\ \updownarrow \text{AT} \\ \_\_ = \_\_ \\ \updownarrow \text{MT} \\ \_\_ = k \end{array}$$

---

a. $2; \frac{2}{3}$ *b.* $7; \frac{7}{5}$ *c.* $5k; 8; \frac{8}{5}$ *d.* $-4; 4k; -1$

---

6 Convert the following equations to the form $k = ?$:

a.
$$\begin{array}{c} 6k - 2 = -3 \\ \updownarrow \text{AT} \\ \_\_ = \_\_ \\ \updownarrow \text{MT} \\ k = \_\_ \end{array}$$

b.
$$\begin{array}{c} -7 = -3 + 2k \\ \updownarrow \text{AT} \\ \_\_ = \_\_ \\ \updownarrow \text{MT} \\ \_\_ = k \end{array}$$

c.
$$\begin{array}{c} 8 = 11 - 6k \\ \updownarrow \text{AT} \\ \_\_ = \_\_ \\ \updownarrow \text{MT} \\ \_\_ = k \end{array}$$

d.
$$\begin{array}{c} -5k + 7 = 12 \\ \updownarrow \text{AT} \\ \_\_ = \_\_ \\ \updownarrow \text{MT} \\ k = \_\_ \end{array}$$

---

a. $6k; -1; -\frac{1}{6}$ *b.* $-4; 2k; -2$ *c.* $-3; -6k; \frac{1}{2}$ *d.* $-5k; 5; -1$

---

7 Solve the following equations:

a. $2y - 3 = -7$ *b.* $-5y + 4 = 1$

c. $4 = 3y - 2$ *d.* $-8 = 2 - 3y$

---

a. -2 *b.* $\frac{3}{5}$ *c.* 2 *d.* $\frac{10}{3}$

---

8 *a.* 1 (is/is not) ______ a solution of $2y + 4 = 8$ because $(2 \times 1) + 4 = 8$ is a (true/false) ______ statement.

b. 2 (is/is not) ______ a solution of $2y + 4 = 8$ because $(2 \times 2) + 4 = 8$ is a (true/false) ______ statement.

c. 3 (is/is not) ______ a solution of $2y + 4 = 8$ because $(2 \times 3) + 4 = 8$ is a (true/false) ______ statement.

---

a. is not; false *b.* is; true *c.* is not; false

---

9 In Frame 8 you found out whether 1, 2, and 3 were solutions of $2y + 4 = 8$ by the process we call ***substitution.***

a. In *a* you substituted __ for y in the equation $2y + 4 = 8$ to obtain the statement $(2 \times 1) + 4 = 8$.

b. In *b* you substituted __ for y in the equation $2y + 4 = 8$ to obtain the statement ____________.

c. Substituting 4 for y in the equation $2y + 4 = 8$ would give you the statement ____________, which is (true/false) ______.

---

a. 1 *b.* 2; $(2 \times 2) + 4 = 8$ *c.* $(2 \times 4) + 4 = 8$; false

---

10 *a.* Substituting 5 for y in the equation $3y - 2 = 13$ gives the statement ________________, which is (true/false) ______.

b. Is 5 a solution of $3y - 2 = 13$?

c. How would you find out if 6 was a solution of $2y + 3 = 15$?

d. Is 6 a solution of $2y + 3 = 15$?

---

a. $(3 \times 5) - 2 = 13$; true *b.* yes
c. by substituting 6 for y in the equation $2y + 3 = 15$ to get a statement and checking if the statement is true or false. *d.* yes

---

11 When you have solved an equation, you should always check your answer by substituting it in the equation and seeing if it gives a true statement.

a. In Frame 7*a* we deduced that the solution of $2y - 3 = -7$ was -2. Substituting -2 for y, we get the statement ________________, which is (true/false) ______.

b. In Frame 7*c* we deduced that the solution of $4 = 3y - 2$ was 2. Check this answer by substitution.

a. $2(-2) - 3 = -7$; true
b. Substituting 2 for y, we get the statement $4 = (3 \times 2) - 2$, which is true. Therefore 2 is a solution of $4 = 3y - 2$.

---

12 Solve the equation $3y + 5 = 2$ by using the AT followed by the MT. Then check your answer by substitution. Show all your work.

$3y + 5 = 2$
$\updownarrow$ AT with 2
$3y = -3$
$\updownarrow$ MT with $\frac{1}{3}$
$y = -1$

-1 is the solution. Substituting -1 for y, we get $3(-1) + 5 = 2$.

---

13 Solve the equation $4y + 9 = 3$ and check your answer by substitution.

$$4y + 9 = 3$$
$$4y = 3 - 9$$
$$y = -\frac{6}{4} = -\frac{3}{2}$$

$-\frac{3}{2}$ is the solution.

CHECK

$$4\left(-\frac{3}{2}\right) + 9 = 3$$
$$-\frac{12}{2} + 9 = 3$$
$$-6 + 9 = 3 \qquad \text{true}$$

---

14 Use trial and error to find a solution of the equation $2y = y + 1$.

1

---

15 Use trial and error to find solutions of the following equations:

a. $3y = 2y + 1$ *b.* $4y = 3y + 1$
c. $5y = 4y + 1$ *d.* $6y = 5y + 1$

a. 1 *b.* 1 *c.* 1 *d.* 1

16 *a.* Adding y to both sides of the equation $y = 1$ gives $2y = y + 1$, whose solution (Frame 14) is ______.

b. Adding $2y$ to both sides of the equation $y = 1$ gives $3y = 2y + 1$, whose solution (Frame 15) is ______.

c. Adding $3y$ gives $4y = 3y + 1$, whose solution is ______.

d. Adding $4y$ gives $5y =$ ______ $+ 1$, whose solution is ______.

e. Adding $5y$ gives ______ $=$ ______ $+ 1$, whose solution is ______.

---

a. 1 *b.* 1 *c.* 1 *d.* $4y$; 1 *e.* $6y$; $5y$; 1

---

17 If we add $-3y$ to both sides of the equation $y = 1$, we get

a. $-2y =$ ______ $+ 1$

b. Is 1 a solution of this new equation?

c. Is this new equation equivalent to $y = 1$?

---

a. $-3y$ *b.* yes *c.* yes

---

18 If we add a multiple of y to both sides of an equation, we obtain a(n) ______ equation. Thus $2y = 5 \leftrightarrow 2y + ny =$ ______ where n is any real number.

---

equivalent; $5 + ny$

---

19 To solve equations of the form $7y = 5y + 3$, first use the ______ transformation with $-5y$ and simplify the resultant expression:

$$7y = 5y + 3 \leftrightarrow 2y = 3$$

then use the ______ transformation with $\frac{1}{2}$:

$$2y = 3 \leftrightarrow y = \_\_\_\_\_\_$$

---

addition; multiplication; $\frac{3}{2}$

---

20 Convert the following equations to the form $k = ?$ by using the AT followed by the MT:

a. $$\begin{array}{c} 8k = 3k + 7 \\ \updownarrow \text{AT} \\ \_\_ = 7 \\ \updownarrow \text{MT} \\ k = \_\_ \end{array}$$

b. $$\begin{array}{c} -4k + 9 = 3k \\ \updownarrow \text{AT} \\ 9 = \_\_ \\ \updownarrow \text{MT} \\ \_\_ = k \end{array}$$

c. $$\begin{array}{c} -2k - 4 = 5k \\ \updownarrow \text{AT} \\ \_\_ = \_\_ \\ \updownarrow \text{MT} \\ \_\_ = k \end{array}$$

d. $$\begin{array}{c} k = 5k + 5 \\ \updownarrow \text{AT} \\ \_\_ = \_\_ \\ \updownarrow \text{MT} \\ k = \_\_ \end{array}$$

---

a. $5k$; $\frac{7}{5}$ *b.* $7k$; $\frac{9}{7}$ *c.* -4; $7k$; $-\frac{4}{7}$ *d.* $-4k$; 5; $-\frac{5}{4}$

---

21 Convert the following equations to the form $k = ?$

a. $$\begin{array}{c} 5k = 2k + 2 \\ \updownarrow \text{AT} \\ \_\_ = \_\_ \\ \updownarrow \text{MT} \\ k = \_\_ \end{array}$$

b. $$\begin{array}{c} -3k = 2 - 2k \\ \updownarrow \text{AT} \\ \_\_ = \_\_ \\ \updownarrow \text{MT} \\ k = \_\_ \end{array}$$

c. $$\begin{array}{c} 7k + 3 = 3k \\ \updownarrow \text{AT} \\ \_\_ = \_\_ \\ \updownarrow \text{MT} \\ \_\_ = k \end{array}$$

d. $$\begin{array}{c} 2k - 5 = 6k \\ \updownarrow \text{AT} \\ \_\_ = \_\_ \\ \updownarrow \text{MT} \\ \_\_ = k \end{array}$$

---

a. $3k$; 2; $\frac{2}{3}$ *b.* $-k$; 2; -2 *c.* 3; $-4k$; $-\frac{3}{4}$ *d.* -5; $4k$; $-\frac{5}{4}$

---

22 Solve the following equations and then check your answers by substitution:

a. $3k - 4 = k$ *b.* $6k + 7 = -2k$ *c.* $-3k = 2k - 6$

a. 2

CHECK

$(3 \times 2) - 4 = 2$ true

b. $-\frac{7}{8}$

CHECK

$6\left(-\frac{7}{8}\right) + 7 = -2\left(-\frac{7}{8}\right)$

$-\frac{42}{8} + 7 = \frac{14}{8}$

$-\frac{42}{8} + \frac{56}{8} = \frac{14}{8}$ true

c. $\frac{6}{5}$

CHECK

$-3\left(\frac{6}{5}\right) = 2\left(\frac{6}{5}\right) - 6$

$-\frac{18}{5} = \frac{12}{5} - 6$

$-\frac{18}{5} = \frac{12}{5} - \frac{30}{5}$ true

23 Solve the following equations and then check your answers by substitution:

a. $k = -4k - 7$ *b.* $5k + 2 = 6$ *c.* $-5k = 7k + 2$

a. $-\frac{7}{5}$

CHECK

$-\frac{7}{5} = -4\left(-\frac{7}{5}\right) - 7$

$-\frac{7}{5} = \frac{28}{5} - 7$

$-\frac{7}{5} = \frac{28}{5} - \frac{35}{5}$ true

b. $\frac{4}{5}$

CHECK

$5\left(\frac{4}{5}\right) + 2 = 6$

$4 + 2 = 6$ true

c. $-\frac{1}{6}$

CHECK

$-5\left(-\frac{1}{6}\right) = 7\left(-\frac{1}{6}\right) + 2$

$\frac{5}{6} = -\frac{7}{6} + 2$

$\frac{5}{6} = -\frac{7}{6} + \frac{12}{6}$ true

You have finished the program for Sec. 11-3. Now take the posttest to see how much you have learned.

POSTTEST 11-3

Solve the following equations and check your answers by substitution:

1. $-6y + 2 = 5$
2. $3y - 6 = -7$
3. $-5 = 2 - 8y$
4. $z = -4z + 5$
5. $2p + 9 = 4p$
6. $3p - 7 = -5p$

When you have finished the posttest, correct it yourself using the answers on page 451. Then check all the questions you got wrong on the list below:

1. ______ 3. ______ 5. ______

2. ______ 4. ______ 6. ______

If you compare your performance with that on Pretest 11-3, you will see how much you have improved. If you made more than one mistake on this posttest, review the material and take the pretest again. Don't move on to the next section until you can get all the questions right.

Now proceed to Sec. 11-4.

Section 11-4 The Solution of Equations with Integral Coefficients

Do this section if you made a mistake in Questions 8 to 10 of Diagnostic Test 11. Begin by taking the pretest below, which will show you how much of Sec. 11-4 you already know and what you will be able to do by the time you have finished it.

PRETEST 11-4

Solve the following equations and check your answers by substitution:

1. $9y + 7 - 2y - 3 = 3y - 10 + 2$
2. $4y + 8 - 12y = 2y - 11 + 5$
3. $3y - 5y + 1 = 2y - 5$
4. $4y - 2y + 2 = 12 - 3y$
5. $7y + 3 - 4y = 8 - 9y + 2$

When you have finished the pretest, correct it yourself using the answers on page 451. Then check all the questions you got wrong on the list below:

1. ______ 3. ______ 5. ______

2. ______ 4. ______

If you made any mistakes (even if it was only *one* mistake), read the introduction and go through the program.

If you made no mistakes at all, skip Sec. 11-4, because you already know its material, and go on to take Final Test 11 (page 449).

Introduction

This section extends the methods of the previous sections to the solution of equations with more than three terms.

FRAMES FOR SECTION 11-4

1 To solve equations of the type

$5y + 6 + 2y = 3 + y + 1$

apply the addition transformation several times until all the terms containing variables are on one side of the equation and all the terms without variables are on the other side:

$$5y + 6 + 2y = 3 + y + 1$$

$\updownarrow$ Adding ______ to both sides

$$5y + 6 + 2y + (-y) = 3 + 1$$

$\updownarrow$ Adding ______ to both sides

$$5y + 2y + (-y) = 3 + 1 + (-6)$$

$-y$ -6

---

2 Then simplify the expressions on either side

$$5y + 2y + (-y) = 3 + 1 + (-6)$$

$$\updownarrow$$

$$\_\_\_\_\_\_ = \_\_\_\_\_\_$$

and use the multiplication transformation to obtain the solution.

$6y = -2$ $-\frac{1}{3}$

---

3 Check your answer by substituting $-\frac{1}{3}$ for y in the original equation:

$5y + 6 + 2y = 3 + y + 1$

$$5\left(-\frac{1}{3}\right) + 6 + 2\left(-\frac{1}{3}\right) = 3 + \left(-\frac{1}{3}\right) + 1$$

$$-\frac{5}{6} + 6 - \frac{2}{3} = 3 - \frac{1}{3} + 1$$

$$6 - \frac{7}{3} = 4 - \frac{1}{3}$$

$$\frac{18 - 7}{3} = \frac{12 - 1}{3}$$

$$\frac{11}{3} = \frac{11}{3} \qquad \text{true}$$

---

4 To solve $7y + 8 = 4y + 2y + 3$ apply the AT several times to get all the terms containing the variable on one side and all the terms without variables on the other side.

$7y + 8 = 4y + 2y + 3$

↕ AT using $-7y$

______ = ______

↕ AT using -3

______ = ______

$8 = 4y + 2y + 3 + (-7y)$ $\qquad$ $8 + (-3) = 4y + 2y + (-7y)$

---

5 $8 + (-3) = 4y + 2y + (-7y)$

↕ Simplify both expressions

______ = ______

↕ MT with -1

______ $= y$

Checking this answer by substitution in the original equation, we get the statement ______, which is (true/false) ______.

$5 = -y$

$-5 = y$

CHECK

$7(-5) + 8 = 4(-5) + 2(-5) + 3$

$-35 + 8 = -20 - 10 + 3$

$-27 = -27 \qquad \text{true}$

---

6 $$6y + 4 + 2y = 5y + 3 + 2$$

$\updownarrow$ AT with $-5y$ and with -4

__________ = __________

(only terms containing the variable) (only terms not containing the variable)

$\updownarrow$ Simplify both expressions

_____ = _____

$\updownarrow$ MT with $\frac{1}{3}$

$y =$ _____

Check your answer by substitution.

$$6y + 2y + (-5y) = 3 + 2 + (-4)$$
$$3y = 1$$
$$y = \frac{1}{3}$$

CHECK

$$6\left(\frac{1}{3}\right) + 4 + 2\left(\frac{1}{3}\right) = 5\left(\frac{1}{3}\right) + 3 + 2$$
$$2 + 4 + \frac{2}{3} = \frac{5}{3} + 3 + 2$$
$$\frac{20}{3} = \frac{20}{3} \quad \text{true}$$

---

7 $$4y - 5 + 3 = -7y + 2y + 6$$

$\updownarrow$ AT with $-4y$ and -6

__________ = __________

(only terms *not* containing the variable) (only terms containing the variable)

$\updownarrow$ Simplify

_____ = _____

$\updownarrow$ MT

_____ $= y$

Check your answer.

$$-5 + 3 + (-6) = -7y + 2y + (-4y)$$
$$-8 = -9y$$
$$\frac{8}{9} = y$$

CHECK

$$4\left(\frac{8}{9}\right) - 5 + 3 = -7\left(\frac{8}{9}\right) + 2\left(\frac{8}{9}\right) + 6$$
$$\frac{32}{9} - 2 = -\frac{56}{9} + \frac{16}{9} + 6$$
$$\frac{32}{9} - \frac{18}{9} = -\frac{40}{9} + \frac{54}{9}$$
$$\frac{14}{9} = \frac{14}{9} \quad \text{true}$$

---

8 $4y + 3y - 1 = 2y - 8 + 3$

$\updownarrow$ AT with ______ and with ______

______ = ______

(only terms containing the variable) = (only terms *not* containing the variable)

$\updownarrow$ Simplify

______ = ______

$\updownarrow$ MT

______ = ______

Check your answer.

$-2y \qquad 1$

$$4y + 3y + (-2y) = -8 + 3 + 1$$
$$5y = -4$$
$$y = -\frac{4}{5}$$

CHECK

$$4\left(-\frac{4}{5}\right) + 3\left(-\frac{4}{5}\right) - 1 = 2\left(-\frac{4}{5}\right) - 8 + 3$$
$$-\frac{16}{5} - \frac{12}{5} - \frac{5}{5} = -\frac{8}{5} - \frac{40}{5} + \frac{15}{5}$$
$$-\frac{33}{5} = -\frac{33}{5} \qquad \text{true}$$

---

9 $3y - 5y + 2 + 4 = 6y - 2y + 8$

$\updownarrow$ AT with ______ and ______ and ______

______ = ______

(only terms *not* containing the variable) = (only terms containing the variable)

$\updownarrow$ Simplify

______ = ______

$\updownarrow$ MT

______ =

Check your answer.

$-3y \qquad 5y \qquad -8$

$$2 + 4 + (-8) = 6y - 2y + (-3y) + 5y$$
$$-2 = 6y$$
$$-\frac{1}{3} = y$$

CHECK

$$3\left(-\frac{1}{3}\right) - 5\left(-\frac{1}{3}\right) + 2 + 4 = 6\left(-\frac{1}{3}\right) - 2\left(-\frac{1}{3}\right) + 8$$
$$-1 + \frac{5}{3} + 2 + 4 = -2 + \frac{2}{3} + 8$$
$$6\frac{2}{3} = 6\frac{2}{3} \qquad \text{true}$$

---

10 Solve $10y + 6 - 2y = 3y + 8 - 4$ and check your answer. Show all your work.

$$10y + 6 - 2y = 3y + 8 - 4$$
$$10y - 2y + (-3y) = 8 - 4 + (-6)$$
$$5y = -2$$
$$y = -\frac{2}{5}$$

CHECK

$$10\left(-\frac{2}{5}\right) + 6 - 2\left(-\frac{2}{5}\right) = 3\left(-\frac{2}{5}\right) + 8 - 4$$
$$-4 + 6 + \frac{4}{5} = -\frac{6}{5} + 8 - 4$$
$$2 + \frac{4}{5} = -\frac{6}{5} + 4$$
$$\frac{14}{5} = \frac{-6 + 20}{5} \qquad \text{true}$$

11 Solve $8y + 3 - y = 4y - 5 + 7$ and check your answer. Show all your work.

$$8y - y + (-4y) = -5 + 7 + (-3)$$
$$3y = -1$$
$$y = -\frac{1}{3}$$

CHECK

$$8\left(-\frac{1}{3}\right) + 3 - \left(-\frac{1}{3}\right) = 4\left(-\frac{1}{3}\right) - 5 + 7$$
$$-\frac{8}{3} + 3 + \frac{1}{3} = -\frac{4}{3} + 2$$
$$3 - \frac{7}{3} = 2 - \frac{4}{3}$$
$$\frac{2}{3} = \frac{2}{3} \qquad \text{true}$$

12 Solve $4y + 2y - 4 + 2 = 8y - 5 + 7 - y$ and check your answer.

$y = -4$

CHECK

$$4(-4) + 2(-4) - 4 + 2 = 8(-4) - 5 + 7 - (-4)$$
$$-16 - 8 - 4 + 2 = -32 - 5 + 7 + 4$$
$$-28 + 2 = -37 + 11$$
$$-26 = -26 \qquad \text{true}$$

13 Solve $8y + 5 - 3y + 2 = 6y + 8 - 7$ and check your answer.

$y = 6$ CHECK

$$8(6) + 5 - 3(6) + 2 = 6(6) + 8 - 7$$
$$48 + 5 - 18 + 2 = 36 + 8 - 7$$
$$55 - 18 = 44 - 7$$
$$37 = 37 \quad \text{true}$$

You have finished the program for Sec. 11-4. Now take the posttest to see how much you have learned.

POSTTEST 11-4

Solve the following equations and check your answers by substitution:

1. $3y + 8 - y - 5 = 5y - 4 + 6$
2. $6y + 20 - 3 - 25 = 12y - 2y - 5$
3. $5y - 7y + 1 = 4y - 1$
4. $y - 11y + 2 = 3 - 2y$
5. $4y + 2 - 6y = 10 - 3y + 5$

When you have finished the posttest, correct it yourself using the answers on page 451. Then check all the questions you got wrong on the list below:

1. ______ 3. ______ 5. ______

2. ______ 4. ______

If you compare your performance with that on Pretest 11-4, you will see how much you have improved. If you made more than one mistake on this posttest, review the material and take the pretest again. Don't move on to the next section until you can get all the questions right.

Now proceed to Final Test 11.

FINAL TEST 11

Solve the following equations and check your answers:

1. $b - 9 = 14$
2. $y + 7 = -3$
3. $a - 5 = -6$
4. $-24k = -8$
5. $4m = \frac{8}{11}$

6. $\frac{3}{4}a = -\frac{5}{6}$
7. $-\frac{4}{9}n = \frac{3}{7}$
8. $7y - 9 - 4y = 9 - 3y - 6$
9. $6d + 5 + 2d = 7 - 5d - 16 + 8d$
10. $11 - 4z + 7 = 4z + 5 - 3z + 1$

The answers to this test are on page 451. Check the questions you got wrong on the list below and compare your results with those on Diagnostic Test 11 to see how much you have improved.

1.	______	4.	______	8.	______
2.	______	5.	______	9.	______
3.	______	6.	______	10.	______
		7.	______		

If you made more than one mistake, review the material and then take the diagnostic test again. Make sure that you can get all the questions right before you finish the chapter. Now proceed to Chap. 12.

ANSWERS TO TESTS FOR CHAPTER 11

DIAGNOSTIC TEST 11

1. $a = 14$
2. $y = -12$
3. $z = -2$
4. $\frac{1}{4}$
5. $\frac{1}{16}$
6. $b = -\frac{5}{6}$
7. $n = -\frac{35}{24}$
8. $y = \frac{11}{5}$ or $2\frac{1}{5}$
9. $d = -\frac{13}{3}$ or $-4\frac{1}{3}$
10. $d = 2$

PRETEST 11-1

1. $z = -13$
2. $z = 9$
3. $z = -2$
4. -4
5. 9
6. -7

POSTTEST 11-1

1. $z = -16$
2. $z = 15$
3. $z = 2$
4. 2
5. 13
6. -13

PRETEST 11-2

1. $k = -\frac{1}{2}$
2. $k = -\frac{8}{5}$
3. $k = \frac{1}{4}$
4. $\frac{17}{12}$ or $1\frac{5}{12}$
5. $\frac{35}{4}$ or $8\frac{3}{4}$
6. $\frac{44}{3}$ or $14\frac{2}{3}$

POSTTEST 11-2

1. $p = -4$
2. $p = -\frac{45}{44}$
3. $p = \frac{1}{4}$
4. $-\frac{13}{15}$
5. $\frac{24}{5}$ or $4\frac{4}{5}$
6. $\frac{77}{6}$ or $12\frac{5}{6}$

PRETEST 11-3

1. $\frac{3}{4}$
2. -2
3. $\frac{10}{3}$ or $3\frac{1}{3}$
4. 1
5. 4
6. $\frac{3}{2}$ or $1\frac{1}{2}$

POSTTEST 11-3

1. $-\frac{1}{2}$
2. $-\frac{1}{3}$
3. $\frac{7}{8}$
4. 1
5. $\frac{9}{2}$ or $4\frac{1}{2}$
6. $\frac{7}{8}$

PRETEST 11-4

1. -3
2. $\frac{7}{5}$
3. $\frac{3}{2}$
4. 2
5. $\frac{7}{12}$

POSTTEST 11-4

1. $\frac{1}{3}$
2. $-\frac{3}{4}$
3. $\frac{1}{3}$
4. $-\frac{1}{8}$
5. 13

FINAL TEST 11

1. $b = 23$
2. $y = -10$
3. $a = -1$
4. $k = \frac{1}{3}$
5. $m = \frac{2}{11}$
6. $a = -\frac{10}{9}$ or $-1\frac{1}{9}$
7. $m = -\frac{27}{28}$
8. $y = 2$
9. $d = -\frac{14}{5}$ or $-2\frac{4}{5}$
10. $z - \frac{12}{5}$

Chapter 12
Ratio and Proportion

In this chapter you will learn a number of methods for expressing the relationship between two quantities and how to use these methods to solve a wide range of problems. You will also get some very useful experience in handling tabulated data and making deductions and inferences from them.

Section 12-1 introduces the concept of proportion for relating a part to its corresponding whole and applies it to the problem of making deductions from tabulated data. Section 12-2 extends the handling of data to the comparison of any two quantities and introduces the concept of ratio. Section 12-3 introduces the colon form of representing ratios and a number of problems which involve multiple ratios comprising more than two quantities at a time. Section 12-4 deals with problems involving directly proportional quantities, and Sec. 12-5 those which are indirectly proportional.

This chapter requires completion of Chaps. 8, 9, and 11.

DIAGNOSTIC TEST 12

Use Tables 12-1 and 12-2 (page 453) as a basis for Questions 1 to 4. *Give your answers in decimal or integer form correct to two figures:*

1. What was the proportion of United States government expenditure in 1957 that went on civil expenditure?
2. If the proportion of consumer expenditure in the United Kingdom in 1965 that went on food was 0.26 and expenditure on food in 1965 was 5,800 million pounds, what was the total consumer expenditure in 1965?
3. What was the ratio of expenditure on food to expenditure on durables in the United States in 1966?
4. The ratio of expenditure on clothing in the United Kingdom in 1966 to that in 1957 was $\frac{3}{2}$. If this proportionate rate of increase remains the same for the period 1966–1975, what will be the expenditure on clothing in 1975?

5. A box is $5\frac{1}{4}$ feet wide, $2\frac{2}{5}$ feet deep, and $1\frac{1}{2}$ feet high. What is the ratio of its width to its depth to its height?
6. A sum of \$900 is to be divided into four parts in the ratio 2:4:5:7. Find the largest part.
7. If I travel from A to B at 33 mph and take a minutes, how fast will someone have to go to do the same journey in b minutes?

TABLE 12-1

United States Expenditure in Millions of Dollars *

	1957	*1966*
Total consumer expenditure:	280,000	450,000
Food	63,000	90,000
Clothing	26,000	42,000
Rent	38,000	64,000
Durables	33,000	54,000
Other	120,000	200,000
Total government expenditure:	80,000	144,000
Defense	45,000	63,000
Civil	35,000	81,000

* These figures have been taken from published Organisation for Economic Co-operation and Development statistics but have been slightly altered to make your calculations easier.

TABLE 12-2

United Kingdom Expenditure in Millions of Pounds *

	1957	*1966*
Total consumer expenditure:	14,400	24,000
Food	4,500	6,000
Clothing	1,600	2,400
Rent	1,200	2,800
Durables	1,200	2,000
Other	5,900	10,000
Total government expenditure:	3,600	6,400
Defense	1,500	2,200
Civil	2,100	4,200

* These figures have been taken from published Organisation for Economic Co-operation and Development statistics but have been slightly altered to make your calculations easier.

8. If the scale of a map is x inches to y miles and two points are n miles apart on the ground, how many inches apart are they on the map?

When you have finished the diagnostic test, correct it yourself using the answers on page 491. Then check all the questions you got wrong on the list below:

1. _____	5. _____	7. _____
2. _____	6. _____	8. _____
3. _____		
4. _____		

If you made no mistakes at all, you need not do any work in this chapter, so you should go straight on to Chap. 13. But if you made any mistakes (even if it was only *one* mistake), follow these instructions:

	section	page
If you made any mistakes in Questions 1 to 4, circle Secs. 12-1 and 12-2.	12-1 12-2	455 462
If you made any mistakes in Question 5 or 6, circle Sec. 12-3.	12-3	469
If you made any mistakes in Question 7 or 8, circle Secs. 12-4 and 12-5.	12-4 12-5	475 482

You should now work through the sections whose numbers you have circled, taking them in numerical order. The sections begin on the pages indicated.

Section 12-1
The Use of Proportion in Interpreting Tabulated Data

Do this section if you made a mistake in Questions 1 to 4 of Diagnostic Test 12. Begin by taking the pretest below, which will show you how much of Sec. 12-1 you already know and what you will be able to do by the time you have finished it.

PRETEST 12-1

Use Tables 12-1 and 12-2 (page 453) to make the following calculations. *Unless otherwise stated, give your answers in decimal or integer form, correct to two figures:*

1. What was the proportion of consumer expenditure in the United States in 1966 that went on durables?
2. What was the proportion of government expenditure in the United Kingdom in 1957 that went on defense? Give your answer in *fractional* form.
3. If the proportion of consumer expenditure in the United States in 1959 that went on food was 0.22 and the expenditure on food was 68,000 million dollars, what was the total consumer expenditure in 1959?
4. If the proportion of consumer expenditure in the United Kingdom in 1963 that went on durables was $\frac{1}{12}$ and the total consumer expenditure in 1963 was 20,000 million pounds, how much was spent on durables?
5. If the expenditure on clothing in the United Kingdom in 1960 was 1,900 million pounds and the proportion of consumer expenditure that went on clothing was $\frac{1}{9}$, what was the total consumer expenditure in 1960?
6. If the total consumer expenditure in the United States in 1963 was 377,000 million dollars and the proportion which went on food was 0.21, what was the expenditure on food?

When you have finished the pretest, correct it yourself using the answers on page 491. Then check all the questions you got wrong on the list below:

1. _____	3. _____	5. _____
2. _____	4. _____	6. _____

If you made any mistakes (even if it was only *one* mistake), read the introduction and go through the program.

If you made no mistakes at all, skip Sec. 12-1, because you already know its material, and go on to Sec. 12-2.

Introduction

In this section you will learn how the concept of proportion can be used to describe the relationship between a part and its corresponding whole. You will learn to express proportions in fractional or decimal form and to use proportions to calculate the part when given the whole and the whole when given the part. What you learn will be applied to the handling of tabulated data.

FRAMES FOR SECTION 12-1

1 Table 12-1 (page 453) shows how the United States spent most of its money in 1957 and in 1966. The money was spent either by consumers, i.e., individual people like you and me, or by the government.

a. What categories does the table use for consumer expenditure?

b. What categories does the table use for government expenditure?

---

a. food; clothing; rent; durables; other *b.* defense; civil

---

2 Since expenditure rose between 1957 and 1966 for every category, we can examine changes in the pattern of expenditure only by seeing if the *proportions* changed. For example, in 1957 the proportion of government expenditure which went on defense was:

a. $\dfrac{45{,}000}{80{,}000} = \dfrac{?}{16}$ (fractional form)

b. ______ (decimal form correct to two figures)

---

a. $\dfrac{9}{16}$ *b.* 0.56

---

3 But in 1966 the proportion of government expenditure which went on defense was:

a. $\dfrac{?}{?} = \dfrac{?}{16}$ (fractional form)

b. ______ (decimal form correct to two figures)

---

a. $\dfrac{63{,}000}{144{,}000} = \dfrac{7}{16}$ *b.* 0.44

---

4 So we can deduce from Table 12-1 that in 1966 the government spent (more/less) ______ money on defense *but* a (smaller/larger) ______ proportion of its money on defense than in 1957. (*Think carefully.*)

---

more (63,000 as opposed to 45,000 million dollars); smaller ($\frac{7}{16}$ as opposed to $\frac{9}{16}$)

---

5 Now let us look at the pattern of consumer spending. In 1957 the proportion of consumer expenditure which went on food was $\frac{63{,}000}{280{,}000} =$ _____ (decimal form), and in 1966 the proportion of consumer expenditure which went on food was $\frac{90{,}000}{450{,}000} =$ _____ (decimal form).

0.225 0.2

---

6 So although (more/less) _____ money was spent on food in 1966, the proportion of consumer expenditure which went on food was (larger/smaller) _____ than in 1957.

---

more; smaller

---

7 *a.* What proportion of consumer expenditure went on clothing in 1957?
b. What proportion of consumer expenditure went on clothing in 1966?
Give your answers in decimal form correct to two figures.

---

a. 0.093 *b.* 0.093

---

8 *a.* What proportion of consumer expenditure went on rent in 1957?
b. What proportion of consumer expenditure went on rent in 1966?
Give your answers in decimal form correct to two figures.

---

a. 0.14 *b.* 0.14

---

9 In Frames 1 to 8 you have been using the equation

$$\frac{\text{Part}}{\text{Whole}} = \text{proportion}$$

You were given the value of a particular *part* of the expenditure and the value of the *whole* of the expenditure and had to calculate the *proportion*. Thus in Frame 8*a* the value of the *part* was 38,000 million dollars, the value of the *whole* was __________, and the *proportion* was _____.

280,000 million dollars; 0.14

---

10 In this frame you are given the proportion and the part, and you have to find the whole. Thus (1) we assume that the proportion of consumer expenditure which went on rent stayed constant at 0.14 for the whole period 1957–1966, and (2) we know that 48,000 million dollars were spent on rent in 1961. If we have to calculate T, the total consumer expenditure in millions of dollars in 1961, which of these equations is correct?

a. $\dfrac{0.14}{48{,}000} = T$

b. $\dfrac{T}{48{,}000} = 0.14$

c. $\dfrac{48{,}000}{T} = 0.14$

---

c $\left(\dfrac{\text{part}}{\text{whole}} = \text{proportion}\right)$

---

11 Now use this equation to find T, the total consumer expenditure in 1961, correct to two figures.

340,000 million dollars **Note:** This calculation depends on our assumption about the proportion 0.14 staying constant being correct.

---

12 In this frame you are given the proportion and the whole and you have to find the part. Thus (1) we assume that the proportion of consumer expenditure which went on clothing stayed constant at 0.093 during the period 1957–1966, and (2) we assume that the total consumer expenditure in 1961 was 340,000 million dollars as we have just calculated. If we have to calculate B, the expenditure on clothing in 1961, which of these equations is correct?

a. $\dfrac{340{,}000}{B} = 0.093$

b. $\dfrac{B}{340{,}000} = 0.093$

c. $\dfrac{0.093}{B} = 340{,}000$

---

b $\left(\dfrac{\text{part}}{\text{whole}} = \text{proportion}\right)$

---

13 Now use this equation to find B, the total expenditure on clothing in 1961 correct to two figures. On what does the accuracy of your result chiefly depend?

32,000 million dollars; the validity of the assumption made in Frame 12.

---

14 In 1962 expenditure on food amounted to 74,000 million dollars. If we assume that the proportion of consumer expenditure which went on food in 1962 was 0.21 (it was 0.225 in 1957 and 0.2 in 1966), calculate the value of T, the total consumer expenditure in 1962 correct to two figures. **Hint:** Are you being asked to find the proportion, the part, or the whole?

$\frac{74,000}{T} = 0.21 \qquad T \approx 350,000$ million dollars

---

15 If we assume that this figure of 350,000 million dollars is the correct value of the consumer expenditure in 1962 and that the proportion of it which went on rent was 0.14, how much was spent on rent in 1962? **Hint:** Are you being asked to find the proportion, the part, or the whole?

$\frac{C}{350,000} = 0.14 \qquad C = 49,000$ million dollars

---

16 Now let us look at Table 12-2, which gives similar figures (in millions of pounds) for the United Kingdom.

a. What was the proportion of consumer expenditure that went on food in 1957?

b. What was the proportion of consumer expenditure that went on food in 1966?

Give your answers correct to two figures.

---

a. 0.31 *b.* 0.25 You may want to compare these figures with your answers for the United States in Frame 5.

---

17 *a.* What proportion of consumer expenditure in the United Kingdom went on clothing in 1957?

b. What proportion of consumer expenditure in the United Kingdom went on clothing in 1966?

Give your answers correct to two figures.

---

a. 0.11 *b.* 0.10 You may want to compare these figures with your answers for the United States in Frame 7.

---

18 If we assume that the proportion of consumer expenditure in the United Kingdom which went on clothing in 1964 was 0.10 and we know that the expenditure on clothing in 1964 was 2,200 million pounds, calculate the total consumer expenditure in 1964 correct to two figures. Look up the figures in Table 12-2 for 1957 and 1966 to check that your answer is about right.

22,000 million pounds $\left(\frac{2{,}200}{T} = 0.10\right)$

---

19 If the total consumer expenditure in the United Kingdom in 1958 was 15,000 million pounds and the proportion spent on food was 0.30, how much was spent on food? Look up the figures in Table 12-2 for 1957 and 1966 to check that your answer is about right.

4,500 million pounds $\left(\frac{A}{15{,}000} = 0.30\right)$

---

20 *a.* What proportion of consumer expenditure in the United Kingdom went on durables in 1957?

b. What proportion of consumer expenditure in the United Kingdom went on durables in 1966?

Give your answers in *fractional* form.

---

a. $\frac{1}{12}$ *b.* $\frac{1}{12}$

---

21 If this proportion was still $\frac{1}{12}$ in 1960 and the expenditure on durables was 1,600 million pounds, what was the total consumer expenditure in 1960 (correct to two figures)? Look up the figures in Table 12-2 for 1957 and 1966 to check that your answer is about right.

19,000 million pounds $\left(\frac{1{,}600}{T} = \frac{1}{12}\right)$

---

22 Assuming a total consumer expenditure in 1964 of 22,000 million pounds and that the proportion spent on durables was still $\frac{1}{12}$, how much was spent on durables in 1964 (correct to two figures)? Look up the figures in Table 12-2 for 1957 and 1966 to check that your answer is about right.

1,800 million pounds $\left(\frac{D}{22,000} = \frac{1}{12}\right)$

You have finished the program for Sec. 12-1. Now take the posttest to see how much you have learned.

POSTTEST 12-1

Use Tables 12-1 and 12-2 (page 453) to make the following calculations. *Unless otherwise stated, give your answers in decimal or integer form, correct to two figures:*

1. What was the proportion of consumer expenditure in the United States in 1957 that went on durables?
2. What was the proportion of government expenditure in the United Kingdom in 1957 that went on civil expenditure? Give your answer in *fractional* form.
3. If the proportion of consumer expenditure in the United Kingdom in 1964 that went on food was 0.27 and expenditure on food in 1964 was 5,600 million pounds, what was the total consumer expenditure in 1964?
4. If the proportion of government expenditure in the United States that went on defense in 1964 was $\frac{3}{7}$ and the total government expenditure was 119,000 million dollars, how much was spent on defense?
5. If expenditure on rent in the United Kingdom in 1958 was 1,400 pounds and the proportion of consumer expenditure that went on rent was $\frac{1}{11}$, what was the total consumer expenditure in 1958?
6. If the total consumer expenditure in the United Kingdom was 17,000 pounds in 1960 and the proportion spent on food was 0.29, how much was spent on food?

When you have finished the posttest, correct it yourself using the answers on page 491. Then check all the questions you got wrong on the list below:

1. ______	3. ______	5. ______
2. ______	4. ______	6. ______

If you compare your performance with that on Pretest 12-1, you will see how much you have improved. If you made more than one mistake on this post-test, review the material and take the pretest again. Don't move on to the next section until you can get all the questions right.

Now proceed to Sec. 12-2.

Section 12-2 The Use of Ratios in Interpreting Tabulated Data

Do this section if you made a mistake in Questions 1 to 4 of Diagnostic Test 12. Begin by taking the pretest below, which will show you how much of Sec. 12-2 you already know and what you will be able to do by the time you have finished it.

PRETEST 12-2

Use Tables 12-1 and 12-2 (page 453) to make the following calculations. *Unless otherwise stated, give your answers in decimal or integer form correct to two figures:*

1. What was the ratio of expenditure on food to expenditure on clothing in the United Kingdom in 1966? *Give your answer in fractional form.*
2. What was the ratio of expenditure in the United Kingdom on rent in 1966 to that in 1957?
3. Which two categories of expenditure increased by the greatest proportion between 1957 and 1966 in the United States?

   a. Food *b.* Clothing
   c. Rent *d.* Durables
   e. Other
4. The ratio of expenditure on defense in 1966 to that in 1957 was 1.4 in the United States. If this proportionate rate of increase remains the same for the period 1966–1975, what will be the expenditure on defense in 1975?
5. The ratio of United States government civil expenditure in 1966 to that in 1957 was 2.3. If this proportionate rate of increase was the same during the period 1948–1957, what would United States government civil expenditure have been in 1948?

When you have finished the pretest, correct it yourself using the answers on page 491. Then check all the questions you got wrong on the list below:

1. ______ 3. ______ 5. ______

2. ______ 4. ______

If you made any mistakes (even if it was only *one* mistake), read the introduction and go through the program.

If you made no mistakes at all, skip Sec. 12-2, because you already know its material, and go on to the next section whose number you circled on page 454. If no further numbers were circled, take Final Test 12 (page 490).

Introduction

In this section you will learn how the relationship between two quantities can be expressed by a ratio and that ratios can be written in fractional or decimal forms. You will also learn how ratios can be used to express the increase in value of a quantity over a given period of time and how, by assuming a constant rate of increase, predictions can be made about the future value of a quantity.

FRAMES FOR SECTION 12-2

1 In Sec. 12-1 we used the word proportion to signify the relationship between a part and its corresponding whole.

$$\frac{\text{Part}}{\text{Whole}} = \text{proportion}$$

But often we want to compare numbers which are not part or whole. We might for example want to use Table 12-1 (page 453) to compare total consumer expenditure and total government expenditure in the United States. In this case we write

$$\frac{\text{Consumer expenditure}}{\text{Government expenditure}} = \textit{ratio} \text{ of consumer expenditure to government expenditure}$$

or else

$$\frac{\text{Government expenditure}}{\text{Consumer expenditure}} = \textit{ratio} \text{ of government expenditure to consumer expenditure}$$

a. What was the ratio of consumer expenditure to government expenditure in the United States in 1957?

b. What was the ratio of government expenditure to consumer expenditure in the United States in 1957?

Give your answers in fractional form.

---

a. $\frac{7}{2}$ *b.* $\frac{2}{7}$

---

2 *a.* What was the ratio of consumer expenditure to government expenditure in 1966?

b. What was the ratio of government expenditure to consumer expenditure in 1966?

---

a. $\frac{25}{8}$ *b.* $\frac{8}{25}$

---

3 *a.* Convert the ratios of Frames 1*a* and 2*a* to decimal form.

b. Was the ratio of consumer expenditure to government expenditure greatest in 1957 or 1966?

---

a. 3.5; 3.125 *b.* 1957

---

4 *a.* Use Table 12-2 (page 453) to find the ratio of consumer expenditure to government expenditure in the United Kingdom in 1957.

b. What was the ratio of consumer expenditure to government expenditure in the United Kingdom in 1966?

Give your answers in decimal form.

---

a. 4 *b.* 3.75

---

5 *a.* Was the ratio of consumer expenditure to government expenditure greatest in the United Kingdom in 1957 or 1966?

b. Was the ratio of consumer expenditure to government expenditure greatest in the United States or the United Kingdom during the period 1957–1966?

---

a. 1957 *b.* United Kingdom

---

6 Now we might well want to exclude defense spending from our comparison.

a. What was the ratio of consumer expenditure to government *civil* expenditure in the United States in 1957?

b. What was the ratio of consumer expenditure to government *civil* expenditure in the United States in 1966?

Give your answers in integer or fractional form.

a. 8 *b.* $\frac{50}{9}$

7 *a.* What was the ratio of consumer expenditure to government *civil* expenditure in the United Kingdom in 1957?

b. What was the ratio of consumer expenditure to government *civil* expenditure in the United Kingdom in 1966?

Give your answers in fractional form.

a. $\frac{48}{7}$ *b.* $\frac{40}{7}$

8 Was the ratio of consumer expenditure to government *civil* expenditure:

a. Higher in the United States in 1957 or 1966?

b. Higher in the United Kingdom in 1957 or 1966?

c. Higher in 1957 in the United Kingdom or the United States?

d. Higher in 1966 in the United Kingdom or the United States?

Hint: You may want to convert ratios to their decimal forms.

a. 1957 *b.* 1957 *c.* United States *d.* United Kingdom

9 If we want to see what form of expenditure increased most in the United States between 1957 and 1966, we can find the ratio of expenditure in 1966 to expenditure in 1957 for each category. Thus the ratio of expenditure on food in 1966 to expenditure on food in 1957 in the United States is _____ (decimal form correct to two figures).

1.4

10 The ratio of expenditure on clothing in the United States in 1966 to expenditure on clothing in 1957 is _____ (decimal form correct to two figures).

1.6

11 Similar ratios for rent, durables and *other* consumer expenditure in the United States are _____, _____, and _____.

1.7 1.6 1.7

12 Similar ratios for defense expenditure and civil expenditure are _____ and _____.

1.4 2.3

---

13 *a.* Which category of expenditure increased by the greatest proportion between 1957 and 1966?

b. Which categories of expenditure increased least?

---

a. government civil expenditure *b.* food and defense

---

14 One way of predicting the pattern of expenditure for 1975 is to assume that each category will expand by the same proportion during the period 1966–1975 as it did during the period 1957–1966 (both are 9-year periods). For example, if the ratio of food expenditure in 1966 to food expenditure in 1957 is 1.4 (Frame 9), we assume that the ratio of food expenditure in 1975 to food expenditure in 1966 is also ______, i.e., that

$$\frac{\text{Food expenditure in 1975}}{\text{Food expenditure in \_\_\_\_}} = \_\_\_\_\_\_$$

1.4 1966 1.4

---

15 Use this equation and the information in Table 12-1 to predict how much money will be spent on food in the United States in 1975.

$$\frac{\text{Food expenditure in 1975}}{90{,}000} = 1.4$$

Food expenditure in 1975 = 126,000 million dollars

---

16 Frame 10 showed that expenditure on clothing increased by a factor of 1.6 between 1957 and 1966. If this proportionate rate of expansion was continued for the period 1966–1975, what would be the expenditure on clothing in 1975 (correct to two figures)?

$$\frac{\text{Clothing expenditure in 1975}}{42{,}000} = 1.6$$

Clothing expenditure in 1975 ≈ 67,000 million dollars

---

17 If clothing expenditure was expanding at this same proportionate rate between 1948 and 1957, which of the following would be true?

a. $\dfrac{\text{Expenditure in 1948}}{26{,}000} = 1.6$ *b.* $\dfrac{26{,}000}{\text{Expenditure in 1948}} = 1.6$

---

b; remember that expenditure in 1948 would have been less than in 1957.

---

18 What would you predict to have been the expenditure on clothing in 1948?

---

about 17,000 million dollars

---

19 Use Table 12-1 and your answer to Frame 11 to predict expenditure on rent:

a. In 1948 *b.* In 1975

---

a. $\dfrac{38{,}000}{\text{Expenditure in 1948}} = 1.7$

Expenditure in 1948 ≈ 22,000 million dollars

b. $\dfrac{\text{Expenditure in 1975}}{64{,}000} = 1.7$

Expenditure in 1975 ≈ 109,000 million dollars

---

20 Predict the expenditure on food in the United States in 1948.

$\dfrac{63{,}000}{\text{Food expenditure in 1948}} = 1.4$

Food expenditure in 1948 ≈ 45,000 million dollars

---

21 Use Table 12-1 and your predictions of Frames 15, 19, and 20 to find the ratio (correct to two figures) of expenditure on food to expenditure on rent in the United States for:

a. 1948 *b.* 1957 *c.* 1966 *d.* 1975

---

a. $\dfrac{45{,}000}{22{,}000} \approx 2.0$ *b.* $\dfrac{63{,}000}{38{,}000} \approx 1.7$ *c.* $\dfrac{90{,}000}{64{,}000} \approx 1.4$

d. $\dfrac{126{,}000}{109{,}000} \approx 1.2$

---

22 Using the same method as Frames 14 to 20, predict:

a. Expenditure on food in 1984.

b. Expenditure on rent in 1984.

c. The food-to-rent ratio in 1984.

a. 176,000 million dollars *b.* 185,000 million dollars *c.* 0.95

You have finished the program for Sec. 12-2. Now take the posttest to see how much you have learned.

POSTTEST 12-2

Use Tables 12-1 and 12-2 (page 453) to make the following calculations. *Unless otherwise stated give your answers in decimal or integer form, correct to two figures:*

1. What was the ratio of expenditure on food to expenditure on rent in the United Kingdom in 1966? *Give your answer in fractional form.*
2. What was the ratio of expenditure in the United Kingdom on clothing in 1966 to that in 1957?
3. Which expenditure increased by the greatest proportion between 1957 and 1966 in the United Kingdom:
   a. Food *b.* Clothing
   c. Rent *d.* Durables
   e. Other
4. The ratio of United States government civil expenditure in 1966 to that in 1957 was 2.3. If this rate of increase is assumed to be the same for the period 1966–1975, what will be the United States government civil expenditure in 1975?
5. The ratio of expenditure in the United States on defense in 1966 to that in 1957 was 1.4. If this rate of increase was the same during the period 1948–1957, what would defense expenditure have been in 1948?

When you have finished the posttest, correct it yourself using the answers on page 491. Then check all the questions you got wrong on the list below:

1. _____ 3. _____ 5. _____

2. _____ 4. _____

If you compare your performance with that on Pretest 12-2, you will see how much you have improved. If you made more than one mistake on this posttest, review the material and take the pretest again. Don't move on to the next section until you can get all the questions right.

Now proceed to the next section whose number you circled on page 454. If no other sections were circled, take Final Test 12 (page 490).

Section 12-3
Multiple Ratios

Do this section if you made a mistake in Question 5 or 6 of Diagnostic Test 12. Begin by taking the pretest below, which will show you how much of Sec. 12-3 you already know and what you will be able to do by the time you have finished it.

PRETEST 12-3

1.

State	*Population, millions*
Kentucky	3.2
Tennessee	4.0
Alabama	3.6
Mississippi	2.4

Given the above populations, what is the population ratio of Kentucky: Tennessee: Alabama: Mississippi in its simplest form?

2. The population ratio of Indiana:New Jersey:Oregon is 8:11:3. If the population of New Jersey is 6.6 million, what is the population of Oregon?
3. A box is $5\frac{1}{4}$ feet wide, $2\frac{1}{3}$ feet deep, and $3\frac{1}{2}$ feet high. What is the ratio (in its simplest form) of its width to its depth to its height?
4. A sum of $700 is to be divided into three parts in the ratio of 3:5:6. Find the largest part.
5. Amy and Joan collect records. For every four records that Amy has, Joan has seven. If they have a total of 88 records, how many does Amy have?
6. You need 40 ounces of a solution containing 2 parts of alcohol to 3 parts of water. How many ounces of alcohol should you use?

When you have finished the pretest, correct it yourself using the answers on page 491. Then check all the questions you got wrong on the list below:

1. ______ 3. ______ 5. ______

2. ______ 4. ______ 6. ______

If you made any mistakes (even if it was only *one* mistake), read the introduction and go through the program.

If you made no mistakes at all, skip Sec. 12-3, because you already know its material, and go on to the next section whose number you circled on page 454. If no further numbers were circled, take Final Test 12 (page 490).

Introduction

In this section you will learn that ratios can also be expressed in colon form, that is, 8:9, and that this form allows more than two quantities to be compared at a time. You learn how to convert ratios in colon form to their simplest colon form and how to use multiple ratios to solve problems.

FRAMES FOR SECTION 12-3

1 So far we have used ratios in fractional form and in decimal form. But there is another common form, the *colon form*. Just as we write

$$\frac{\text{Population of West Virginia}}{\text{Population of North Carolina}} = \frac{3}{8}$$

we can also write that the ratio of the population of West Virginia to the population of North Carolina is 3:8. Similarly, if we are told that the ratio of the population of North Carolina to the population of Georgia is 8:7, we can deduce that

$$\frac{\text{Population of North Carolina}}{\text{Population of Georgia}} = \frac{?}{?}$$

$\frac{8}{7}$

---

2 The colon form has the special advantage of allowing more than two numbers to be compared at a time. Thus we can write that the ratio of the population of West Virginia to that of North Carolina to that of Georgia is 3:8:7, and know that

a. $\dfrac{\text{Population of West Virginia}}{\text{Population of Georgia}} = \dfrac{?}{?}$

b. $\dfrac{\text{Population of North Carolina}}{\text{Population of West Virginia}} = \dfrac{?}{?}$

c. $\dfrac{\text{Population of Georgia}}{\text{Population of North Carolina}} = \dfrac{?}{?}$

$\frac{3}{7}$ $\frac{8}{3}$ $\frac{7}{8}$

---

3 If there are 1.8 million people in West Virginia, how many are there in Georgia? **Hint:** Do you use Eq. 2*a*, 2*b*, or 2*c*?

Eq. $2a$ becomes $\dfrac{1.8 \text{ million}}{\text{population of Georgia}} = \dfrac{3}{7}$, so the population of Georgia is 4.2 million.

---

4 How many people are there in North Carolina?

Using Eq. $2b$: $\dfrac{\text{population of North Carolina}}{1.8 \text{ million}} = \dfrac{8}{3}$

or using Eq. $2c$: $\dfrac{4.2 \text{ million}}{\text{population of North Carolina}} = \dfrac{7}{8}$

gives 4.8 million as the population of North Carolina.

---

5 The ratio of the population of North Carolina to that of Florida is 6:7. What is the population of Florida?

$\dfrac{4.8 \text{ million}}{\text{Population of Florida}} = \dfrac{6}{7}$; population of Florida = 5.6 million

---

6 The ratio of the population of Florida to that of Oklahoma to that of Michigan is 7:3:10. What is the population of Oklahoma, and what is the population of Michigan?

$\dfrac{5.6 \text{ million}}{\text{Population of Oklahoma}} = \dfrac{7}{3}$, so population of Oklahoma = 2.4 million.

$\dfrac{\text{Population of Michigan}}{5.6 \text{ million}} = \dfrac{10}{7}$; so population of Michigan = 8.0 million.

---

7 Ratios in colon form can be simplified like fractions by dividing through by a common factor. Thus

a. $10:12 = 5:\_$ *b.* $12:16:20 = 3:\_:\_$

---

6 4 5

---

8 Express the following ratios in their simplest forms:

a. 35:45:55 *b.* 42:14:35

---

a. 7:9:11 *b.* 6:2:5

9

State	*Population, millions*
Connecticut	2.8
Minnesota	3.5
Georgia	4.2

Given the above populations, what is the ratio (in simplest form) of the population of Connecticut to that of Minnesota to that of Georgia?

$28:35:42 = 4:5:6$

10

State	*Population, millions*
Indiana	4.8
New Jersey	6.6
Oregon	1.8

Given the above populations, what is the ratio (in simplest form) of the population of Indiana to that of New Jersey to that of Oregon?

$48:66:18 = 8:11:3$

11 Converting a ratio to its simplest form may also involve multiplying by a common denominator. Thus

a. $5\frac{1}{2}:4\frac{1}{2} = \frac{11}{2}:\frac{?}{2} =$ ______:______

b. $3\frac{1}{3}:2\frac{1}{2} = \frac{20}{6}:\frac{?}{6} =$ ______:______ $= 4:$______

a. 9; 11; 9 *b.* 15; 20; 15; 3

12 Suppose I have a picture which is $4\frac{1}{2}$ feet high and $3\frac{1}{3}$ feet wide. What is the ratio of its height to its width in the simplest colon form?

$4\frac{1}{2}:3\frac{1}{3} = \frac{9}{2}:\frac{10}{3} = \frac{27}{6}:\frac{20}{6} = 27:20$

13 A box is $2\frac{2}{3}$ feet high, $3\frac{3}{7}$ feet wide, and $1\frac{1}{3}$ feet deep. What is the ratio of its height to its width to its depth?

$2\frac{2}{3}:3\frac{3}{7}:1\frac{1}{3} = \frac{8}{3}:\frac{24}{7}:\frac{4}{3} = \frac{2}{3}:\frac{6}{7}:\frac{1}{3}$; multiplying by 21 ($3 \times 7$) gives height:width:depth $= 14:18:7$

14 A box is $3\frac{2}{3}$ feet high, $2\frac{3}{4}$ feet deep, and $4\frac{2}{5}$ feet wide. What is the ratio of its height to its depth to its width?

$3\frac{2}{3}:2\frac{3}{4}:4\frac{2}{5} = \frac{11}{3}:\frac{11}{4}:\frac{22}{5} = \frac{1}{3}:\frac{1}{4}:\frac{2}{5}$; multiplying by 60 $(3 \times 4 \times 5)$ gives height : depth : width = 20 : 15 : 24

---

15 Suppose you need 50 ounces of a solution consisting of 7 parts of alcohol to 3 parts water and want to find out how much water and alcohol to use. You can use a multiple ratio. Amount of alcohol : amount of water : amount of solution = 7 : _____ : _____.

---

3 10

---

16 $\dfrac{\text{Amount of alcohol needed}}{\text{Amount of solution needed}} = \dfrac{7}{?}$

and $\dfrac{\text{Amount of water needed}}{\text{Amount of solution needed}} = \dfrac{?}{?}$

Amount of alcohol needed = _____

Amount of water needed = _____

---

10 $\frac{3}{10}$ 35 ounces 15 ounces

---

17 If you need 20 ounces of a weed killer solution containing 1 part weed killer to 4 parts water, how much weed killer and water do you need? Use the method of Frames 15 and 16.

---

amount of weed killer : amount of water : amount of solution = 1 : 4 : 5

$\dfrac{\text{Weed killer needed}}{20} = \dfrac{1}{5}$

Weed killer needed = 4 ounces

Water needed = 16 ounces

---

18 Joe and Ben receive a total of \$1.50 for delivering newspapers. If Joe delivers 32 and Ben delivers 48, how much should Joe get?

Joe's papers : Ben's papers : total number of papers = 32 : 48 : 80

$\dfrac{\text{Joe's pay}}{\$1.50} = \dfrac{32}{80}$

Joe should get 60 cents.

---

19 A sum of \$600 is to be divided into three parts in the ratio of 3:5:7. Find the largest part. You need to use a quadruple ratio!

---

smallest part : middle part : largest part : total = 3:5:7:15

$\frac{\text{Largest part}}{\$600} = \frac{7}{15}$ Largest part = \$280

---

20 A sum of \$500 is to be divided into four parts in the ratio of 2:5:6:7. Find the smallest part.

2:5:6:7 : total = 2:5:6:7:20

$\frac{\text{Smallest part}}{500} = \frac{2}{20}$ Smallest part = \$50

You have finished the program for Sec. 12-3. Now take the posttest to see how much you have learned.

POSTTEST 12-3

1.

State	*Population*
Delaware	600,000
Arkansas	2,100,000
New Mexico	1,200,000
Louisiana	3,900,000

Given the above populations, what is the population ratio of Delaware: Arkansas : New Mexico : Louisiana?

2. The population ratio of Arizona : Nevada : Hawaii is 8:2:3. If the population of Arizona is 2 million, what is the population of Hawaii?
3. A box is $4\frac{1}{6}$ feet wide, $3\frac{3}{4}$ feet deep, and $1\frac{2}{3}$ feet high. What is the ratio (in simplest form) of its width to its depth to its height?
4. A sum of \$800 is to be divided into 3 parts in the ratio of 3:6:7. Find the largest part.
5. Amy and Joan collect records. For every 3 records that Amy has, Joan has 8. If they have a total of 99 records, how many does Amy have?
6. A certain solution was made up of 2 parts drug to 3 parts water. How many ounces of drug are there in 60 ounces of solution?

When you have finished the posttest, correct it yourself using the answers on page 491. Then check all the questions you got wrong on the list below:

1. _____ 3. _____ 5. _____

2. _____ 4. _____ 6. _____

If you compare your performance with that on Pretest 12-3, you will see how much you have improved. If you made more than one mistake on this posttest, review the material and take the pretest again. Don't move on to the next section until you can get all the questions right.

Now proceed to the next section whose number you circled on page 454. If no other sections were circled, take Final Test 12 (page 490).

Section 12-4 Direct Proportionality

Do this section if you made a mistake in Question 7 or 8 of Diagnostic Test 12. Begin by taking the pretest below, which will show you how much of Sec. 12-4 you already know and what you will be able to do by the time you have finished it.

PRETEST 12-4

Give your answers as fractions or as integers:

1. If the scale of a map is 3 inches to 1 mile and the distance between *A* and *B* on the map is 2 inches, what is the distance between *A* and *B* on the ground?
2. If the scale of a map is 3 inches to 4 miles and the distance between *A* and *B* on the ground is 6 miles, what is the distance between *A* and *B* on the map?
3. If 8 cans of fruit cost \$3.10, what will 20 cans cost?
4. A painting measures 24 by 30 inches. If a reproduction of the painting is to have a shorter dimension of 6 inches, what should be its larger dimension?
5. If a jet plane can travel 4,500 kilometers in 5 hours, how many kilometers will it travel in 25 minutes?
6. Calcium and chlorine combine in the weight ratio of 36:64. How much chlorine will combine with 18 grams of calcium?

When you have finished the pretest, correct it yourself using the answers on page 492. Then check all the questions you got wrong on the list below:

1. ______ 3. ______ 5. ______

2. ______ 4. ______ 6. ______

If you made any mistakes (even if it was only *one* mistake), read the introduction and go through the program.

If you made no mistakes at all, skip Sec. 12-4, because you already know its material, and go on to Sec. 12-5.

Introduction

After learning the meaning of the term *directly proportional,* you will solve a wide variety of problems involving directly proportional quantities.

FRAMES FOR SECTION 12-4

1 Another common example of the use of ratios is a map. If a map has a scale of 1 inch to 1 mile, this means that if two points on the ground are ______ (what distance?) apart, then the two points on the map which represent those points on the ground will be ______ (what distance?) apart.

---

1 mile 1 inch

---

2 So if two points are 3 inches apart on the map, they will be ______ miles apart on the ground, and if they are 5 miles apart on the ground, they will be ______ inches apart on the map.

---

3 5

---

3 If, however, the scale of the map is 2 inches to 1 mile, two points which are 2 inches apart on the map will be ______ mile(s) apart on the ground, and points which are 3 inches apart on the map will be ______ mile(s) apart on the ground.

---

1 $\frac{3}{2}$

---

4 If the scale of a map is 2 inches to 1 mile:

a. Points which are 5 inches apart on the map will be ______ miles apart on the ground.

b. Points which are 5 miles apart on the ground will be ______ inches apart on the map.

---

a. $\frac{5}{2}$ *b.* 10

---

5 If the scale of a map is 4 inches to 1 mile:

a. Points which are 8 inches apart on the map will be ______ miles apart on the ground.

b. Points which are 8 miles apart on the ground will be ______ inches apart on the map.

---

a. 2 *b.* 32

---

6 If the scale of a map is $\frac{1}{4}$ inch to 1 mile:

a. Points which are 12 inches apart on the map will be ______ miles apart on the ground.

b. Points which are 12 miles apart on the ground will be ______ inches apart on the map.

---

a. 48 *b.* 3

---

7 If two points are 6 miles apart on the ground, how far apart will they be on a map:

a. With a scale of 3 inches to 1 mile?

b. With a scale of $\frac{1}{2}$ inch to 1 mile?

---

a. 18 inches *b.* 3 inches

---

8 If two points are 18 inches apart on a map, how far apart will they be on the ground if the map has a scale:

a. Of 2 inches to 1 mile?

b. Of $\frac{1}{3}$ inch to 1 mile?

---

a. 9 miles *b.* 54 miles

---

9 The scale of a map is just one example of a situation in which two quantities are said to be *directly proportional.* Because for any given scale the ratio

$$\frac{\text{Distance between } A \text{ and } B \text{ on map}}{\text{Distance between } A \text{ and } B \text{ on ground}}$$

is constant. What is this ratio if the scale is 1 inch to 1 mile? (5,280 feet = 1 mile)

$\frac{1}{63,360}$

10 Did you have to know this ratio in order to answer Frames 1 to 9?

No. You just had to assume that the ratio of the distance on the map to the distance on the ground was constant, i.e., that the two distances were directly proportional.

11 The answer to Frame 8*a*, for example, is based on the equation

$$\frac{18 \text{ inches}}{n \text{ miles}} = \frac{2 \text{ inches}}{1 \text{ mile}}$$

$$n = \_\_\_\_\_\_ \text{ miles}$$

What equation was the answer to Frame 8*b* based on?

9 $\qquad \frac{18 \text{ inches}}{n \text{ miles}} = \frac{\frac{1}{3} \text{ inch}}{1 \text{ mile}}$

12 If two points are 5 inches apart on a map whose scale is 3 inches to 4 miles, how many miles apart are they on the ground? (Give your answer as a fraction.)

$\frac{5 \text{ inches}}{n \text{ miles}} = \frac{3 \text{ inches}}{4 \text{ miles}} \qquad n = \frac{20}{3} = 6\frac{2}{3} \text{ miles}$

13 If two points are 7 miles apart on the ground, how far apart will they be on a map whose scale is 2 inches to 5 miles?

$\frac{n \text{ inches}}{7 \text{ miles}} = \frac{2 \text{ inches}}{5 \text{ miles}} \qquad n = \frac{14}{5} = 2\frac{4}{5} \text{ inches}$

14 Another example of two quantities being directly proportional is the number of articles you buy and the price you have to pay. For example, if you pay $1.20 for 3 cans, you can infer what you have to pay for 7 cans. If *n* cents is the cost of 7 cans, then

$$\frac{\text{Price in cents}}{\text{Number of cans}} = \frac{n}{7} = \frac{?}{3}$$

$$n = \_\_\_\_\_\_$$

120 $\qquad$ 280 cents = $2.80

15 **Problem:** If 6 cans cost 45 cents, what will 16 cans cost?

Solution: Let n cents be the cost of 16 cans. Then

$$\frac{\text{Cost}}{\text{Number of cans}} = \frac{n}{?} = \frac{?}{?}$$

$$n = \_\_\_\_$$

$\frac{n}{16} = \frac{45}{6}$ 120 cents or \$1.20

---

16 Solve the following problem by the same method:

Problem: If 4 yards of cloth cost \$13, what will 14 yards cost?

Let n dollars be the cost of 14 yards. Then

$\frac{n}{14} = \frac{13}{4}$ $n = \frac{91}{2} = \$45.50$

---

17 **Problem:** A piece of cloth 26 inches long shrinks to a length of 24 inches in washing. How many yards long should a roll of cloth be if it is required to be 36 yards long after washing?

Solution: Let n ________.

$$\frac{\text{Length before shrinking}}{\text{Length after shrinking}} = \frac{n}{?} = \frac{?}{?}$$

$$n = \_\_\_\_$$

Let n yards be the length of the roll of cloth before washing. Then

$\frac{n}{36} = \frac{26}{24}$ $n = 39$ yards

---

18 **Problem:** A piece of metal 1 inch long at 0°C is 1.1 inches long at 100°C. Another piece of the same metal is 30 inches long at 0°C. What length will it be at 100°C?

Let n be the length at 100°C.

$$\frac{\text{Length at }100°}{\text{Length at }0°} = \frac{n}{30} = \frac{1.1}{1}$$

$$n = \frac{30 \times 1.1}{1} = 33 \text{ inches}$$

---

19 Problem: A person weighing 150 pounds weighed only 120 pounds after a period of dieting. If someone weighing 175 pounds went on the same diet for the same length of time and lost the same proportion of weight, what would his new weight be?

Solution: Let n pounds be his new weight.

$$\frac{\text{New weight}}{\text{Old weight}} = \frac{n}{175} = \frac{120}{150}$$

$$n = \frac{175 \times 120}{150} = 140 \text{ pounds}$$

---

20 Problem: A snapshot measures $1\frac{7}{8}$ by $2\frac{1}{2}$ inches. It is enlarged so that the longer dimension becomes 4 inches. Find the shorter dimension.

Solution: Let n inches be __________.

$$\frac{\text{Width}}{\text{Length}} = \frac{n}{?} = \_\_\_\_$$

$$n = \_\_\_\_$$

Let n inches be new width.

$$\frac{\text{Width}}{\text{Length}} = \frac{n}{4} = \frac{1\frac{7}{8}}{2\frac{1}{2}}$$

$$n = \frac{\frac{15}{8} \times 4}{\frac{5}{2}} = \frac{\frac{15}{2}}{\frac{5}{2}} = 3 \text{ inches}$$

---

21 Problem: A painting measures 11 by 14 inches. A reproduction of the painting has a shorter dimension of $5\frac{1}{2}$ inches; what is its longer dimension?

Let n inches be the length of the reproduction.

$$\frac{\text{Length}}{\text{Width}} = \frac{n}{5\frac{1}{2}} = \frac{14}{11}$$

$$n = \frac{14}{11} \times 5\frac{1}{2} = \frac{14}{11} \times \frac{11}{2} = 7 \text{ inches}$$

---

22 Problem: An ammeter incorrectly reads 25 instead of 30 amperes. If it reads 40 amperes, what should the correct reading be?

Let n be the correct reading.

$$\frac{\text{Correct reading}}{\text{Observed reading}} = \frac{n}{40} = \frac{30}{25}$$

$$n = \frac{30 \times 40}{25} = 48 \text{ amperes}$$

23 If four balls of string give 90 feet of string, how many feet of string will three balls give?

$67\frac{1}{2}$

24 A certain drink blends 5 ounces of orange juice for every 4 ounces of grapefruit juice. How many ounces of orange juice are required to go with 6 ounces of grapefruit juice?

$7\frac{1}{2}$

25 If the scale of a map is $\frac{1}{4}$ inch to 1 mile and the distance between A and B on the map is $\frac{3}{8}$ inch, what is the distance between A and B on the ground?

$1\frac{1}{2}$ miles

You have finished the program for Sec. 12-4. Now take the posttest to see how much you have learned.

POSTTEST 12-4

Give your answers as fractions or as integers:

1. If the scale of a map is 4 inches to 1 mile and the distance between A and B on the map is 3 inches, what is the distance between A and B on the ground?
2. If the scale of a map is 3 inches to 5 miles and the distance between A and B on the ground is 7 miles, what is the distance between A and B on the map?
3. If 6 cans of fruit cost \$2.80, what will 15 cans cost?
4. A painting measures 25 by 32 inches. If a reproduction of the painting is to have a shorter dimension of 5 inches, what should be its larger dimension?
5. If a jet plane can travel 3,600 kilometers in 4 hours, how many kilometers will it travel in 35 minutes?

6. Analysis of carbon disulfide shows that every 19 grams contains 16 grams of sulfur. How much carbon disulfide contains 96 grams of sulfur?

When you have finished the posttest, correct it yourself using the answers on page 492. Then check all the questions you got wrong on the list below:

1. ______ 3. ______ 5. ______

2. ______ 4. ______ 6. ______

If you compare your performance with that on Pretest 12-4, you will see how much you have improved. If you made more than one mistake on this posttest, review the material and take the pretest again. Don't move on to the next section until you can get all the questions right.

Now proceed to Sec. 12-5.

Section 12-5 Inverse Proportionality

Do this section if you made a mistake in Question 7 or 8 of Diagnostic Test 12. Begin by taking the pretest below, which will show you how much of Sec. 12-5 you already know and what you will be able to do by the time you have finished it.

PRETEST 12-5

1. If 9 men take 28 hours to do a certain job, how long will it take 13 men to do the same job? Give your answer in fractional form.
2. If m pounds of steak costs y dollars, what will t pounds of steak cost?
3. If a scale of a map is m inches to q miles and two points are 6 inches apart on the map, how many miles apart are they on the ground?
4. If I travel from A to B at k mph and take f minutes, how many minutes will I take if I travel at q mph?

When you have finished the pretest, correct it yourself using the answers on page 492. Then check all the questions you got wrong on the list below:

1. ______ 3. ______

2. ______ 4. ______

If you made any mistakes (even if it was only *one* mistake), read the introduction and go through the program.

If you made no mistakes at all, skip Sec. 12-5, because you already know its material, and go on to take Final Test 12 (page 490).

Introduction

After learning the meaning of the term *inversely proportional*, you learn how to distinguish between quantities which are directly proportional, those which are inversely proportional, and those which are neither. You then practice solving problems containing inversely proportional quantities, including some which involve variables.

FRAMES FOR SECTION 12-5

1 In Sec. 12-4 we called two quantities A and B for which the ratio A/B is constant *directly proportional*. If A and B are directly proportional and you double A, what happens to B?

B is also doubled

---

2 Which of the following pairs of quantities are directly proportional?

a. The number of articles you buy and the total cost of those articles.

b. The distance between two places on the ground and the distance between the points representing those places on a map.

c. The speed at which you travel a certain distance and the time it takes you.

---

a and *b*

---

3 Which of the following pairs of quantities are directly proportional?

a. The width of any enlargement of a photograph and the length of the same enlargement.

b. The number of men working on a job and the time it takes them to do the job.

c. The mass of oxygen in a given mass of water and the mass of hydrogen in the same mass of water.

---

a and *c*

---

4 *a.* If two men take 2 hours to do a job, how long will four men take? (Think carefully.)

b. How long will one man take?

---

a. 1 hour *b.* 4 hours

---

5 *a.* If you double the number of men on the job, you (double/halve) ______ the time they take.

b. If you halve the number of men on the job, you (double/halve) ______ the time they take.

---

a. halve *b.* double

---

6 The job takes a given number of man-hours, thus

No. of men $\times$ No. of hours = No. of man-hours

which is constant.

a. How many man-hours is a job which takes six men 3 hours to complete?

b. How long will the job take three men to complete?

c. How long will the job take nine men to complete?

---

a. 18 man-hours *b.* 6 hours *c.* 2 hours

---

7 *a.* How many man-days is a job which takes four men 3 days?

b. How long will the job take two men?

c. How many men would be needed to complete the job in a day?

---

a. 12 man-days *b.* 6 days *c.* 12

---

8 If A = number of men working on a job and B = time taken to do the job, which of the following is constant?

$\frac{A}{B}$ $A \times B$ $A + B$

$A \times B$

---

9 Two quantities A and B are called *inversely proportional* if $A \times B$ is constant. If A and B are inversely proportional and you double A, what happens to B?

B is halved

---

10 Which of the following pairs of quantities are inversely proportional?

a. The number of men working on a job and the time it takes them to do the job.

b. The distance between two places on the ground and the distance between the points representing those places on a map.

c. The speed at which you travel a certain distance and the time it takes you.

---

a and *c*

---

11 Which of the following pairs of quantities are directly proportional, which are inversely proportional, and which are neither?

a. The amount of flour you use and the amount of sugar you use in making cookies according to a certain recipe.

b. The population of Chicago and the population of New York.

c. Your present age and the number of years you still have to live.

d. The number of men on a desert island with a limited amount of water and the number of days they are likely to survive if they each have the same amount of water every day.

---

a. directly proportional *b.* neither *c.* neither
d. inversely proportional

---

12 Suppose we use K to represent a constant.

a. If $\frac{A}{B} = K$, A and B are ________ proportional.

b. If $A \times B = K$, A and B are ________ proportional.

---

a. directly *b.* inversely

---

13 By transforming these equations we find that:

a. When A and B are directly proportional, $A =$ ______.

b. When A and B are inversely proportional, $A =$ ______.

---

a. $K \times B$ *b.* $\frac{K}{B}$

---

14 If $A = \frac{K}{B}$, then $A = K \times$ ______.

$\frac{1}{B}$

---

15 $\frac{1}{B}$ is called the *reciprocal* or *inverse* of B, which is why we use the term *inversely proportional* to describe the situation where $A = K \times$ ______ or $K =$ ________.

$\frac{1}{B}$ $\quad A \times B$

---

16 In solving problems involving proportionality it is useful to base your solution on the constant factor, as shown below.

Problem: If 11 men take 26 hours to do a certain job, how long will it take 13 men to do the same job?

Solution: Let n be the time it takes 13 men to do the job. Constant factor (number of man-hours to do the job) = __ × __ = __ × __. Therefore $n =$ ________.

$11 \times 26 = 13 \times n \qquad n = 22$ hours

---

17 Use the same method to solve the following problem.

Problem: If 10 men take 6 hours to do a certain job, how long will it take 12 men to do the same job?

Solution: Let n be ________________________________.

Constant factor is ____________________ = ________ and ________.

________ = ________

$n =$ ________

Let n be the time it takes 12 men to do the job. Constant factor is number of man-hours to do the job $= 10 \times 6$ and $12 \times n$.

$10 \times 6 = 12 \times n$

$n = 5$ hours

---

18 Are the speed at which you travel a certain distance and the time it takes you related by formula *a* or formula *b*?

a. $\dfrac{\text{Speed}}{\text{Time}} = \text{constant}$

b. $\text{Speed} \times \text{time} = \text{constant}$

---

b

---

19 If a car traveling at 35 mph covers a certain distance in 54 minutes, how fast would it have to travel to cover the same distance in 42 minutes?

Let n be the speed required to cover the distance in 42 minutes.

Constant factor is $35 \times 54 = 42 \times n$

$$n = \frac{35 \times 54}{42} = 45 \text{ mph}$$

---

20 The fact that speed is measured in miles per hour or $\dfrac{\text{miles}}{\text{hour}}$ should remind you of the formula

$$\text{Speed} = \frac{\text{distance}}{\text{time}}$$

a. This formula can also be written as distance = ________________.

b. What was the constant factor in Frames 18 and 19?

---

speed × time distance

---

21 **Problem:** An airplane traveling at 400 mph takes 3 hours to get from *A* to *B*. If it takes 4 hours to get back to *A*, what was its speed for the return trip?

Solution: Let n be ________________.

Distance is ________ = ________

$n =$ ________

Let n be the speed for the return trip.

$400 \times 3 = n \times 4$

$n = 300$ mph

---

22 Ratio problems stated in terms of variables can be solved in exactly the same way because the ratio is constant.

Problem: If the scale of a map is p inches to q miles and two points are 5 miles apart on the ground, how many inches apart are they on the map?

Solution: Let n be ______________________________.

Constant $= \frac{?}{?}$ and $\frac{?}{?}$

$n =$ ______

Let n be the distance between the two points on the map.

Constant $= \frac{q}{p}$ and $\frac{5}{n}$ $\left(\text{or } \frac{p}{q} \text{ and } \frac{n}{5}\right)$

$n = \frac{5p}{q}$ inches

---

23 Problem: If I travel from A to B at 35 mph and take a minutes, how fast will someone have to go to do the same distance in b minutes?

Solution: Let n be ______________________________.

Constant = distance = ______________ and ______________

$n =$ ______________

Let n be the speed required to do the distance in b minutes.

$35 \times a$ and $n \times b$

$n = \frac{35a}{b}$ mph

---

24 If a car can travel d miles in t minutes, how many minutes will it take to travel e miles if it still goes at the same speed? (Think carefully about the constant factor. Are the distance and the time taken directly or inversely proportional?)

Let n be the time it takes to travel e miles.

Constant = speed $= \frac{d}{t}$ and $\frac{e}{n}$

$n = \frac{et}{d}$ minutes

---

25 If m pounds of coffee costs t dollars, how much does n pounds cost?

Let x be the cost of n pounds of coffee.

Constant = price of 1 pound = $\frac{t}{m}$ and $\frac{x}{n}$

$$x = \frac{nt}{m} \text{ dollars}$$

---

26 If c men can do a job of work in q days, how many men are needed to do the same job in r days?

Let n be the number of men to do the job in r days.
Constant = number of man-days for job = cq and rn

$$n = \frac{cq}{r} \text{ men}$$

You have finished the program for Sec. 12-5. Now take the posttest to see how much you have learned.

POSTTEST 12-5

1. If 7 men take 27 hours to do a certain job, how long will it take 11 men to do the same job? Give your answer in fractional form.
2. If t pounds of potatoes costs b dollars, what will n pounds cost?
3. If a scale of a map is z inches to x miles and two points are 7 miles apart on the ground, how many inches apart are they on the map?
4. If I travel from A to B at y mph and take c minutes, how fast will someone have to go to complete the same journey in d minutes?

When you have finished the posttest, correct it yourself using the answers on page 492. Then check all the questions you got wrong on the list below:

1. ______ 3. ______

2. ______ 4. ______

If you compare your performance with that on Pretest 12-5, you will see how much you have improved. If you made more than one mistake on the posttest, review the material and take the pretest again.

Now proceed to Final Test 12.

FINAL TEST 12

Use Tables 12-1 and 12-2 (page 453) as a basis for Questions 1 to 4. *Give your answers in decimal or integer form correct to two figures.*

1. What was the proportion of United States government expenditure in 1966 that went on civil expenditure?
2. If the proportion of consumer expenditure in the United Kingdom in 1964 that went on rent was 0.11 and the expenditure on rent was 2,300 million pounds, what was the total consumer expenditure in 1964?
3. What was the ratio of expenditure on food to expenditure on durables in the United States in 1957?
4. The ratio of expenditure on food in the United Kingdom in 1966 to that in 1957 was $\frac{4}{3}$. If this proportionate rate of increase remains the same for the period 1966–1975, what will be the expenditure on food in 1975?

5. A box is $3\frac{3}{4}$ feet wide, $2\frac{4}{7}$ feet deep, and $4\frac{1}{2}$ feet high. What is the ratio of its width to its depth to its height?
6. A sum of $750 is to be divided into four parts in the ratio 1:3:5:6. Find the largest part.
7. If I travel from A to B at 26 mph and take a minutes, how long will someone take who travels at t mph?
8. If the scale of a map is x inches to y miles, and two points are n inches apart on the map, how many miles apart are they on the ground?

The answers to this test are on page 492. Check all the questions you got wrong on the list below and compare your results with those on Diagnostic Test 12 to see how much you have improved.

1. _____	5. _____	7. _____
2. _____	6. _____	8. _____
3. _____		
4. _____		

If you made more than one mistake, review the material and then take the diagnostic test again. Make sure that you can get all the questions right before you finish.

Now proceed to Chap. 13.

ANSWERS TO TESTS FOR CHAPTER 12

DIAGNOSTIC TEST 12

1. 0.44
2. 22,000 million pounds
3. 1.7
4. 3,600 million pounds
5. 35:16:10
6. $350
7. $\frac{33a}{b}$
8. $\frac{nx}{y}$

PRETEST 12-1

1. 0.12
2. $\frac{5}{12}$
3. 310,000 million dollars
4. 1,700 million pounds
5. 17,000 million pounds
6. 79,000 million dollars

POSTTEST 12-1

1. 0.12
2. $\frac{7}{12}$
3. 21,000 million dollars
4. 51,000 million dollars
5. 15,000 million pounds
6. 4,900 million pounds

PRETEST 12-2

1. $\frac{5}{2}$
2. 2.3
3. rent; other consumer expenditure
4. 88,000 million dollars
5. 15,000 million dollars

POSTTEST 12-2

1. $\frac{15}{7}$
2. 1.5
3. rent
4. 190,000 million dollars
5. 32,000 million dollars

PRETEST 12-3

1. 8:10:9:6
2. 1.8 million
3. 9:4:6
4. $300
5. 32 records
6. 16 ounces

POSTTEST 12-3

1. 2:7:4:13
2. 750,000
3. 10:9:4
4. $350
5. 27 records
6. 24 ounces

PRETEST 12-4

1. $\frac{2}{3}$ mile
2. $4\frac{1}{2}$ inches
3. $7.75
4. $7\frac{1}{2}$ inches
5. 375 kilometers
6. 32 grams

POSTTEST 12-4

1. $\frac{3}{4}$ mile
2. $4\frac{1}{5}$ inches
3. $7.00
4. $6\frac{2}{5}$ inches
5. 525 kilometers
6. 114 grams

PRETEST 12-5

1. $19\frac{5}{13}$ hours
2. $\dfrac{yt}{m}$ dollars
3. $\dfrac{6q}{m}$ miles
4. $\dfrac{fk}{q}$ minutes

POSTTEST 12-5

1. $17\frac{2}{11}$ hours
2. $\dfrac{bn}{t}$ dollars
3. $\dfrac{7z}{x}$ inches
4. $\dfrac{cy}{d}$ minutes

FINAL TEST 12

1. 0.56
2. 21,000 million pounds
3. 1.9
4. 8,000 million pounds
5. 35:24:42
6. $300
7. $\dfrac{26a}{t}$
8. $\dfrac{ny}{x}$ miles

Chapter 13
Percentages

In this chapter you will learn about percentages as a method for describing proportions and how to convert information into and out of percentage form. Section 13-1 introduces percentages and shows how they can be converted into decimal or fractional form. Section 13-2 deals with the use of information in percentage form. Section 13-3 shows how information can be converted into percentage form and introduces a variety of problems involving percentages. Then Sec. 13-4 deals specifically with percent change and problems which involve it.

For this chapter you will need a knowledge of Chaps. 8, 9, 11, and 12.

DIAGNOSTIC TEST 13

1. Convert $\frac{2}{5}\%$ to a proportion expressed in decimal form.
2. Convert $4\frac{3}{8}\%$ to a proportion expressed in fractional form.
3. Convert 0.008% to a proportion expressed in fractional form.
4. If the sales tax is $4\frac{1}{2}\%$, how much tax do you pay when you buy goods costing $38?
5. Use Table 13-1 (page 495) to find the number of craftsmen and foremen who have completed college (including those who went on to do graduate work).
6. A sample of hydrogen chloride is decomposed into 35.96 grams of chlorine and 1.04 grams of hydrogen. Calculate the percentages of hydrogen and chlorine in the sample correct to one decimal place.
7. Use Table 13-2 (page 496) to calculate (correct to one decimal place) the percentage of personal expenditure in 1965 that went on private education and research.
8. If a rock contains about 0.006% scandium and you need 0.005 gram of scandium for an accurate weighing, how big a sample of rock do you need in order to find the "exact" amount of scandium (correct to three figures)?

Use Table 13-5 (page 524) to find the percentage changes in the populations of the following cities between 1950 and 1960 (correct to three figures):

9. Boston
10. Fort Lauderdale

11. The volume of a certain gas was 720 milliliters. If this represents an increase of 8% on its previous volume under different conditions, what was its previous volume correct to three figures?
12. What is the percentage error involved in using 60 as an approximation for 57 (correct to one decimal place)?

When you have finished the diagnostic test, correct it yourself using the answers on page 526. Then check all the questions you got wrong on the list below:

1. _____	4. _____	6. _____	9. _____
2. _____	5. _____	7. _____	10. _____
3. _____		8. _____	11. _____
			12. _____

If you made no mistakes at all, you need not do any work in this chapter, so you have finished the book. But if you made any mistakes (even if it was only *one* mistake), follow these instructions:

	section	page
If you made any mistakes in Questions 1 to 3, circle Sec. 13-1.	13-1	497
If you made any mistakes in Question 4 or 5, circle Sec. 13-2.	13-2	504
If you made any mistakes in Questions 6 to 8, circle Sec. 13-3.	13-3	509
If you made any mistakes in Questions 9 to 12, circle Sec. 13-4.	13-4	517

You should now work through the sections whose numbers you have circled, taking them in numerical order. The sections begin on the pages indicated.

TABLE 13-1
Educational Levels in Different Types of Occupations (Male Workers Only) *

Occupation	Total employed	*Highest educational level completed, %* Elementary	High school	College	Graduate
All occupations (male workers only)	44,200,000	24.7	51.2	18.1	5.4
Professional and technical workers	5,620,000	2.1	20.8	43.5	33.6
Farmers and farm managers	2,090,000	54.0	38.2	6.2	0.4
Managers, proprietors (except farm)	6,370,000	15.3	49.0	31.2	4.2
Clerical workers	3,190,000	10.0	62.0	26.6	1.4
Sales workers	2,440,000	9.3	51.9	35.9	2.7
Craftsmen and foremen	8,330,000	28.1	62.1	9.0	0.5
Operatives	9,170,000	32.1	61.7	5.2	0.2
Service workers	2,960,000	31.7	56.8	9.9	0.7
Farm laborers and foremen	880,000	59.7	29.2	4.1	
Laborers (except farm and mine)	3,150,000	44.2	49.2	4.9	0.3

* Department of Labor, 1965.

TABLE 13-2

Personal Expenditure in Millions of Dollars in 1962 and 1965 *

Type of expenditure	Expenditure in millions of dollars	
	1965	*1962*
Clothing, jewelry	43,400	35,200
Food and tobacco	106,700	91,700
Foreign travel	3,200	2,500
Household operation	61,800	50,500
Housing	63,100	51,200
Medical care	28,100	21,700
Personal business	22,000	16,200
Personal care	7,500	6,200
Private education, research	5,500	4,300
Recreation	26,300	20,100
Religious and welfare activities	5,600	5,000
Transportation	57,800	45,300
Total	430,000	350,000

* Adapted from figures provided by Department of Commerce. Some of these figures have been adjusted to make your calculations easier, but none have been changed by more than 2%.

Section 13-1
Definition and Conversion to Decimals or Fractions

Do this section if you made a mistake in Questions 1 to 3 of Diagnostic Test 13. Begin by taking the pretest below, which will show you how much of Sec. 13-1 you already know and what you will be able to do by the time you have finished it.

PRETEST 13-1

Convert the following percentages to decimals:

1. 35% 2. 2.37% 3. $12\frac{1}{2}\%$ 4. 0.71% 5. $\frac{1}{4}\%$

Convert the following percentages to fractions (simplest forms):

6. 32% 7. 0.95% 8. $1\frac{2}{3}\%$ 9. 0.003% 10. $\frac{1}{2}\%$

When you have finished the pretest, correct it yourself using the answers on page 526. Then check all the questions you got wrong on the list below:

1. ______	5. ______	8. ______
2. ______	6. ______	9. ______
3. ______	7. ______	10. ______
4. ______		

If you made any mistakes (even if it was only *one* mistake), read the introduction and go through the program.

If you made no mistakes at all, skip Sec. 13-1, because you already know its material, and go on to the next section whose number you circled on page 494. If no further numbers were circled, take Final Test 13 (page 524).

Introduction

After learning that proportions can be expressed in percentage form as well as in decimal or fractional form, you will find out how to convert proportions in percentage form to decimals or to fractions.

FRAMES FOR SECTION 13-1

1 In Chap. 12 we discussed proportion and defined it by the equation

$$\text{Proportion} = \frac{\text{part}}{\text{whole}}$$

Thus in a pack of playing cards, the proportion of red cards is $\frac{26}{52} = \frac{1}{?}$ and the proportion of diamond cards is $\frac{13}{52} = \frac{1}{?}$.

$\frac{1}{2}$ $\frac{1}{4}$

---

2 We also showed in Chap. 12 that proportion could be expressed in fractional form as above or in decimal form. The proportion of red cards in a pack expressed in decimal form is _____, and the proportion of diamond cards is _____.

0.5 0.25

---

3 A third form of representing a proportion is the *percentage form. Percent* means per hundred (a century is 100 years), so

$$25 \text{ percent} = \frac{25}{100} = \_\_\_\_\_ \text{ (decimal form)}$$

$$50 \text{ percent} = \frac{?}{100} = \_\_\_\_\_ \text{ (decimal form)}$$

$$63 \text{ percent} = \_\_\_\_\_ \text{ (decimal form)}$$

0.25 $\frac{50}{100}$ 0.50 0.63

---

4 Percentages are in such common use that we use a special symbol % to denote percent. Thus

$$30\% = 30 \text{ percent} = \frac{?}{100} = \_\_\_\_\_ \text{ (decimal form)}$$

$$45\% = \frac{?}{100} = \_\_\_\_\_ \text{ (decimal form)}$$

$$62\% = \_\_\_\_\_ \text{ (decimal form)}$$

$\frac{30}{100}$ 0.30 $\frac{45}{100}$ 0.45 0.62

---

5 Convert the following percentages to decimals:

73% 70% 7%

0.73 0.70 0.07

---

6 To convert an integral percentage to decimal form, you divide by ______.

100

---

7 The percentage form may itself contain a fraction or a decimal point. A half of 25% is $12\frac{1}{2}$%, or 12.5%, and a half of $12\frac{1}{2}$% is ______% or ______%.

$6\frac{1}{4}$% 6.25%

---

8 Convert the following percentages to decimals:

a. 13% *b.* 12% *c.* $12\frac{1}{2}$% *d.* $6\frac{1}{4}$%

---

a. 0.13 *b.* 0.12 *c.* 0.125 *d.* 0.0625

---

9 To convert a percentage containing fractions to a decimal:

a. Change the fractions to decimals.

b. Divide by ______.

c. $5\frac{1}{2}$% = ______% = ______ (decimal form).

---

b. 100 *c.* 5.5% = 0.055

---

10 Convert the following percentages to decimals:

a. $13\frac{2}{5}$% *b.* $3\frac{2}{5}$% *c.* $\frac{2}{5}$%

---

a. 13.4% = 0.134 *b.* 3.4% = 0.034 *c.* 0.4% = 0.004

---

11 Convert the following percentages to decimals:

a. $\frac{3}{4}$% *b.* $1\frac{5}{8}$% *c.* $2\frac{1}{4}$%

---

a. 0.75% = 0.0075 *b.* 1.625% = 0.01625 *c.* 2.25% = 0.0225

---

12 Convert the following percentages to decimals (correct to four places):

a. $3\frac{1}{3}$% *b.* $\frac{5}{7}$% *c.* $1\frac{5}{6}$%

---

a. 3.33% = 0.0333 *b.* 0.71% = 0.0071 *c.* 1.83% = 0.0183

---

13 Convert the following percentages to decimals:

a. 12.5% *b.* 2.5% *c.* 0.5%

---

a. 0.125 *b.* 0.025 *c.* 0.005

---

14 Convert the following percentages to decimals:

a. 0.7% *b.* 5.43% *c.* 0.86%

---

a. 0.007 *b.* 0.0543 *c.* 0.0086

---

15 Convert the following percentages to decimals (correct to three figures):

a. 0.62% *b.* $\frac{2}{3}$% *c.* 2.78% *d.* $1\frac{2}{7}$%

---

a. 0.0062 *b.* 0.00667 *c.* 0.0278 *d.* 0.0129

---

16 Integral percentages can be considered as fractions with denominator 100, and conversion to fractional form is merely a matter of simplification. Thus

$$30\% = \frac{?}{100} = \frac{?}{10}$$

$$25\% = \frac{?}{100} = \underline{\qquad} \text{ (simplest fractional form)}$$

$$20\% = \underline{\qquad} \text{ (simplest fractional form)}$$

$\frac{30}{100} = \frac{3}{10}$ $\frac{25}{100} = \frac{1}{4}$ $\frac{1}{5}$

---

17 Give the simplest fraction forms of the following:

a. 15% *b.* 16% *c.* 5%

---

a. $\frac{15}{100} = \frac{3}{20}$ *b.* $\frac{16}{100} = \frac{4}{25}$ *c.* $\frac{5}{100} = \frac{1}{20}$

---

18 If the percentage is a mixed numeral, then the procedure is:

a. Change the mixed numeral to a fraction.

b. Divide by 100.

c. Simplify.

Thus

$$12\tfrac{1}{2}\% = \frac{?}{2}\% = \frac{?}{200} = \underline{\qquad} \text{ (simplest fraction form)}$$

and

$$3\tfrac{1}{3}\% = \frac{?}{3}\% = \frac{?}{300} = \underline{\qquad} \text{ (simplest fraction form)}$$

$\frac{25}{2}\% = \frac{25}{200} = \frac{1}{8}$ $\quad \frac{10}{3}\% = \frac{10}{300} = \frac{1}{30}$

---

19 Convert the following percentages to fractions (simplest forms):

a. $6\tfrac{1}{4}\% = \frac{?}{4}\% = \underline{\qquad}$

b. $3\tfrac{2}{5}\% = \underline{\qquad}\% = \underline{\qquad}$

c. $7\tfrac{1}{2}\% = \underline{\qquad}\% = \underline{\qquad}$

---

a. $\frac{25}{4}\% = \frac{25}{400} = \frac{1}{16}$ *b.* $\frac{17}{5}\% = \frac{17}{500}$ *c.* $\frac{15}{2}\% = \frac{15}{200} = \frac{3}{40}$

---

20 Convert the following percentages to fractions (simplest forms):

a. $2\tfrac{2}{3}\% = \underline{\qquad}\% = \underline{\qquad}$

b. $\tfrac{2}{3}\% = \underline{\qquad}$

c. $12\tfrac{2}{3}\% = \underline{\qquad}$

---

a. $\frac{8}{3}\% = \frac{8}{300} = \frac{2}{75}$ *b.* $\frac{2}{300} = \frac{1}{150}$ *c.* $\frac{38}{300} = \frac{19}{150}$

---

21 Convert the following percentages to fractions (simplest forms):

a. $\tfrac{2}{7}\%$ *b.* $3\tfrac{5}{8}\%$ *c.* $7\tfrac{2}{3}\%$

---

a. $\frac{2}{700} = \frac{1}{350}$ *b.* $\frac{29}{800}$ *c.* $\frac{23}{300}$

---

22 If the percentage is in decimal form, the procedure is essentially similar:

a. Divide by 100.

b. Eliminate decimals.

Thus

$$2.8\% = \frac{2.8}{100} = \frac{28}{?}$$

$$2.98\% = \frac{2.98}{100} = \frac{298}{?}$$

$$3.6\% = \frac{36}{?}$$

$\frac{28}{1,000}$ $\frac{298}{10,000}$ $\frac{36}{1,000}$

---

23 Convert the following percentages to fractions without decimal points. Do not bother to convert the fractions to their simplest forms.

a. 3.56% *b.* 11.4% *c.* 0.78%

---

a. $\frac{356}{10,000}$ *b.* $\frac{114}{1,000}$ *c.* $\frac{78}{10,000}$

---

24 Convert the following percentages to fractions without decimal points. Do not bother to convert the fractions to their simplest forms.

a. 0.83% *b.* 6.792% *c.* 0.0035%

---

a. $\frac{83}{10,000}$ *b.* $\frac{6,792}{100,000}$ *c.* $\frac{35}{1,000,000}$

---

25 Convert the following percentages to fractions (simplest forms):

a. 0.0025% *b.* 4.5% *c.* 0.8%

---

a. $\frac{25}{1,000,000} = \frac{1}{40,000}$ *b.* $\frac{45}{1,000} = \frac{9}{200}$ *c.* $\frac{8}{1,000} = \frac{1}{125}$

---

26 Convert the following percentages to fractions (simplest forms):

a. $22\frac{2}{3}\%$ *b.* 0.07% *c.* $\frac{3}{8}\%$ *d.* 0.195%

---

a. $\frac{68}{300} = \frac{17}{75}$ *b.* $\frac{7}{10,000}$ *c.* $\frac{3}{800}$ *d.* $\frac{195}{100,000} = \frac{39}{20,000}$

---

27 Convert the following percentages to decimals (correct to four places):

a. 0.38% *b.* $\frac{5}{6}$% *c.* $2\frac{5}{9}$% *d.* 6.49%

---

a. 0.0038 *b.* 0.0083 *c.* 0.0256 *d.* 0.0649

You have finished the program for Sec. 13-1. Now take the posttest to see how much you have learned.

POSTTEST 13-1

Convert the following percentages to decimals:

1. 47% 2. 6.95% 3. $16\frac{1}{4}$% 4. 0.42% 5. $\frac{2}{5}$%

Convert the following percentages to fractions:

6. 24% 7. 0.45% 8. $1\frac{1}{4}$% 9. 0.007% 10. $\frac{1}{8}$%

When you have finished the posttest, correct it yourself using the answers on page 526. Then check all the questions you got wrong on the list below:

1. ______ 5. ______ 8. ______
2. ______ 6. ______ 9. ______
3. ______ 7. ______ 10. ______
4. ______

Now proceed to the next section whose number you circled on page 494. If no other sections were circled, take Final Test 13 (page 524).

Section 13-2
The Use of Information in Percentage Form

Do this section if you made a mistake in Question 4 or 5 of Diagnostic Test 13. Begin by taking the pretest below, which will show you how much of Sec. 13-2 you already know and what you will be able to do by the time you have finished it.

PRETEST 13-2

1. If the sales tax is 4%, how much tax do you pay when you buy goods costing $37?
2. If the sales tax is $3\frac{1}{2}$%, how much tax do you pay when you buy goods costing $64?
3. If you are employed, 4.9% of your earnings goes on social security. So if you earn $3,800 a year, how much social security will you pay each year?
4. The nickel contains only 25% nickel, the other 75% being copper. If it weighs 77.16 grains, how many grains of nickel does it contain?

Look at Table 13-1 (page 495) and answer the following questions. *Give your answers correct to three figures.*

5. What number of clerical workers have completed high school but not completed college?
6. What number of service workers have completed college but not completed any graduate work?

When you have finished the pretest, correct it yourself using the answers on page 526. Then check all the questions you got wrong on the list below:

1. ______	3. ______	5. ______
2. ______	4. ______	6. ______

If you made any mistakes (even if it was only *one* mistake), read the introduction and go through the program.

If you made no mistakes at all, skip Sec. 13-2, because you already know its material, and go on to the next section whose number you circled on page 494. If no further numbers were circled, take Final Test 13 (page 524).

Introduction

In this section you learn how to interpret and use information presented in percentage form. In all the examples you are given proportions in percentage form and use these data to obtain other information.

FRAMES FOR SECTION 13-2

1 One use of percentages which you must surely know well is *sales tax.* In most states all goods sold are taxed at a standard rate, which is expressed as a percentage. If the sales tax is 4%, the tax on \$100 worth of goods will be

$$4\% \text{ of } \$400 = \frac{4}{100} \times \$100 = \_\_\_\_$$

What will be the tax on \$50 worth of goods?

\$4 \$2

---

2 If the sales tax is 3%, the tax on \$50 worth of goods will be

$$\frac{?}{100} \times \$\_\_\_\_ = \$\_\_\_\_$$

What will be the tax on \$25 worth of goods?

$\frac{3}{100} \times \$50$ \$1.50 \$0.75

---

3 *a.* If the sales tax is 5%, what will be the tax on \$20 worth of goods?

b. What will be the tax on \$5 worth of goods?

---

a. \$1 *b.* \$0.25

---

4 *a.* If the sales tax is 6%, what will be the tax on \$3 worth of goods?

b. What will be the tax on \$13 worth of goods?

---

a. \$0.18 *b.* \$0.78

---

5 *a.* If the sales tax is 2%, what will be the tax on \$7 worth of goods?

b. What will be the tax on \$7.50 worth of goods?

---

a. \$0.14 *b.* \$0.15

---

6 *a.* If the sales tax is 4%, what will be the tax on \$3.50 worth of goods?

b. What will be the tax on \$2.25 worth of goods?

---

a. \$0.14 *b.* \$0.09

---

7 If the sales tax is $3\frac{1}{2}$%, the tax on \$4 worth of goods will be

$$\frac{3\frac{1}{2}}{100} \times 4 = \frac{?}{200} \times 4 = \_\_\_\_$$

What will be the tax on \$14 worth of goods?

$\frac{7}{200}$ \$0.14 \$0.49

---

8 *a.* If the sales tax is $4\frac{1}{2}$%, what will be the tax on \$6 worth of goods?

b. What will be the tax on \$22 worth of goods?

---

a. \$0.27 *b.* \$0.99

---

9 *a.* If the sales tax is $2\frac{1}{4}$%, what will be the tax on \$8 worth of goods?

b. What will be the tax on \$20 worth of goods?

---

a. \$0.18 *b.* \$0.45

---

10 The composition of alloys and mixtures is often described in terms of percentages. For example the silver coinage minted in the United States before 1965 contained 90% silver and 10% copper. If a newly minted silver quarter dollar weighed 96.45 grains, what weight of silver did it contain? Give your answer correct to two decimal places in grains (a grain is the unit of weight used for precious metals).

86.81 grains $\left(96.45 \times \frac{90}{100}\right)$

---

11 If a newly minted silver dime weighed 38.58 grains, what weight of silver did it contain? Give your answer correct to two decimal places in grains.

34.72 grains

---

12 Since 1965 the new dime has weighed 35 grains and contained 91.67% copper and 8.33% nickel. What weight of copper does it contain? Give your answer in grains correct to two decimal places.

32.08 grains

---

13 Since 1965 the new quarter dollar also has contained 91.67% copper and 8.33% nickel. It weighs 87.50 grains. What weight of nickel does it contain? Give your answer in grains correct to two decimal places.

7.29 grains

---

14 Percentages are often used to tabulate information as they make comparisons easy. Table 13-1 (page 495), for example, shows how educational level varies with type of occupation for male workers.

a. What type of occupation has a majority of workers who have completed college (this includes those who go on to complete graduate work)?

b. Which types of occupation have a majority of workers who have not completed high school?

a. professional and technical work
b. farming and farm managing; farm laboring

---

15 Now suppose we want to know how many of those who have studied at the graduate level are professional and technical workers. Table 13-1 tells us that there are ______ professional and technical workers altogether and that ______% of them have completed some education at the graduate level. So how many professional and technical workers have completed some education at the graduate level? Give your answer correct to three figures.

5,620,000 33.6% 1,890,000

---

16 *a.* What is the total number of male workers who have completed some education at the graduate level?

b. Are the majority of them engaged in professional and technical work? Give your answers correct to three figures.

a. 2,390,000 *b.* yes, 1.89 out of 2.39 million is a substantial majority

---

17 *a.* How many farmers and farm managers did not complete high school?

b. How many farm laborers and foremen did not complete high school? Give your answers correct to three figures.

a. 1,130,000 *b.* 525,000

18 *a.* What is the total number of male workers who did not complete high school?

b. Did the majority of them go into farming?

a. 10,900,000 *b.* no

19 Unless you are self-employed, 4.9% of your income goes on social security payments and your employer pays an equivalent contribution. So if you are earning $400 a month, you pay social security of $_____ a month, and your employer pays $_____ a month.

$19.60 $19.60

20 *a.* If you are earning $3,400 a year, how much social security do you pay?

b. How much social security do your employers pay on your behalf?

a. $166.60 a year *b.* $166.60 a year

You have finished the program for Sec. 13-2. Now take the posttest to see how much you have learned.

POSTTEST 13-2

1. If the sales tax is 3%, how much tax do you pay when you buy $55 worth of goods?
2. If the sales tax is $4\frac{1}{2}$%, how much do you pay when you buy $64 worth of goods?
3. If you are self-employed, 7.1% of your earnings goes on social security. So if you earn $6,200 a year, how much social security will you pay each year?
4. The half-dollar coin contains 40% silver and 60% copper. If it weighs 177.72 grains, how much silver does it contain?

Look at Table 13-1 (page 495) and answer the following questions. *Give your answers correct to three figures.*

5. How many craftsmen and foremen have completed high school but not completed college?
6. How many operatives have completed college but not completed any graduate work?

When you have finished the posttest, correct it yourself using the answers on page 526. Then check all the questions you got wrong on the list below:

1. ______ 3. ______ 5. ______

2. ______ 4. ______ 6. ______

If you compare your performance with that on Pretest 13-2, you will see how much you have improved. If you made more than one mistake on this posttest, review the material and take the pretest again. Don't move on to the next section until you can get all the questions right.

Now proceed to the next section whose number you circled on page 494. If no other sections were circled, take Final Test 13 (page 524).

Section 13-3 Conversion into Percentage Form and Problems with Percentages

Do this section if you made a mistake in Questions 6 to 8 of Diagnostic Test 13. Begin by taking the pretest below, which will show you how much of Sec. 13-3 you already know and what you will be able to do by the time you have finished it.

PRETEST 13-3

1. A sample of water is decomposed into 2.846 grams of oxygen and 0.354 gram of hydrogen. Calculate the percentages of oxygen and hydrogen in the sample of water (correct to one decimal place).

Use Table 13-2 (page 496) to calculate the following (correct to one decimal place):

2. The percentage of personal expenditure in 1962 which went on medical care.
3. The percentage of personal expenditure in 1962 which went on foreign travel.

4. If a rock contains about 0.07% cobalt and you need 0.005 gram of cobalt for an accurate weighing, how big a sample of rock do you need in order to find the "exact" amount of cobalt (correct to two figures)?
5. Lead sulfide contains 86.6% lead. How much lead is there in 800 grams of lead sulfide?

6. A 400-gram sample of moon rock contains 0.73 gram of copper. What is the percentage of copper in the sample?

When you have finished the pretest, correct it yourself using the answers on page 526. Then check all the questions you got wrong on the list below:

1. ______ 3. ______ 5. ______

2. ______ 4. ______ 6. ______

If you made any mistakes (even if it was only *one* mistake), read the introduction and go through the program.

If you made no mistakes at all, skip Sec. 13-3, because you already know its material, and go on to the next section whose number you circled on page 494. If no further numbers were circled, take Final Test 13 (page 524).

Introduction

In this section you learn how to convert information into percentage form and how to recognize and solve different kinds of problems involving percentages.

FRAMES FOR SECTION 13-3

1 So far in this chapter we have been concerned with interpreting and using information presented in percentage form. We now have to tackle the problem of getting information into percentage form. If the data are already in decimal form, this is a simple procedure.

$0.5 = \frac{?}{100} =$ ______% and $0.1 =$ ______%

$\frac{50}{100}$ 50% 10%

---

2 Convert the following proportions from decimal form to percentage form:

a. 0.36 *b.* 0.19 *c.* 0.07

---

a. 36% *b.* 19% *c.* 7%

---

3 To convert a proportion from decimal form to percentage form, you (multiply/divide) ______ by 100. So

a. 0.364 = ______% *b.* 0.075 = ______% *c.* 0.007 = ______%

---

multiply *a.* 36.4% *b.* 7.5% *c.* 0.7%

---

4 In the last chapter we calculated proportions in decimal form, and we can now convert these to percentage form.

Problem: A 5.000-gram sample of iron ore contains 3.105 grams of iron. Express the iron content of the sample in percentage form.

Solution:

$$\text{Proportion of iron in sample} = \frac{?}{?} = \_\_\_\_\_ \text{ (decimal form)}$$

$$\text{Percentage of iron in sample} = \_\_\_\_\_\%$$

$\frac{3.105}{5}$ 0.621 62.1%

---

5 In the last chapter we used the equation

$$\text{Proportion} = \frac{\text{part}}{\text{whole}}$$

In this chapter we can use the equation

$$\text{Percentage} = \frac{\text{part}}{\text{whole}} \times \_\_\_\_\_$$

100

---

6 **Problem:** An 8.0-gram sample of iron sulfide contains 5.1 grams of iron by weight. What is the percentage of iron in the iron sulfide?

Solution:

$$\text{Percentage} = \frac{?}{?} \times 100 = \_\_\_\_\_$$

$\frac{5.1}{8.0}$ 63.75%

---

7 A 29-gram sample of sodium chloride contains 17.58 grams of chlorine. What is the percentage of chlorine in sodium chloride? (Give your answer correct to three figures.)

$\frac{17.58}{29} \times 100 = 60.6\%$

---

8 Problem: A sample of water is decomposed into 0.96 gram of oxygen and 0.120 gram of hydrogen. Calculate the percentage weights of oxygen and hydrogen in the sample (correct to three figures). This problem is slightly more complicated because you have to begin by finding the weight of the sample.

Solution:

Weight of sample of water = ______ grams

$$\text{Percentage of oxygen} = \frac{?}{?} \times \_\_\_\_\_\_ = \_\_\_\_\_\_$$

$$\text{Percentage of hydrogen} = 100 - \_\_\_\_\_\_ = \_\_\_\_\_\_$$

1.08 $\frac{0.96}{1.08}$ 100 88.9 88.9 11.1

Answer: 88.9% oxygen, 11.1% hydrogen

---

9 A sample of hydrogen chloride is decomposed into 72.90 grams of chlorine and 2.10 grams of hydrogen. Calculate the percentages of hydrogen and chlorine in the sample.

Weight of sample = 75.00 grams

$$\text{Percentage of hydrogen} = \frac{2.10}{75.00} \times 100 = 2.8$$

Answer: 2.8% hydrogen, 97.2% chlorine

---

10 Tabulated data are often more meaningful in percentage form. Table 13-2, for example, is probably more easily interpreted in percentage form. The table, which is on page 496, gives 12 major categories of personal expenditure and shows how expenditure was divided among them in 1962 and in 1965. In 1962 the percentage of consumer expenditure that went on food and tobacco was $\frac{?}{?} \times 100 =$ ______________.

$$\frac{91{,}700}{350{,}000} \times 100 = \frac{917}{35} = \frac{1{,}834}{70} = 26.2\%$$

---

11 In 1962 the proportion of consumer expenditure that went on housing was ______%, and the proportion that went on personal care was ______% (give your answers correct to one decimal place).

$$\frac{51{,}200}{350{,}000} \times 100 = \frac{512}{35} = \frac{1{,}024}{70} \approx 14.6\%$$

$$\frac{6{,}200}{350{,}000} \times 100 = \frac{62}{35} = \frac{124}{70} \approx 1.8\%$$

12 What percentage of consumer expenditure in 1962 went:

a. On transportation? *b.* On foreign travel?

(Give your answer correct to one decimal place.)

a. $\frac{45{,}300}{350{,}000} \times 100 \approx 12.9\%$ *b.* $\frac{2{,}500}{350{,}000} \times 100 \approx 0.7\%$

13 We now want to construct a table to show the allocation of personal expenditure in 1962 on a percentage basis. The table below includes all the figures you have already calculated. Now complete the table.

Type of expenditure	*Percentage of expenditure in 1962*
Clothing, jewelry	*a.* ____
Food and tobacco	26.2
Foreign travel	0.7
Household operation	*b.* ____
Housing	14.6
Medical care	6.2
Personal business	*c.* ____
Personal care	1.8
Private education, research	*d.* ____
Recreation	*e.* ____
Religious and welfare activities	*f.* ____
Transportation	12.9
Total	100.0

a. 10.1 *b.* 14.4 *c.* 4.6 *d.* 1.2 *e.* 5.7 *f.* 1.4

14 So far we have tackled two main kinds of problem which involve percentages. In Sec. 13-2 we were given the whole and the percentage and had to find the part. The equation we used was:

a. $\text{Part} = \text{whole} \times \text{percentage} \times 100$

b. $\text{Part} = \text{whole} \times \dfrac{\text{percentage}}{100}$

c. $\text{Part} = \text{whole} \times \dfrac{100}{\text{percentage}}$

---

b

---

15 In this section we have been doing problems in which we have been given the part and the whole and have had to find the percentage. The equation we have used has been:

a. $\text{Percentage} = \dfrac{\text{part}}{\text{whole}} \times 100$

b. $\text{Percentage} = \dfrac{\text{whole}}{\text{part}} \times 100$

c. $\text{Percentage} = \dfrac{\text{part}}{\text{whole}} \div 100$

d. $\text{Percentage} = \dfrac{\text{whole}}{\text{part}} \div 100$

---

a

---

16 The equations of Frames 14 and 15 are both forms of the basic percentage equation.

$$\frac{\text{Part}}{\text{Whole}} = \text{proportion} = \frac{?}{?} \quad \left(\frac{\text{percentage}}{100} \quad \text{or} \quad \frac{100}{\text{percentage}}\right)$$

$\dfrac{\text{percentage}}{100}$

---

17 A third kind of problem is as follows.

Problem: Silver solder contains 63.00% silver. How much of the solder can be made with 720.0 grams of silver?

Solution: Does this problem ask for the part, the whole, or the percentage? Substitute the figures you are given in the equation of Frame 16.

whole $\quad \dfrac{720.0}{\text{whole}} = \dfrac{63.00}{100}$

18 Now find the answer to the problem correct to four figures.

$$\text{Whole} = \frac{720.0}{63.00} \times 100 = \frac{72{,}000}{63} = \frac{8{,}000}{7} \approx 1{,}143 \text{ grams}$$

19 **Problem:** A copper compound contains 80.0% copper by weight. How much copper is there in a 1.95-gram sample of the compound?

Solution: Does this problem ask for the part, the whole, or the percentage? Substitute the figures you are given in the equation of Frame 16.

part $\quad \dfrac{\text{part}}{1.95} = \dfrac{80.0}{100}$

20 Now find the answer to the problem correct to three figures.

$$\text{Part} = 1.95 \times \frac{80.0}{100} = 1.56 \text{ grams}$$

21 You want to make up a 20% solution of salt in water. You have 700 grams of salt. How much solution can you make?

$$\frac{700}{\text{Whole}} = \frac{20}{100}$$

$$\text{Whole} = 700 \times \frac{100}{20} = 3{,}500 \text{ grams}$$

22 An analysis of 700 grams of moon rock showed that it contained 0.04 gram of chromium. What was the percentage of chromium correct to two figures?

$$\frac{0.04}{700} = \frac{\text{percentage}}{100}$$

$$\text{Percentage} = \frac{0.04 \times 100}{700} = \frac{4}{700} \approx 0.0057\%$$

23 A typical earth rock contains 0.007% chromium. How much chromium would 700 grams of this rock contain?

$$\frac{\text{Part}}{700} = \frac{0.007}{100}$$

$$\text{Part} = \frac{700 \times 0.007}{100} = 0.049 \text{ gram}$$

24 You believe that moon rock is likely to contain about 0.09% manganese, and you need 0.005 gram of manganese for an accurate weighing. How big a sample do you need if you leave no margin of error?

$$\frac{0.005}{\text{Whole}} = \frac{0.09}{100}$$

$$\text{Whole} = \frac{0.005 \times 100}{0.09} \approx 5.5 \text{ grams}$$

You have finished the program for Sec. 13-3. Now take the posttest to see how much you have learned.

POSTTEST 13-3

1. A sample of water is decomposed into 5.68 grams of oxygen and 0.72 gram of hydrogen. Calculate the percentages of oxygen and hydrogen in the sample.

Use Table 13-2 (page 496) to calculate the following correct to one decimal place:

2. The percentage of personal expenditure in 1965 which went on housing.
3. The percentage of personal expenditure in 1965 which went on foreign travel.

4. If a rock contains about 0.04% vanadium and you need 0.005 gram for an accurate weighing, how big a sample of rock do you need in order to find the "exact" percentage of vanadium?
5. Iron oxide contains 70.02% iron. How much iron is there in 600.0 grams of iron oxide? Give your answer correct to four figures.
6. A 200-gram sample of moon rock contains 0.68 gram of nickel. What is the percentage of nickel in the sample?

When you have finished the posttest, correct it yourself using the answers on page 526. Then check all the questions you got wrong on the list below:

1. _____ 3. _____ 5. _____

2. _____ 4. _____ 6. _____

If you compare your performance with that on Pretest 13-3, you will see how much you have improved. If you made more than one mistake on this posttest, review the material and take the pretest again. Don't move on to the next section until you can get all the questions right.

Now proceed to the next section whose number you circled on page 494. If no other sections were circled, take Final Test 13 (page 524).

Section 13-4 Percent Change

Do this section if you made a mistake in Questions 9 to 12 of Diagnostic Test 13. Begin by taking the pretest below, which will show you how much of Sec. 13-4 you already know and what you will be able to do by the time you have finished it.

PRETEST 13-4

Use Table 13-5 (page 524) to find the percentage changes in the populations of the following cities between 1950 and 1960. (*Give all answers correct to one decimal place.*)

1. Anchorage 2. Atlanta 3. Kansas City

4. An electric resistance has decreased by 4%. If the present value of the resistance is 36.3 ohms, what was the value before the change took place?
5. If we assume that water increases in volume by 9% on freezing, how many cubic feet of water would be required to produce 654 cubic feet of ice?
6. What is the percentage error involved in using 80 as an approximation for 84?

When you have finished the pretest, correct it yourself using the answers on page 527. Then check all the questions you got wrong on the list below:

1. ______	3. ______	5. ______
2. ______	4. ______	6. ______

If you made any mistakes (even if it was only *one* mistake), read the introduction and go through the program.

If you made no mistakes at all, skip Sec. 13-4, because you already know its material, and go on to take Final Test 13 (page 524).

Introduction

In this section you learn how changes are described in terms of percent change and how to recognize and solve different kinds of problems involving present changes.

TABLE 13-3

Population of Cities in Illinois in 1956 and in 1960 and Percent Changes over the Period 1950–1960 *

City	*Population*		*Percent change 1950–1960*
	1960	*1950*	
Arlington Heights	27,878	8,768	218.0
Aurora	63,715	50,576	26.0
Chicago	3,550,404	3,620,962	−1.9
Peoria	103,162	111,856	−7.8
Skokie	59,364	14,832	300.2

* Figures taken from those supplied by U.S. Bureau of the Census.

FRAMES FOR SECTION 13-4

1 Another common use of percentages is in the measurement of change. Table 13-3 above, for example, gives the population of a number of cities in Illinois in 1950 and in 1960 and the *percent change* in population between 1950 and 1960. In calculating *percent change* we are concerned with *change* as a *proportion* of the *original value.* So if we look at the figures for Aurora in Table 13-3, we can see that:

a. Change in population between 1950 and 1960 is

________ − ________ = ________

b. Percent change $= \dfrac{26}{100} = \dfrac{\text{change in population}}{\text{population in 19\_\_}} = \dfrac{?}{?}$

---

a. $63{,}715 - 50{,}576 = 13{,}139$ *b.* 1950; $\dfrac{13{,}139}{50{,}576}$

---

2 As a check, find 26% of 50,000.

$50{,}000 \times \dfrac{26}{100} = 13{,}000$

---

3 If 13,139 is 26% of 50,576 and 50,576 is 100% of 50.576, then 63,715 will be ______% of 50,576.

126%

---

4 *a.* What was the change in the population of Chicago between 1950 and 1960?

b. Was this an increase or a decrease?

c. $\frac{1.9}{100} = \frac{\text{change in population}}{\text{population in 19\_\_}} = \frac{?}{?}$

---

a. 3,620,962 − 3,550,404 = 70,558 *b.* decrease *c.* 1950; $\frac{70{,}558}{3{,}620{,}962}$

---

5 What percentage of 3,620,962 is 3,550,404?

---

100 − 1.9 = 98.1%

---

6 Look at the figures for Peoria. What percentage of 111,856 is 103,162?

---

100 − 7.8 = 92.2%

---

7 Now look at the figures for Arlington Heights.

a. What was the change in population between 1950 and 1960?

b. What was the percentage change in population between 1950 and 1960?

c. What percentage of 8,768 is 27,878?

---

a. 27,878 − 8,768 = 19,110 *b.* 218% *c.* 100 + 218 = 318%

---

8 *a.* What was the percentage change in the population of Skokie between 1950 and 1960?

b. What percentage of 14,832 is 59,364?

---

a. 300.2 *b.* 400.2

---

9 Table 13-4 on page 520 shows the populations of some California cities in 1950 and in 1960. What was the percent change in the population of Los Angeles between 1950 and 1960?

Change = 500,000 percent change = $\frac{500{,}000}{2{,}000{,}000} \times 100 = 25$

---

10 The population of Los Angeles in 1960 was _____% of what it was in 1950.

125%

---

11 *a.* What was the percent change in the population of Sacramento between 1950 and 1960?

b. The population of Sacramento in 1960 was _____% of what it was in 1950. (Give your answers correct to one decimal place.)

a. $\frac{55{,}000}{140{,}000} \times 100 = \frac{550}{14} \approx 39.3$ *b.* 139.3%

---

12 *a.* What was the percent change in the population of Santa Clara between 1950 and 1960?

b. The population of Santa Clara in 1960 was _____% of what it was in 1950. (Give your answers correct to three figures.)

a. $\frac{47{,}000}{12{,}000} \times 100 \approx 392$ *b.* 492%

TABLE 13-4

Population of Cities in California in 1950 and in 1960*

City	*Population*	
	1960	*1950*
Garden Grove	84,000	3,800
Los Angeles	2,500,000	2,000,000
Oakland	360,000	380,000
Sacramento	195,000	140,000
San Francisco	740,000	780,000
Santa Clara	59,000	12,000

* Adapted from figures supplied by U.S. Bureau of the Census. Some of these figures have been adjusted to make your calculations easier.

13 *a.* What was the percent change in the population of San Francisco between 1950 and 1960? (Remember to indicate that it was a decrease and give your answer correct to one decimal place.)

b. The population of San Francisco in 1960 was ______% of what it was in 1950.

---

a. $\frac{40{,}000}{780{,}000} \times 100 = \frac{400}{78} \approx 5.1$; percent change $= -5.1\%$ *b.* 94.9%

---

14 *a.* What was the percent change in the population of Oakland between 1950 and 1960?

b. The population of Oakland in 1960 was ______% of what it was in 1950. (Give your answers correct to one decimal place.)

---

a. $\frac{20{,}000}{380{,}000} \times 100 = \frac{200}{38} \approx 5.3\%$; percent change $= -5.3\%$ *b.* 94.7%

---

15 *a.* What was the percent change in the population of Garden Grove between 1950 and 1960?

b. The population of Garden Grove in 1960 was ______% of what it was in 1950. (Give your answers correct to three figures.)

---

a. $\frac{80{,}200}{3{,}800} \times 100 \approx 2{,}110$ *b.* 2,210%

---

16 In general:

$$\frac{\text{Percentage change}}{100} = \frac{\text{change}}{\text{(old/new) \_\_\_\_\_\_ value}} = \frac{\text{new value} - \text{old value}}{\text{(old/new) \_\_\_\_\_\_ value}}$$

---

old old

---

17 A gas contracts from a volume of 32 milliliters to a volume of 30.5 milliliters. What is the percent change in volume correct to two figures?

$\frac{30.5 - 32}{32} = \frac{-1.5}{32} = -0.047$; percent change in volume $= -4.7\%$

---

18 The length of a brass rod is 3.594 meters. If the rod were to expand by 0.2%, what would be its new length (to three decimal places)?

Change $= 3.594 \times \frac{0.2}{100} \approx 0.007$ meter new length $= 3.601$ meters

---

19 Problem: An electric resistance decreased 2%. If the present value of the resistance is 74.8 ohms, what was it before the change took place? (Give your answer correct to one decimal place.)

Solution:

$$74.8 \text{ ohms} = \_\_\_\_\_\% \text{ of original value} = \text{original value} \times \frac{?}{100}$$

$$\text{Original value} = \frac{74.8 \times 100}{?} = \_\_\_\_\_\_\_$$

98% 98 $\frac{74.8 \times 100}{98} \approx 76.3$ ohms

---

20 Problem: Water increases in volume by 9% on freezing. How many cubic feet of water are needed to produce 545 cubic feet of ice?

Solution: 545 cubic feet is _____% of the required volume of water. Required volume = _______.

109% $\frac{545 \times 100}{109} = 500$ cubic feet

---

21 The volume of a certain gas is 46 milliliters. If this represents a decrease of 8% from its previous volume under different conditions, what was its previous volume?

---

46 is 92% of previous volume; previous volume $= \frac{46 \times 100}{92} = 50$ milliliters

---

22 A voltmeter reads 120 volts. It is known that this reading is 6% too high. What should the correct reading have been? (Give your answer correct to one decimal place.)

---

120 volts is 106% of correct reading

$$\text{correct reading} = \frac{120 \times 100}{106} \approx 113.2 \text{ volts}$$

---

23 Frame 22 was an example of the use of percentages for representing error. In Chap. 9 we defined proportional error as $\dfrac{\text{error}}{\text{true value}}$ and in percentage form we can write

$$\frac{\text{Percentage error}}{100} = \frac{\text{error}}{\text{true value}}$$

Thus if the dry weight of a pure sample is 60 grams and after exposure to moisture it weighs 60.75 grams, what is the percentage error in recording its moist weight?

$$\text{Percentage error} = \frac{0.75 \times 100}{60} = 1.25\%$$

---

24 What is the percentage error involved in making the following approximations? Give your answers correct to three figures.

a. 60 for 64 *b.* 20 for 24 *c.* 140 for 144

---

a. $\dfrac{4}{64} \times 100 \approx 6.25\%$ *b.* $\dfrac{4}{24} \times 100 \approx 16.7\%$ *c.* $\dfrac{4}{144} \times 100 \approx 2.78\%$

You have finished the program for Sec. 13-4. Now take the posttest to see how much you have learned.

POSTTEST 13-4

Use Table 13-5 (page 524) to find the percentage changes in the populations of the following cities between 1950 and 1960. Give all answers correct to one decimal place.

1. Washington 2. Miami 3. Las Vegas

4. An electric resistance has decreased by 3%. If the present value of the resistance is 32.4 ohms, what was the value before the change took place?
5. If we assume that water increases in volume by 9% on freezing, how many cubic feet of water would be required to produce 763 cubic feet of ice?
6. What is the percentage error involved in using 30 as an approximation for 28?

When you have finished the posttest, correct it yourself using the answers on page 527. Then check all the questions you got wrong on the list below:

1. ______ 3. ______ 5. ______

2. ______ 4. ______ 6. ______

If you compare your performance with that on Pretest 13-4, you will see how much you have improved. If you made more than one mistake on this posttest, review the material and take the pretest again.

Now proceed to Final Test 13.

TABLE 13-5

Population of Some United States Cities in 1950 and in 1960 *

City	*Population* 1960	1950
Anchorage	44,000	11,000
Atlanta	487,000	330,000
Atlantic City	60,000	62,000
Boston	697,000	800,000
El Paso	276,000	130,000
Fort Lauderdale	83,000	36,000
Kansas City	122,000	130,000
Las Vegas	64,000	24,000
Miami	292,000	250,000
Washington	760,000	800,000

* Adapted from figures supplied by U.S. Bureau of the Census. Some of these figures have been adjusted to make your calculations easier.

FINAL TEST 13

1. Convert $\frac{3}{4}\%$ to a proportion expressed in decimal form.
2. Convert $3\frac{4}{5}\%$ to a proportion expressed in fractional form.
3. Convert 0.006% to a proportion expressed in fractional form.
4. If the sales tax is $2\frac{1}{2}\%$, how much tax do you pay when you buy $44 worth of goods?
5. Use Table 13-1 (page 495) to find the number of managers and proprietors who have not completed high school.
6. A sample of hydrogen chloride is decomposed into 106.91 grams of chlorine and 3.09 grams of hydrogen. Calculate to one decimal place the percentage of chlorine and hydrogen in the sample.
7. Use Table 13-2 (page 496) to calculate (correct to one decimal place) the percentage of personal expenditure in 1965 that went on religious and welfare activities.

8. If a rock contains 0.007% titanium and you need 0.005 gram of titanium for an accurate weighing, how big a sample of rock do you need in order to find the "exact" amount of titanium (correct to three figures)?

Use Table 13-5 (page 524) to find the percentage changes in the populations of the following cities between 1950 and 1960 (correct to three figures):

9. Atlantic City
10. El Paso

11. The volume of a certain gas was 63 milliliters. If this represents an increase of 7% from its previous volume under different conditions, what was its previous volume (correct to three figures)?
12. What is the percentage error involved in using 80 as an approximation for 77 (correct to one decimal place)?

The answers to this test are on page 527. Check the questions you got wrong on the list below and compare your results with those on Diagnostic Test 13 to see how much you have improved.

1. ______	4. ______	6. ______	9. ______
2. ______	5. ______	7. ______	10. ______
3. ______		8. ______	11. ______
			12. ______

If you made more than one mistake, review the material and then take the diagnostic test again. Make sure that you can get all the questions right before you finish.

You have now completed this volume. I hope you have found it useful. You can get further help with your algebra, if you want it, from *Fundamentals of Elementary Algebra*.

ANSWERS TO TESTS FOR CHAPTER 13

DIAGNOSTIC TEST 13

1. 0.004
2. $\frac{35}{800}$ or $\frac{7}{160}$
3. $\frac{8}{100,000}$ or $\frac{1}{12,500}$
4. \$1.71
5. 791,350
6. 2.8% hydrogen, 97.2% chlorine
7. 1.3%
8. 83.3 grams
9. −12.9%
10. 131%
11. 66.7 milliliters
12. 5.3%

PRETEST 13-1

1. 0.35
2. 0.0237
3. 0.125
4. 0.0071
5. 0.0025
6. $\frac{8}{25}$
7. $\frac{19}{2,000}$
8. $\frac{1}{60}$
9. $\frac{3}{100,000}$
10. $\frac{1}{200}$

POSTTEST 13-1

1. 0.47
2. 0.0695
3. 0.1625
4. 0.0042
5. 0.004
6. $\frac{6}{25}$
7. $\frac{9}{2,000}$
8. $\frac{1}{80}$
9. $\frac{7}{100,000}$
10. $\frac{1}{800}$

PRETEST 13-2

1. \$1.48
2. \$2.24
3. \$186.20
4. 19.29 grains
5. 1,977,800
6. 293,040

POSTTEST 13-2

1. \$1.65
2. \$2.88
3. \$440.20
4. 71.088 grains
5. 5,172,930
6. 476,840

PRETEST 13-3

1. 88.9% oxygen, 11.1% hydrogen
2. 6.2%
3. 0.7%
4. 7.1 grams
5. 692.8 grams
6. 0.1825%

POSTTEST 13-3

1. 88.75% oxygen, 11.25% hydrogen
2. 14.7%
3. 0.7%
4. 12.5 grams
5. 420.1 grams
6. 0.34%

PRETEST 13-4

1. 300%
2. 47.6%
3. −6.2%
4. 37.8 ohms
5. 600 cubic feet
6. 4.8%

POSTTEST 13-4

1. −5.0%
2. 16.8%
3. 166.7%
4. 33.4 ohms
5. 700 cubic feet
6. 7.1%

FINAL TEST 13

1. 0.0075
2. $\frac{19}{500}$
3. $\frac{6}{100,000}$
4. \$1.10
5. 974,610
6. 97.2% chlorine, 2.8% hydrogen
7. 1.3%
8. 71.4 grams
9. −3.23%
10. 112%
11. 58.9 milliliters
12. 3.9%